Principles and Applications of Digital Electronics

Larry D. Jones

Oklahoma State University

MACMILLAN PUBLISHING COMPANY New York

Collier Macmillan Publishers London

MACMILLAN PUBLISHING COMPANY
866 Third Avenue, New York, New York 10022

Collier Macmillan Canada, Ltd.

Library of Congress Cataloging in Publication Data

Jones, Larry D.
   Principles and applications of digital electronics.

   Includes index.
   1. Digital electronics.   I. Title.
TK7868.D5J65 1986   621.381   84-15457
ISBN 0-02-361320-3

Printing: 1 2 3 4 5 6 7 8   Year: 6 7 8 9 0 1 2 3 4

ISBN 0-02-361320-3

Preface

Few, if any, areas of technology have touched more people in recent years than digital electronics. Areas of electronics that were formerly the sole domain of analog systems are rapidly giving way to digital techniques. At the same time a wide range of industrial and consumer products that did not even exist a few years ago are commonplace today. Because of the tremendous interest in digital electronics, one of the most sought-after skills by employers of technical personnel is in this area.

The objective of this textbook is to provide an introduction to digital concepts and to present a wide range of practical applications of these concepts. The book is intended for use in introductory courses in digital electronics and was written with one overriding thought in mind—textbooks are primarily for students. Therefore, considerable effort was exerted to make the material as readable as possible.

There are several good-quality textbooks for digital electronics already available, each with their own strengths and weaknesses. However, as with most other things, not everyone likes the same textbook. It is my hope that the topic coverage, readability, problem selection, practical examples, and practical applications in this textbook will make it a worthy contender.

Although the entire book reflects modern digital technology, its particular technical strengths include the following:

1. Up-to-date coverage of digital IC families including TTL, MOS, CMOS, ECL, I^2L, and G_AA_S.
2. The use of both general block diagrams and detailed schematic diagrams.
3. Coverage of tri-state devices and bus organization.
4. Thorough discussion of multiplexers, demultiplexers, encoders, decoders, and code converters.
5. In-depth discussion of analog-to-digital and digital-to-analog converters.
6. Coverage of MSI and LSI semiconductor memory devices and programmable logic arrays.
7. Introduction to microprocessors and computers.

Each chapter begins with a list of instructional objectives and self-evaluation questions. Examples, block diagrams, schematics, tables, and graphs are used throughout the book.

Any textbook is the result of the combined efforts of many people, all of whom I wish to thank. In particular, I wish to thank Violeta Revere and Jamie Hadden for skillfully and cheerfully typing the manuscript and Judy Green and her editorial staff at Macmillan, as well as the entire production staff, for their splendid work.

I also wish to express appreciation to the very knowledgeable and thorough reviewers of the manuscript for their valuable and constructive suggestions and to the many companies that provided technical data and photographs of their products. Finally, I want to say ''thanks'' to my wife Kay for her patience and encouragement.

L.D.J.

Contents

CHAPTER 6   NUMBER SYSTEMS AND ARITHMETIC OPERATIONS   144

CHAPTER 9   COUNTERS   **242**

CHAPTER 14   D/A AND A/D CONVERSION   **371**

CHAPTER 15   MEMORY AND MEMORY DEVICES   **401**

Principles and Applications of Digital Electronics

1

Introduction to Digital Principles

INSTRUCTIONAL OBJECTIVES

In this introductory chapter we discuss the development of digital electronics and compare digital and analog systems. After completing the chapter, you should be able to

1. Define the terms *analog* and *digital*.
2. Associate names with significant events in the evolution of digital computers.
3. List examples of analog and digital devices or signals.
4. List advantages of digital instruments over analog instruments.
5. List chronologically several major events in the history of computers.
6. List applications, other than computers, for digital circuits.

SELF-EVALUATION QUESTIONS

At the beginning of each chapter in this textbook is a set of self-evaluation questions related to the material in the chapter. Read the questions prior to studying the chapter and, as you read the chapter, watch for the answers to the questions. After you have completed the chapter, return to this section and evaluate your comprehension of the material by answering the questions again.

1. What is the root word of *digital,* and what does it mean?
2. What is the difference between an analog signal and a digital signal?
3. What important device related to digital circuits was invented at Bell Laboratories?
4. Who developed logic algebra?
5. What are five broad categories in which digital electronics find applications?
6. Who started the company that eventually became IBM?
7. What company developed a basic computer on a single logic chip?
8. What contribution did Charles Babbage make toward the development of the modern computer?

1-1
INTRODUCTION

The integrated circuit revolution that began in the 1960s has remarkably transformed every area of electronics. The fantastic reduction in size and the increased reliability of integrated circuits have made it possible to construct complex electronic circuits on a single, incredibly small, semiconductor chip. It has been said that large-scale integrated circuits are creating a second industrial revolution. One of the primary benefactors of integrated-circuit technology has been digital electronics, which has experienced phenomenal growth in the last two decades.

An understanding of digital electronics is just as fundamental and important to the technician, technologist, or engineer as an understanding of ac and dc circuits. This introductory chapter looks briefly at the historical development of digital electronics, compares digital and analog techniques, and begins to lay a foundation on which to build an understanding of digital electronics.

1-2
BRIEF HISTORY OF DIGITAL COMPUTATIONS

Counting in discrete units has been practiced throughout recorded history. Early man computed by making marks in the dust or by using fingers and toes. Both of these techniques introduced terms that are still associated with counting. The terms *digit* and *digital* both stem from the Latin word *digitus*, which means "finger" or "toe." The earliest calculating device, the abacus, is mentioned in both European and Asian historical records dating back to about 2000 B.C. (The word *abacus* originally meant "dust" and referred to the practice of counting by making marks in the dust.) This simple yet effective device, which is shown in Fig. 1-1, is still widely used in Japan and China. It represents one of the first attempts at mechanized calculation and reflects an early refinement in man's approach to the concept of counting and arithmetic.

(a)

(b)

FIGURE 1-1   Abacus: (a) Japanese model; (b) Chinese model

Very little improvement was made in computational techniques until the seventeenth century, when Scottish mathematician John Napier made several very significant contributions. He is credited with inventing logarithms, introducing the decimal point in decimal notations, using tables of trigonometric functions, and developing a multiplication process that utilized numbering rods. The numbering rods were often called ''Napier's Bones.''

The first mechanical calculating machine was invented in 1642 by Blaise Pascal, a French scientist. Pascal's machine, shown in Fig. 1-2, had geared wheels numbered from zero to nine. He mechanized the carry by means of a rachet so that a gear wheel passing from nine to zero caused the next higher-order gear wheel to move forward one unit. The machine could add and subtract directly; however, multiplication and division had to be done by repeated addition or subtraction.

Baron von Leibnitz, a German mathematician, modified Pascal's machine in 1670 by adding several additional gear wheels. His machine, which he called a ''stepped reckoner,'' was capable of adding, subtracting, multiplying, dividing, and extracting roots. This development was the forerunner of twentieth-century mechanical calculators.

The next significant development, which was not to have a direct influence on computing until later, was the development in 1870 by Frenchman Joseph Jacquard of the automatic weaving loom. The loom, shown in Fig. 1-3, could be programmed by punched cards to weave intricate design into fabric. The punched cards he used were very similar to those used with modern computers, where a hole represents a 1 and no hole represents a 0.

The British mathematician Charles Babbage was the first to conceive the idea of a calculator that would not only solve problems, but would also print out the solutions. He developed a mechanical calculator that would accept a complete problem and proceed to a complete solution in printed form without additional

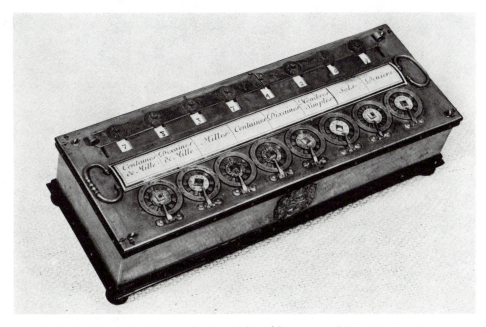

FIGURE 1-2 Pascal's calculating machine. (Courtesy of IBM Archives)

FIGURE 1-3 Original model of the loom invented by Jacquard utilizing punched cards. (Courtesy of The Bettmann Archive)

input from a human operator. He called his machine, which is shown in Fig. 1-4, a ''difference engine.'' Early attempts to build the machine failed, in part because of an inability to obtain precision components. The British government eventually withdrew financial support and the machine never became operational; however, despite this, Babbage established the principles of modern computers, including control, memory, and an arithmetic unit.

At the same time that Babbage was working to build a mechanical calculator, a fellow Englishman named George Boole was applying his mind to the basic concepts of mathematical calculation. Boole reasoned that arithmetic—in fact, all mathematical calculations— were not really ends unto themselves but special cases of a wider logical scheme. He developed a system of logical algebra called Boolean algebra and then showed that arithmetic and other algebra could be derived

FIGURE 1-4 *Babbage difference engine. (Courtesy of IBM Archives)*

from it. Later, Boole's insight was to provide a powerful tool for designers of computing systems.

In the late 1880s Herman Hollerith, a statistician for the U.S. Bureau of the Census, devised a punched card system for coding information, shown in Fig. 1-5. The principle behind this system, which mechanized the processing of information, is still used. In 1911 Hollerith started a manufacturing company called the Computing-Tabulating-Recording Company, which changed its name to International Business Machines in 1924.

In the period between 1924 and 1945, many refinements were made in electric accounting machines. Great strides were made during World War II when, because of the urgency to solve gigantic scientific problems, tremendous efforts were exerted to develop an electronic computing machine. The two primary machines that resulted from this effort were the Mark I, shown in Fig. 1-6, which was

FIGURE 1-5　Hollerith's punched-card sorter. (Courtesy of IBM Archives)

FIGURE 1-6　IBM's Mark I. (Courtesy of IBM Archives)

constructed at Harvard, and the Electronic Numerical Integrator and Computer (ENIAC) shown in Fig. 1-7, which was constructed at the University of Pennsylvania. Early computers used large numbers of vacuum tubes, which were prone to failure because of the heat generated. ENIAC contained 18,000 vacuum tubes.

In 1948 two major events related to computers occurred. International Business Machines (IBM) built the first computer to employ a stored program. It was called the Selective Sequence Electronic Calculator (SSEC). The second major event was the discovery at Bell Laboratories of the principle of the transistor. The increased reliability and reduced power consumption offered by transistors paved the way for the mass production of larger and more complex computers.

Commercial production of data processing computers began in 1951 with the production by Remington Rand of the UNIVAC I, shown in Fig. 1-8. Several

FIGURE 1-7 Electronic Numerical Integrator and Computer (ENIAC). (Courtesy of Sperry Corporation and Eleutherian Mills Historical Library)

FIGURE 1-8 UNIVAC I Computer. (Courtesy of Sperry Corporation and Eleutherian Mills Historical Library)

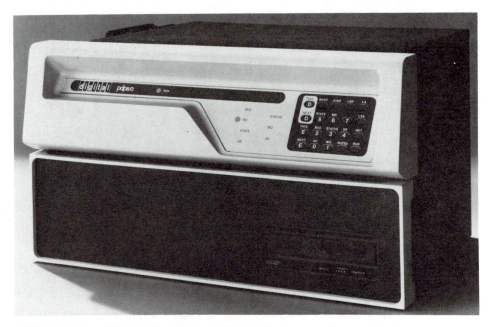

FIGURE 1-9 PDP-8. (Courtesy of Digital Equipment Corporation)

other companies, including IBM, National Cash Register Company, Radio Corporation of America, and General Electric Company, soon followed suit. By the early 1960s virtually all computers were completely transistorized.

The computer industry has grown at a phenomenal rate during the past two decades. During the 1960s the trend was toward large computers for scientific applications. However, as computer size increased, computer cost decreased, due to integrated-circuit technology.

In 1965 Digital Equipment Corporation (DEC) introduced the PDP-8, shown in Fig. 1-9. The relatively low price of the PDP-8—$50,000—helped to move the computer out of sophisticated research laboratories and into manufacturing plants across the country, and thus the minicomputer industry was born.

In 1969 Datapoint Corporation contracted with Intel and Texas Instruments to design an elementary computer on a single logic chip. Intel succeeded, but their design executed instructions too slowly and Datapoint declined to buy the product. Intel was left with a single-chip, computerlike logic device, which they decided to market in 1972 as the Intel 8008—and thus the microcomputer era began. During the last decade the microcomputer explosion has been phenomenal. Figure 1-10 shows only a few of the many moderately priced microcomputer systems presently on the market.

1-3
DIGITAL AND ANALOG TECHNIQUES

There are two basic types of electronic signals, analog and digital. Most of us are more familiar with analog signals, since most of the parameters in our physical world change in a continuous, or *analog,* manner. Examples of analog signals or analog devices are encountered many times daily. Temperature varies in an analog fashion, as does the light associated with the sunrise or sunset. The speedometer on your car is an analog device, as is the volume control on your radio or television set.

Many electrical signals, such as the signals shown in Fig. 1-11, are analog signals. Some devices may be either analog or digital. For example, a watch or clock may be either an analog or a digital device. The same is true of a voltmeter, which may be either an analog-type instrument or a digital instrument. Examples of digital devices are television channel selectors, automobile odometers, and any type of switch.

Digital signals are essentially a series of abruptly changing voltage levels that switch between two fixed values of voltage. Figure 1-12 shows several typical digital signals. Electronic circuits that process digital signals are called digital circuits or logic circuits and are of primary concern in this book.

1-4
APPLICATIONS OF DIGITAL TECHNIQUES

Digital electronics, which until a decade or so ago found application almost exclusively in digital computers, is now common in virtually every area of electronics. You have probably seen or used several things already today that incorporate digital circuits. The following are just a few examples.

(a) IBM PC

(c) Commodore 64

(e) Radio Shack
Model 100

FIGURE 1-10 Modern microcomputers. [Courtesy of (a) International Business Machines Corporation; (b) Texas Instruments, Inc.; (c) Commodore Electronics Limited; (d) Apple Computer, Inc.; (e) Radio Shack, a division of Tandy Corporation; and (f) Hewlett-Packard Company]

(b) Texas Instruments Professional Computer

(d) Apple
Macintosh

(f) HP 75
Portable Computer

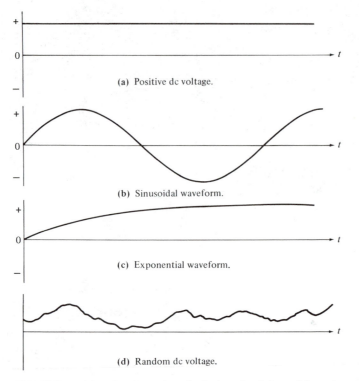

FIGURE 1-11   Analog electrical signals: (a) positive dc voltage; (b) sinusoidal waveform; (c) exponential waveform; (d) random dc voltage

Consumer Electronics.   The list of consumer items that utilize digital circuits is already extensive and will continue to grow. Included are hand-held calculators such as those shown in Fig. 1-13, digital watches and clocks, electronic and video games, digital channel selectors on television sets, digital controls on microwave ovens and washing machines, and digital frequency synthesizers for tuning hi-fi receivers.

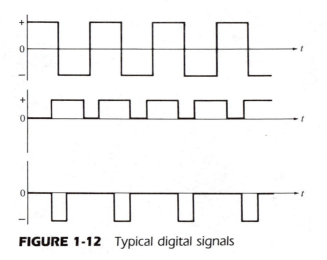

FIGURE 1-12   Typical digital signals

(a) TI55II Hand-Held Calculator.

(b) HP-16C Programmable Calculator.

FIGURE 1-13   Hand-held calculators. [Courtesy of (a) Texas Instruments, Inc.; (b) Hewlett-Packard Company]

Electronic Test Instruments.   A very substantial percentage of electronic test instruments of recent design incorporate some digital circuitry. Digital instruments offer several very attractive advantages over analog instruments, including increased accuracy and resolution, reduction in user errors, and the ability to interface instruments with computers for data acquisition and control purposes. The most common digital test instruments are digital voltmeters (DVMs); digital multimeters (DMMs), such as the one shown in Fig. 1-14; and electronic counters.

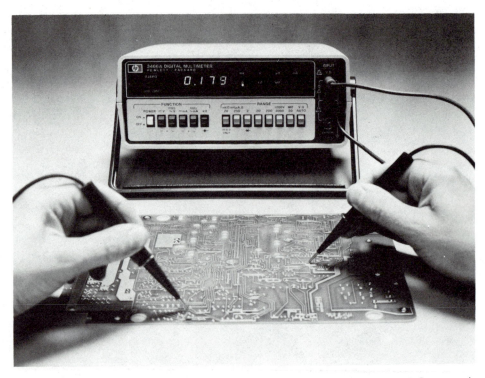

FIGURE 1-14   Digital Multimeter 3466A. (Courtesy of Hewlett-Packard Company)

Aerospace.   Digital techniques are used extensively in missiles and satellites in guidance systems, in on-board computers, and in telemetry systems. Today it is common practice to convert analog signals such as voltage, current, temperature, humidity, light, pressure, and radiation to digital signals prior to transmitting them, to increase accuracy and reliability.

Automotive Electronics.   Digital circuitry is finding applications both within automobiles—in digital clocks, panel meters, and circuitry to monitor gas mileage—and in diagnostic equipment used to analyze and repair automobiles.

Office Equipment.   Significant changes have occurred in office equipment in recent years, including considerable use of digital circuitry in telephone pushbutton dialing and automatic answering systems, reproduction machines, and word processors.

Industrial Users.   Digital circuitry is used extensively in automatic process controls and in the numerical control of machine tools. An industrial application of digital techniques that is growing rapidly is the use of industrial robots such as that shown in Fig. 1-15.

Other Applications.   Digital circuits also find applications in traffic control, biomedical instrumentation, banks, postal service, and point-of-sale terminals in supermarkets and department stores.

FIGURE 1-15 Industrial robot making arc welds. (Courtesy of Cincinnati Milacron)

1-5
SUMMARY

Counting in a digital mode is as old as man himself. From the beginning of counting by making marks in the dust, digital counting techniques eventually progressed to the abacus, then to various types of mechanical computing machines, and finally to a vacuum-tube type of electronic computer. With the invention of the transistor and the development of integrated-circuit technology, computer development has surged forward at a phenomenal rate in the past two decades. Integrated-circuit technology has also spurred the use of digital techniques in virtually every area of electronics.

PROBLEMS

1. Define the term *analog*.

2. Define the term *digital*.

3. Where was the abacus invented, and what does the word *abacus* mean?

4. List three examples of devices or physical parameters that are analog.

5. List three devices that may be either analog or digital.

6. List three examples of devices or occurrences that are digital in nature.

7. What are two advantages of digital instruments over analog instruments?

8. Match the following associated terms:

(a) First minicomputer   A. John Napier
(b) Adding machine   B. Charles Babbage
(c) First program stored electronically   C. Digital Equipment Corp.
(d) Numbering rods or bones   D. George Boole
(e) First microcomputer   E. Blaise Pascal
(f) Programmable loom   F. Intel
(g) Logic algebra   G. Joseph Jacquard
(h) Programmable mechanical calculator H. IBM

REFERENCES

Osborne, Adam, *An Introduction to Microcomputers*. Berkeley, Calif.: Adam Osborne and Associates, Inc., 1977.

Randell, Brian, ed., *The Origins of Digital Computers*. New York: Springer-Verlag, 1973.

Reid-Green, Keith S., "A Short History of Computing," *BYTE*, July 1978, pp. 84–94.

Logic Gates

INSTRUCTIONAL OBJECTIVES

In this chapter we discuss logic gates, which are the fundamental building blocks from which most digital systems are built. After completing the chapter, you should be able to

1. Write a statement that is *obviously true*.
2. Describe the term *two-valued logic*.
3. Define the term *logic gate*.
4. Develop the truth table for an AND gate or an OR gate.
5. Describe the function of the NOT circuit.
6. Describe what is meant by *gating action*.
7. Define the term *current hogging*.
8. Describe the difference between positive and negative logic.
9. Given a problem statement, draw the corresponding logic diagram.

SELF-EVALUATION QUESTIONS

The following questions deal with the material in this chapter. Read the questions prior to studying the material and, as you read the chapter, watch for the answers to the questions. After completing the chapter, return to this section and evaluate your comprehension of the material by answering the questions again.

1. What symbols are used in a two-valued logic system?
2. What is the function of a logic gate?
3. What electronic component functions as a NOT circuit?
4. What does the term *current hogging* mean?
5. In what type of logic system does current hogging show up, and is it always a problem?
6. What does the term *gating action* mean?
7. What type of logic gate does an AND gate in a positive logic system function as in a negative logic system?

2-1
INTRODUCTION

Ancient Greek philosophers were fascinated by the process of logical inference. Aristotle made use of a *two-valued logic system* in devising a method for getting to the truth, given a set of true assumptions. Logic based on Aristotle's system makes no guarantees regarding the accuracy of input statements; however, it does make guarantees about correctly drawn inferences. Statements drawn from correct inferences are called *demonstrably true*. There is another kind of truth that is *obviously true* just by its very nature. A statement such as ''Either Bill is here or he is not'' is obviously true. Bill must either be here or not be here; there are no other alternatives. Such statements are easily dealt with in a two-valued logic system.

We encounter many things every day to which two-valued logic can be applied. Is the light *on* or *off?* Is the answer *true* or *false?* Is the action *right* or *wrong?* Digital logic circuits that respond to obviously true, two-valued-logic, input signals are the topic of interest in this chapter.

2-2
TWO-VALUED LOGIC SYMBOLS

The symbols that are used to represent the two levels of a two-valued logic system are 1 and 0. The symbol 1 may represent a closed switch, a true statement, an ''on'' lamp, a correct action, a high voltage, or many other things. The symbol 0 may represent an open switch, a false statement, an ''off'' lamp, an incorrect action, a low voltage, or many other things.

Since we will be dealing with electronic circuits and signals, henceforth a logic 1 will represent a closed switch, a high voltage, or an ''on'' lamp, and a logic 0 will represent an open switch, a low voltage, or an ''off'' lamp. These describe the only two states that exist in digital logic systems and will be used exclusively to represent the input and output conditions of logic gates.

2-3
LOGIC GATES

A *logic gate* is defined as a circuit with two or more input terminals and one output terminal. The most basic logic circuits are *AND gates, OR gates,* and *inverters*. Strictly speaking, inverters are not logic gates, since they have only one input terminal; however, they are best introduced at the same time as basic gates and will therefore be dealt with in this section.

Circuits that function as logic gates can be fabricated in many ways. This includes switch circuits, the use of diodes and resistors, transistors and resistors, or by several other techniques, which will be discussed in some detail in subsequent chapters.

2-3.1   AND Gates

An *AND gate* is a logic circuit with two or more input terminals and one output terminal. The output signal of an AND gate is *high* (logic 1) only if all inputs are *high* (logic 1). The symbol for the logic AND gate is shown in Fig. 2-1. Because of the extensive use of integrated circuits, detailed knowledge of the actual digital circuitry within the chip is becoming less important for those involved in the development, application, or maintenance of digital equipment. However, some

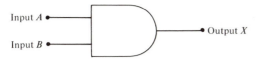

FIGURE 2-1   Logic AND-gate symbol

discussion of circuit implementation for each type of gate is presented to give you a better grasp of what the gate is made of and how it works.

A circuit that will function as an AND gate can be implemented in several ways. A mechanical AND gate can be fabricated by connecting two switches in series as shown in Fig. 2-2. Output X rises to 5 V only when switches A *and* B are closed. Table 2-1 shows this set of conditions together with all other possible combinations of switches A and B and the resulting output. Table 2-1 is called an AND-gate *truth table*.

An AND gate can also be fabricated with resistors and semiconductor diodes. Consider the circuit shown in Fig. 2-3. For the benefit of analysis of the circuit, we shall restrict the values of the input voltages A and B to either 0 V or 5 V. If both inputs are initially set to 0 V, both diodes appear to be forward biased by the +5 V source, V_{CC}. When diodes are forward biased, they present a low-resistance path to ground; therefore, most of the V_{CC} is dropped across resistor R, placing output X near 0 V. This low voltage corresponds to logic 0. In actual practice, if both inputs are set to 0 V at the same instant, one diode or the other will invariably start to conduct before the other. Because of the nonlinear characteristic of diodes, the current through the diode that begins to conduct may be much larger than the current through the other diode. This phenomenon is called *current hogging* and is usually a factor to be considered when diodes are connected in parallel.

If one or the other, but not both input voltages, is set to 5 V, the diode to which the 5 V is applied will be reverse biased; however, output X is still at logic 0 because the other diode is forward biased, thereby holding the output *low*.

If both input voltages are set to 5 V, both diodes will be reverse biased, thus reducing the current that flows from V_{CC} through resistor R to a very low value. Since there is very little current through resistor R, there will be very little voltage dropped across the resistor; therefore, the output voltage X will be close to the value of V_{CC} (logic 1). The truth table for the two-input logic AND gate is shown in Table 2-2.

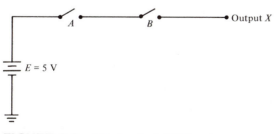

FIGURE 2-2   Mechanical AND gate

TABLE 2-1. / Truth Table for a Switch Circuit Operating as an AND Gate

Switch A	Switch B	Output X (V)
Open	Open	0
Closed	Open	0
Open	Closed	0
Closed	Closed	5

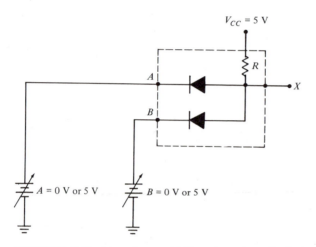

$V_{CC} = 5$ V

R

A

B

X

$A = 0$ V or 5 V $B = 0$ V or 5 V

FIGURE 2-3   Two-input AND gate using diode logic

TABLE 2-2. / Truth Table for a Two-Input AND Gate

A	B	X
0 V	0 V	0 V
0 V	5 V	0 V
5 V	0 V	0 V
5 V	5 V	5 V

(a) Voltage Levels.

A	B	X
0	0	0
0	1	0
1	0	0
1	1	1

(b) Logic Levels.

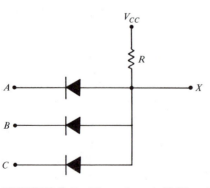

V_{CC}

R

A

B

C

X

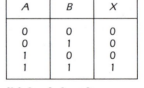

FIGURE 2-4   Three-input AND gate

TABLE 2-3. / Truth Table for a Three-Input AND Gate

A	B	C	X
0	0	0	0
0	0	1	0
0	1	0	0
0	1	1	0
1	0	0	0
1	0	1	0
1	1	0	0
1	1	1	1

An AND gate may have more than two inputs. Figure 2-4 shows a three-input AND gate and Table 2-3 shows the truth table for the circuit. The number of possible combinations of inputs, and therefore entries in the truth table, equals 2 raised to the nth power, where n equals the number of input variables. In the case of a three-input gate, n equals 3; therefore, there are eight possible combinations of 1's and 0's because $2^3 = 8$.

There are other combinations of discrete components that can be used to fabricate AND gates; however, the basic principle of operation remains the same—the output will be *high* only when all inputs are *high*. There are also several integrated-circuit logic families that have AND-gate IC chips. These are discussed in Chapter 5.

2-3.2 OR Gates

An *OR gate* is a logic circuit in which the output signal will be *high* (logic 1) if any single input signal is *high* (logic 1). The symbol for the logic OR gate is shown in Fig. 2-5.

As with the AND gate, a circuit that will function as an OR gate can be implemented in several ways. A mechanical OR gate can be constructed by connecting two switches in parallel as shown in Fig. 2-6. The output X rises to 5 V whenever switch A *or* switch B is closed. Table 2-4 shows this set of conditions together with all other possible combinations of switches A and B and the resulting output. Table 2-4 is called an OR-gate truth table.

An OR gate can also be fabricated with resistors and semiconductor diodes as shown in Fig. 2-7. If both input voltages are initially set to 0 V, output X must be *low* (logic 0), since there is no voltage anywhere in the circuit. If input A is set to 5 V, diode A will be forward biased. Current from the input will flow through diode A and resistor R to ground. Since the forward-biased diode has a

FIGURE 2-5 Logic symbol for an OR gate

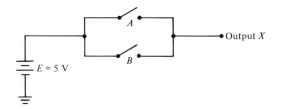

FIGURE 2-6   Mechanical OR gate

TABLE 2-4. / Truth Table for a Switch Circuit Operating as an OR Gate

Switch A	Switch B	Output X (V)
Open	Open	0
Closed	Open	5
Open	Closed	5
Closed	Closed	5

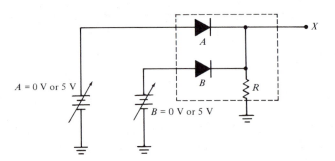

FIGURE 2-7   Two-input OR gate using diode logic

TABLE 2-5. / Truth Table for a Two-Input OR Gate

A	B	X
0 V	0 V	0 V
0 V	5 V	5 V
5 V	0 V	5 V
5 V	5 V	5 V

(a) Voltage Levels.

A	B	X
0	0	0
0	1	1
1	0	1
1	1	1

(b) Logic Levels.

low resistance, most of the input voltage is dropped across R; therefore, the output voltage at X is *high* (logic 1).

If both inputs are set to 5 V, both diodes will be forward biased and the output voltage at X will be *high* (logic 1). The truth table for a two-input OR gate is shown in Table 2-5.

As with the AND gate, an OR gate may have more than two inputs. The circuit

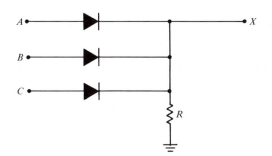

FIGURE 2-8   Three-input OR gate

shown in Fig. 2-8 is a three-input OR gate. The truth table for the circuit is given
in Table 2-6.

**TABLE 2-6. / Truth Table
for a Three-Input OR
Gate**

A	B	C	X
0	0	0	0
0	0	1	1
0	1	0	1
0	1	1	1
1	0	0	1
1	0	1	1
1	1	0	1
1	1	1	1

AND and OR Functions Combined.   Many logic circuit applications require a
combination of logic functions. Consider the following example.

EXAMPLE 2-1 ▬▬▬▬▬▬▬▬▬▬▬▬▬▬▬▬▬▬▬▬▬▬▬▬▬▬▬▬▬

Show how mechanical logic gates (switches) can be constructed to provide an
indication (light or buzzer) of whether a car is not in "park" *or* the hand brake
is not "set" *or* the lights are left "on" *and* the key is removed from the ignition.

SOLUTION

Table 2-7 lists the switch functions.

TABLE 2-7. / Switch Functions

Switch	Open	Closed
A	Car in "park"	Car not in "park"
B	Hand brake "set"	Hand brake not "set"
C	Lights not "on"	Lights "on"
D	Key not removed	Key removed

□

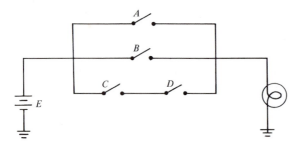

FIGURE 2-9   Circuit for Example 2-1

From the problem statement and using the tabulated switch functions, we can draw the circuit shown in Fig. 2-9.

2-3.3   The NOT Circuit

The NOT circuit performs the basic logic operation of *complementing* or *inverting* a signal and is therefore often called an *inverter*. The purpose of the inverter is to invert the logic level. Since we are working with a two-valued logic system, if either of the two logic values is applied to the input of an inverter, the inverter output will be the other logic value. If we apply a *high* to the inverter input, its output will be *low*.

The logic symbol for the NOT circuit is shown in Fig. 2-10. The circle, or "bubble," at the output indicates inversion. It might also be viewed as a distinction between the symbol for the NOT circuit and the symbol for an operational amplifier or certain types of buffers. Table 2-8 is the truth table for the NOT circuit.

A circuit that functions as a NOT circuit can be constructed with a bipolar transistor as shown in Fig. 2-11. This very basic circuit is called a *resistor–transistor logic (RTL) inverter,* so named because of the resistor used to couple the input signal to the base of the transistor.

The transistor is ideally suited to function as a NOT circuit because of its switching characteristics and because of the inherent signal inversion between the base and collector terminals. Figure 2-12 illustrates these characteristics. When a positive voltage is applied to the base of the transistor, as in Fig. 2-12a, the transistor is driven into saturation. This saturated condition is analogous to a closed switch. The saturated transistor, or closed switch, presents a low resistance from its collector to its emitter; therefore, the output V_o is near 0 V.

When the input is at 0 V as in Fig. 2-12b, the transistor is biased *off,* which is analogous to an open switch. The *off* transistor, or open switch, presents a large

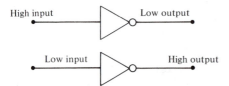

FIGURE 2-10   Symbol for a NOT circuit showing inverting action

**TABLE 2-8. / Truth
Table for a NOT
Circuit**

Input	Output
0	1
1	0

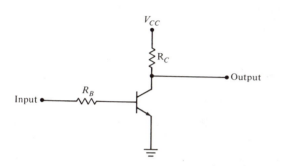

FIGURE 2-11   Basic RTL inverter

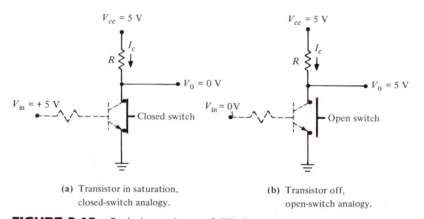

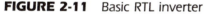

(a) Transistor in saturation,   (b) Transistor off,
closed-switch analogy.   open-switch analogy.

FIGURE 2-12   Switch analogy of RTL inverter operation: (a) transistor in saturation, closed-switch analogy; (b) transistor off, open-switch analogy

collector-to-emitter resistance. Since I_c is approximately zero, there is virtually
no voltage drop across R; therefore, V_o is approximately 5 V.

2-4
THE GATING FUNCTION

The purpose of logic gates is to provide an output signal for a specific input
condition or set of input conditions. An AND gate will provide a *high* at its output
when all inputs are *high*. This characteristic can be used to pass a certain number
of pulses through an AND gate, as shown in Fig. 2-13. A pulse train consisting

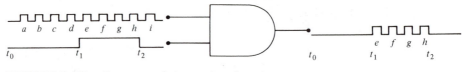

FIGURE 2-13   Concept of the gating function

of pulses *a* through *i* is applied to input *A* of the AND gate. The signal on input *B* goes *high* at time t_1 and remains *high* until t_2. During the time interval $t_2 - t_1$ pulses *e* through *h* will be passed by the AND gate. This is called *gating action*.

2-5
POSITIVE AND NEGATIVE LOGIC

In a two-valued logic system, the two values are generally $+5$ V and 0 V. The relationship between the two voltage levels and the logic symbols 1 and 0 is a matter of definition. By definition, a positive logic system is a system in which a 1 represents the more positive of the two voltage levels and a 0 represents the less positive voltage. In a negative logic system, a 1 represents less positive voltage. We are free to choose whichever relationship between voltage levels and logic symbols we wish; however, once that choice is made we are bound, by definition, to the logic system of our choice. Up to now, we have been using positive logic and will continue to use primarily positive logic.

The distinction between positive and negative logic is important with regard to logic gates. An AND gate in a negative logic system becomes an OR gate in a positive logic system. To see why, consider the logic gate circuit shown in Fig. 2-14. The same circuit is shown in Fig. 2-7 and was identified in Section 2-3.2 as an OR gate. If we apply the various combination of input voltage shown in Table 2-9, we should observe the output voltages shown in Table 2-9. Whether the circuit is functioning as an AND gate or as an OR gate depends on whether it is being used as part of a positive logic system or a negative logic system. Table 2-10 tabulates the relationship between logic levels and logic symbols for positive and negative logic.

Using the data from Tables 2-9 and 2-10, we can construct a positive logic truth table and a negative logic truth table for the circuit in Fig. 2-14. These are given in Table 2-11. Positive and negative logic symbols are substituted in place of the corresponding voltage levels in Table 2-9. As can be seen, the positive logic truth table is an OR-gate truth table, and the negative logic truth table is an AND-gate truth table. The thing to remember is that the circuit behavior is the

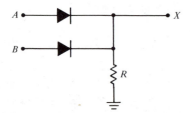

FIGURE 2-14   Basic diode logic gate

TABLE 2-9. / Input and Output Voltages for the Circuit Shown in Fig. 2-14

A	B	X
0 V	0 V	0 V
0 V	5 V	5 V
5 V	0 V	5 V
5 V	5 V	5 V

TABLE 2-10. / Logic Levels and Logic Symbol Relationships for Positive and Negative Logic

Logic Level	Positive Logic	Negative Logic
Low	0	1
High	1	0

TABLE 2-11. / Positive and Negative Logic Truth Tables

A	B	X
0	0	0
0	1	1
1	0	1
1	1	1

A	B	X
1	1	1
1	0	0
0	1	0
0	0	0

same regardless of the type of logic, although the type of logic determines whether we call the circuit an AND gate or an OR gate.

2-6
THE FUNCTION OF LOGIC GATES

Logic gates are used singularly or in combination for one of two purposes: to perform a logic operation or to perform an arithmetic operation. When performing either a logic operation or an arithmetic operation, the symbols 1 and 0 are used. In the case of a logic operation, the symbols represent high or low logic levels. In the case of an arithmetic operation, the same symbols are used to represent binary numbers. Whether the logic gates are interconnected so that the output represents the correct results of an arithmetic operation or the correct response to a logic operation, the logic gates respond in the same way. Any signal applied to the input of a logic gate is responded to as a logic level regardless of whether it represents a logic level or a binary number.

2-7
APPLICATIONS

There are many commercial and industrial applications for logic gates used either singularly or in conjunction with other logic gates. The following examples illustrate a few of these applications.

EXAMPLE 2-2 ━━━━━━━━━━━━━━━━━━━━━━━━━━━━━━━━━━━━

Design a basic circuit that uses a logic gate and that will allow a person to start a car only if the doors are closed and the seat belts are fastened.

SOLUTION

The "and" requirement is satisfied with an AND gate. A basic circuit such as the one shown in Fig. 2-15 could be used.   ❏

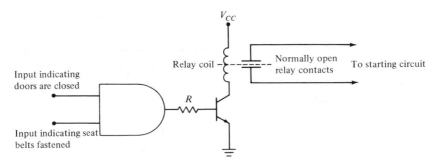

FIGURE 2-15   Circuit for Example 2-2

EXAMPLE 2-3 ━━━━━━━━━━━━━━━━━━━━━━━━━━━━━━━━━━━━

Design a basic logic circuit to monitor two signals that vary randomly and turn *on* a light when both are positive at the same time.

SOLUTION

The problem statement requires the light to be *on* when signal *A and* signal *B* are *high;* therefore, we can use an AND gate as shown in Fig. 2-16.   ❏

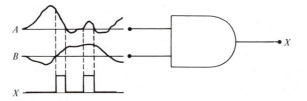

FIGURE 2-16   Waveforms and logic gate for Example 2-3

EXAMPLE 2-4 ▬▬▬▬▬▬▬▬▬▬▬▬▬▬▬▬▬▬▬▬▬▬▬▬▬

There are three elevators in a building. If elevators *A* and *B* are in use and going up, and elevator *C* is not in use, a signal is desired to send *C* to the ground floor. Design such a circuit.

SOLUTION

The condition "*A and B* are in use" implies that an AND gate is needed. The condition "*C* is *not* in use" implies that a NOT circuit is needed to obtain a logic 1 even though *C* is not in use. The fact that the first condition *and* the second condition must both be satisfied implies that we need an AND gate. The circuit can be implemented with either of the circuits shown in Fig. 2-17. ❑

It should be apparent that the reasoning process used to obtain the necessary logic circuit for Example 2-4 would not be practical, and in fact would not work, for complex problems. In Chapter 3 we will investigate the use of Boolean algebra to solve such problems.

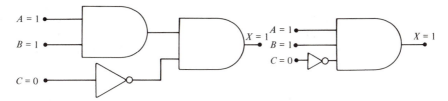

FIGURE 2-17   Logic circuit for Example 2-4

2-8
SUMMARY

Two-valued logic, which can be applied to many things we encounter daily, is the foundation on which digital logic systems are built. The symbols that represent the two levels are 1 and 0. In logic circuits, these logic levels are equated to dc voltage levels, where logic 0 is equated to 0 V and logic 1 is equated to 5 V.

The most elementary logic circuits are AND and OR gates and NOT circuits. Applying logic levels 0 or 1 to the input terminals or these elementary logic circuits causes them to behave in a very predictable manner. A table of all possible combinations of input signals applied to the elementary logic circuits and the resulting output is called a truth table.

Circuits that will function as elementary logic circuits can be constructed with switches or relays or with various combinations of electronic components such as diodes and resistors.

PROBLEMS

1. Write a statement that is obviously true regarding Jack and ownership of a bicycle.

2. List two examples (other than those used in the chapter) of events or conditions to which two-valued logic can be applied.

3. Draw a mechanical switching circuit that will provide a logic 1 output if A is true, or B and C are true, or C and A or B are true.

4. How many combinations of inputs are possible with a four-input logic gate?

5. Redraw the input waveforms and the output waveform at point X if the waveforms shown are applied to inputs A, B, and C in the following figure.

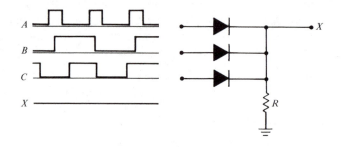

6. Using the principles of positive logic, develop two truth tables for the following circuit. Indicate voltage levels for all positive combinations of input conditions in the first table and the corresponding logic levels in the second table.

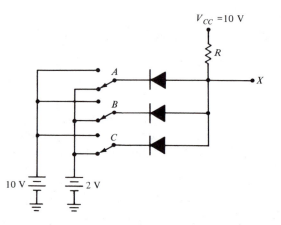

7. What is the logic level at the output of the following circuit?

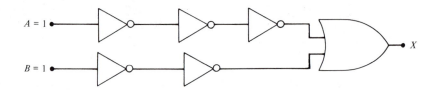

8. Define the term *current hogging*.

9. Develop the truth table for the following circuit by employing the principles of negative logic.

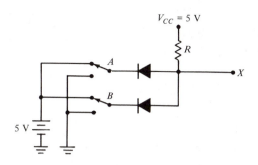

10. Design a logic circuit to provide an indication if a calculator battery is low (near 0) and the charger is not connected.

11. Redraw the input waveforms and the resulting output waveform at X for the following circuit.

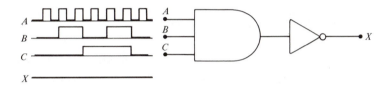

12. Draw a logic diagram for a control circuit that will allow a car to be started only if the transmission is in neutral, the hand brake is set, and the ignition key is turned to the "start" position.

13. Develop the truth table for Problem 12.

14. Just before the launch of a rocket, many systems, such as pressurization, cooling systems, liquid oxygen tanks, electrical systems, and range operations, are checked. Draw a logic diagram that will allow the countdown to continue if range operations are acceptable, the liquid oxygen tanks are full, and the cooling system is working correctly.

15. Each store in a chain of convenience stores places large-denomination bills in a safe that requires two keys to open. One key is kept by the manager of the store and the other by the business manager from the central office. Draw a logic diagram that will permit access to the safe only if both keys are used.

16. At lift-off of a rocket, a signal is needed to energize a relay to change from external power to on-board power. Draw a logic diagram that will operate the relay if the umbilical disconnect is complete, engine firing has occurred, and the liquid hydrogen valve is not open.

17. Four large tanks at a processing plant contain different solutions that are being heated. Level sensors are being used to detect whether the level of the solution in tanks M and N rises above a predetermined level. Temperature sensors are used with tanks P and Q to provide an indication if the temperature of these tanks decreases to less than a predetermined temperature limit. The level-sensor outputs M and N should be *low* when the level is acceptable and *high* when the level is too high. The temperature-sensor outputs P and Q should be *low* when the temperature is acceptable and *high* when the temperature is too low. Design a logic circuit that will provide an indication if the level in tank *M* or tank *N* is too high at the same time that the temperature in either tank *P* or tank *Q* is too low.

REFERENCES

Dempsey, John A., *Basic Digital Electronics with MSI Applications*. Reading, Mass.: Addison-Wesley Publishing Co., Inc., 1977.

Malvino, Albert P., and Donald P. Leach, *Digital Principles and Applications,* 3rd ed. New York: McGraw-Hill Book Company, 1981.

Ward, Dennis M., *Applied Digital Electronics*. Columbus, Ohio: Charles E. Merrill Publishing Company, 1982.

Boolean Algebra

INSTRUCTIONAL OBJECTIVES

In this chapter we discuss Boolean algebra. This ''logical algebra'' provides a tool for systematically analyzing digital circuits and systems. After completing the chapter, you should be able to

1. Define or describe the following terms:
   (a) Perfect induction.
   (b) Universal building block.
   (c) Minimization.
   (d) Implicant.
   (e) Maxterm and minterm.
2. Reduce Boolean expressions by use of laws and theorems.
3. Prove Boolean algebra statements by perfect induction.
4. Convert Boolean expressions from product-of-sums form to sum-of-products form, or vice versa, by use of De Morgan's theorems.
5. Implement logic circuits from Boolean expressions.
6. Write Boolean expressions for given logic circuits.
7. Construct logic circuits using only NAND gates.
8. List the rules for logic multiplication and addition.
9. Construct logic circuits using only NOR gates.
10. Given a problem statement, write a corresponding Boolean algebra expression.

SELF-EVALUATION QUESTIONS

The following questions deal with the material in this chapter. Read the questions prior to studying the material and, as you read the chapter, watch for answers to the questions. After completing the chapter, return to this section and evaluate your comprehension of the material by answering the questions again.

1. What arithmetic operations are permitted in Boolean algebra?
2. What logic gates are called *universal building blocks,* and why?
3. What does the term *minterm* mean?
4. What are the two most useful Boolean algebra theorems?
5. What technique is used to prove Boolean algebra theorems?

6. Define the term *implicant*.
7. What Boolean theorems are used to convert Boolean expressions from product-of-sums to form sum-of-products form?

3-1
INTRODUCTION

George Boole, a nineteenth-century English mathematician, is credited with developing a system of logical algebra by which reasoning can be expressed mathematically. In 1854 Boole published a classic book, *An Investigation of the Laws of Thought on Which Are Founded the Mathematical Theories of Logic and Probabilities*. Boole's stated intention was to perform a mathematical analysis of logic; however, his work bore little fruit until years later.

Boole's system of logical algebra, now called *Boolean algebra,* was investigated as a tool for analyzing and designing relay switching circuits by Claude E. Shannon at the Massachusetts Institute of Technology in 1938. Shannon, a research assistant in the Electrical Engineering Department, wrote a thesis entitled ''A Symbolic Analysis of Relay and Switching Circuits.'' As a result of his work, Boolean algebra is now used extensively in the analysis and design of logic circuits.

There are several advantages in having a mathematical tool that can be used to describe digital logic circuits. First, it is much more convenient to describe logic circuits with a mathematical expression and to make necessary changes mathematically than to make changes on schematic or logic diagrams. Second, Boolean algebra expressions can be simplified by the use of basic theorems in much the same way as is done in ordinary algebra. This enables the logic designer to reduce the circuitry, thus reducing the cost and increasing the reliability of operation. Boolean algebra is an indispensable tool for anyone involved in digital systems.

3-2
FUNDAMENTAL CONCEPTS OF BOOLEAN ALGEBRA

Boolean algebra is a *logical algebra* in which symbols are used to represent logic levels. Any symbol can be used; however, letters of the alphabet are generally used. Since logic levels are generally associated with the symbols 1 and 0, whatever letters are used are viewed as variables that can take on values of 1 or 0. Since any letter used can take on two values, we distinguish between the values by placing a bar above the letter for one of the values. For example, if we use the symbol A to represent logic 1, then $\overline{A}$ (read ''A bar'' or ''A not'') represents logic 0.

In Boolean algebra, any symbol can be used; however, there are only two constants: 1 and 0. There are no negative numbers or fractional numbers; therefore,

$$\text{if } A = 1 \quad \text{then } \overline{A} = 0$$
$$\text{if } A = 0 \quad \text{then } \overline{A} = 1$$

Boolean algebra permits only two mathematical operations, addition and multiplication. These operations are associated with the AND gate and the OR gate, respectively.

3-3
LOGIC MULTIPLICATION

The AND gate, which was introduced in Chapter 2, is capable of performing Boolean algebra multiplication according to the following rules for multiplication in Boolean algebra:

$$0 \cdot 0 = 0$$
$$1 \cdot 0 = 0$$
$$0 \cdot 1 = 0$$
$$1 \cdot 1 = 1$$

Recall that the output signal of an AND gate is 1 only when all input signals to the AND gate are 1, which agrees with the rules above. The Boolean algebra expression for the two-input AND gate shown in Fig. 3-1 is

$$X = A \times B \tag{3-1}$$

or

$$X = A \cdot B \tag{3-2}$$

The times symbol ($\times$ or $\cdot$) is read as "and"; therefore, Eq. 3-1 or Eq. 3-2 is read as

$$X = A \text{ "and" } B \tag{3-3}$$

The Boolean algebra expression for the three-input AND gate shown in Fig. 3-2 is

$$X = A \cdot B \cdot C$$

or

$$X = ABC$$

FIGURE 3-1   Two-input AND gate

FIGURE 3-2   Three-input AND gate

The truth tables for two- and three-input AND gates are shown in Table 3-1.

**TABLE 3-1. / Truth
Tables for Two- and
Three-Input AND Gates**

A	B	X = AB
0	0	0
0	1	0
1	0	0
1	1	1

(a) Two-Input.

A	B	C	X = ABC
0	0	0	0
0	0	1	0
0	1	0	0
0	1	1	0
1	0	0	0
1	0	1	0
1	1	0	0
1	1	1	1

(b) Three-Input.

3-4
LOGIC ADDITION

The OR gate, also introduced in Chapter 2, is capable of performing Boolean algebra addition according to the following rules for addition in Boolean algebra:

$$0 + 0 = 0$$
$$1 + 0 = 1$$
$$0 + 1 = 1$$
$$1 + 1 = 1$$

Recall that the output signal of an OR gate is 1 when any input signal is 1, which agrees with the rules above. The Boolean algebra expression for the two-input OR gate shown in Fig. 3-3 is

$$X = A + B \tag{3-4}$$

The plus symbol ($+$) is read as "or"; therefore, Eq. 3-4 is read as

$$X = A \text{ "or" } B$$

The Boolean algebra expression for the three-input OR gate shown in Fig. 3-4 is

$$X = A + B + C \tag{3-5}$$

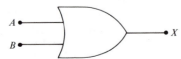

FIGURE 3-3 Two-input OR gate

FIGURE 3-4 Three-input OR gate

The truth tables for two- and three-input OR gates are shown in Table 3-2.

**TABLE 3-2. / Truth Tables
for Two- and Three-Input
OR Gates**

A	B	$X = A + B$
0	0	0
0	1	1
1	0	1
1	1	1

(a) Two-Input.

A	B	C	$X = A + B + C$
0	0	0	0
0	0	1	1
0	1	0	1
0	1	1	1
1	0	0	1
1	0	1	1
1	1	0	1
1	1	1	1

(b) Three-Input.

3-5
COMPLEMENTATION

The logical operation of *complementing* or *inverting* a variable is performed by an inverter, or a NOT circuit, which was introduced in Chapter 2. Since there are only two values that variables can assume in two-value logic systems, the input to an inverter must be either 0 or 1. The purpose of the inverter is to invert, or complement, the input signal; therefore, if the input is 1, the output is 0, and if the input is 0, the output is 1. The symbol used to represent complementation of a variable in a Boolean algebra equation is a bar ($^-$) above the variable. For example, the complement of A is written as $\bar{A}$ and is read as "complement of A" or "A not," as stated in Section 3-2. Since variables can only be equal to 0 or 1, we can say that $\bar{0} = 1$ or $\bar{1} = 0$.

3-6
LAWS AND THEOREMS OF BOOLEAN ALGEBRA

The AND, OR, and NOT functions make up all the basic symbols of digital logic; however, these functions are generally used as part of more complex logic systems rather than individually. As the complexity of the logic system increases, our ability to reason out a solution decreases; therefore, as with any other type of mathematics, a set of rules governing the treatment of Boolean algebra equations must be established.

The fundamental rules, or postulates, of Boolean algebra are those associated with the AND, OR, and NOT functions and are tabulated in Table 3-3. Three of the basic laws of Boolean algebra are the same as in ordinary algebra: the *commutative law*, the *associative law*, and the *distributive law*.

The *commutative law for addition* of two variables is written as

$$A + B = B + A \tag{3-6}$$

TABLE 3-3. / Rules for AND, OR, and NOT Functions

AND	OR	NOT
$0 \cdot 0 = 0$ $0 \cdot 1 = 0$ $1 \cdot 0 = 0$ $1 \cdot 1 = 1$	$0 + 0 = 0$ $0 + 1 = 1$ $1 + 0 = 1$ $1 + 1 = 1$	$\overline{0} = 1$ $\overline{1} = 0$

Equation 3-6 states that the order of addition is immaterial. The physical meaning of Eq. 3-6 is shown in Fig. 3-5, which shows that the order of ORing is immaterial.

The *commutative law for multiplication* of two variables is written as

$$A \cdot B = B \cdot A \tag{3-7}$$

Equation 3-7 states that the order of multiplication of the variables is immaterial. The physical meaning of Eq. 3-7 is shown in Fig. 3-6, which shows that the order of ANDing is immaterial.

The *associative law for addition* of three variables is written as

$$(A + B) + C = A + (B + C) \tag{3-8}$$

Equation 3-8 states that the grouping of variables for the purpose of addition is immaterial. The physical meaning of Eq. 3-8 is shown in Fig. 3-7, which shows that the grouping of variables for ORing is immaterial.

The *associative law for multiplication* of three variables is written as

$$(A \cdot B)C = A(B \cdot C) \tag{3-9}$$

FIGURE 3-5   Physical meaning of the commutative law of addition

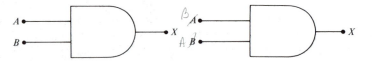

FIGURE 3-6   Physical meaning of the commutative law of multiplication

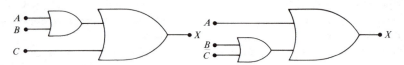

FIGURE 3-7   Physical meaning of the associative law for addition

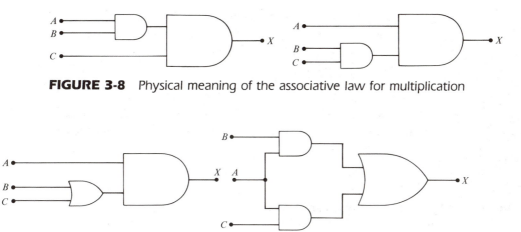

FIGURE 3-8 Physical meaning of the associative law for multiplication

FIGURE 3-9 Physical meaning of the distributive law

Equation 3-9 states that the grouping of variables for the purpose of multiplication is immaterial. The physical meaning of Eq. 3-9 is shown in Fig. 3-8, which shows that the grouping of variables for ANDing is immaterial.

The *distributive law* for three variables involves both addition and multiplication and is written as

$$A(B + C) = AB + AC \qquad (3\text{-}10)$$

TABLE 3-4. / Fundamental Laws and Theorems of Boolean Algebra

1. $A + B = B + A$ 2. $A \cdot B = B \cdot A$	Commutative laws
3. $A + (B + C) = (A + B) + C$ 4. $A \cdot (B \cdot C) = (A \cdot B) \cdot C$	Associative laws
5. $(A + B)(A + C) = A + BC$ 6. $(A \cdot B) + (A \cdot C) = A(B + C)$	Distributive laws
7. $A = A$ 8. $\overline{A} = A$	Identity laws
9. $A + A = A$ 10. $A \cdot A = A$	Idempotent laws
11. $A + \overline{A} = 1$ 12. $A \cdot \overline{A} = 0$ 13. $A = \overline{\overline{A}}$	Complementation laws
14. $A + AB = A$ 15. $A \cdot (A + B) = A$	Laws of absorption
16. $A + 0 = A$ 17. $A + 1 = 1$	Laws of union (Properties of OR operation)
18. $A \cdot 1 = A$ 19. $A \cdot 0 = 0$	Laws of intersection (Properties of AND operation)
20. $A + \overline{A}B = A + B$ 21. $A(\overline{A} + B) = A \cdot B$	Identity theorems
22. $\overline{A + B} = \overline{A} \cdot \overline{B}$ 23. $\overline{A \cdot B} = \overline{A} + \overline{B}$	De Morgan's theorems

Equation 3-10 states that adding several variables and then multiplying the sum by a single variable is equivalent to multiplying each of the variables by the single variable and then adding the products. The physical meaning of Eq. 3-10 is shown in Fig. 3-9, which shows that ORing two variables and then ANDing is equivalent to ANDing the single variable with the two other variables and then ORing.

In addition to the associative, commutative, and distributive laws, there are several other very useful laws for manipulating and simplifying Boolean algebra expressions. The eight fundamental laws of Boolean algebra together with several very useful theorems are listed in Table 3-4.

3-7
PROOF BY PERFECT INDUCTION

Any of the Boolean algebra theorems are easily proved by the method of perfect induction. This is done by using a truth table and evaluating both sides of a theorem for all possible combinations of 1's and 0's.

EXAMPLE 3-1

Prove that $A + AB = A$ by using the technique of perfect induction.

SOLUTION

Since the theorem contains two variables, there are only four combinations of 1's and 0's to be considered. These combinations are

TABLE 3-5. / Combinations of 1's and 0's shown in Table 3-5

A	B	$A + AB = X$
0	0	$0 + 0 \cdot 0 = 0$
0	1	$0 + 0 \cdot 1 = 0$
1	0	$1 + 1 \cdot 0 = 1$
1	1	$1 + 1 \cdot 1 = 1$

Since the entries in the column beneath X, which are the values for $A + AB$, are equal in each case to the corresponding value of A, we have proved the theorem.   ❏

3-8
DE MORGAN'S THEOREMS

Theorems 22 and 23 in Table 3-4 are known as De Morgan's theorems, named for Augustus De Morgan, and English mathematician and friend of George Boole. De Morgan's theorems, which are perhaps the most powerful and useful theorems

in Boolean algebra, provide a designer of logic circuits with two very helpful tools used

1. To facilitate removing individual variables from beneath a complementing bar.

2. To facilitate changing from a *product-of-sums* form or *maxterm* form of a Boolean algebra expression to a *sum-of-products* form or *minterm* form, or vice versa.

Minterm and maxterm are associated with the two common forms of logic equations: the sum-of-products and the product-of-sums forms. The name *minterm* is used in association with the products that are separated by a plus sign in the sum-of-products form. The name *maxterm* refers to the sum of variables that are separated by a multiplication sign in a product-of-sums expression. The names are identified in the following Boolean equations:

Sum-of-products form:
$$X = AB\overline{C} + A\overline{B}C + \overline{A}\overline{B}C$$

minterms

Product-of-sums form:
$$X = (A + \overline{B} + C)(A + \overline{B} + \overline{C})(\overline{A} + \overline{B} + C)$$

maxterms

The following steps allow us to make the preceding transformations readily.

1. Complement the entire function.
2. Change all OR signs to ANDs and all AND signs to ORs.
3. Complement each of the individual variables.

EXAMPLE 3-2

Apply the three steps listed above to remove the complementing bar from the expression

$$X = \overline{\overline{AB} + C}$$

SOLUTION

Complement entire function: $\quad X = \overline{\overline{\overline{AB} + C}} = \overline{AB} + C$

Interchange operators: $\quad X = (\overline{A} + B)(C)$

Complement individual variables: $\quad X = (\overline{\overline{A}} + \overline{B})(\overline{C}) = (A + \overline{B})(\overline{C})$

The procedure outlined above is sometimes referred to as *demorganization*. ❑

EXAMPLE 3-3

Convert the following product-of-sums expression to a sum-of-products expression.

$$(\overline{\overline{A} + B})(A + \overline{B} + C)$$

SOLUTION

Complement entire function:

$$X = \overline{\overline{(\overline{A} + B)(A + \overline{B} + C)}}$$
$$= (\overline{A} + B)(A + \overline{B} + C)$$

Interchange operators:

$$X = \overline{A} \cdot B + A \cdot \overline{B} \cdot C$$

Complement individual variables: $X = \overline{\overline{A}} \cdot \overline{B} + \overline{A} \cdot \overline{\overline{B}} \cdot \overline{C} = A\overline{B} + \overline{A}B\overline{C}$ ❏

The physical significance of De Morgan's theorems is that any Boolean function can be implemented using either AND gates and NOT circuits or OR gates and NOT circuits. De Morgan's theorems also provide us with a convenient way of describing NAND and NOR gates.

3-9
UNIVERSAL BUILDING BLOCKS

To this point, three fundamental building blocks—AND gates, OR gates, and NOT circuits—have been discussed. Boolean algebra theorems 22 and 23 in Table 3-4 mathematically describe combinations of these building blocks to form two new gates, NAND and NOR gates. These gates are the most popular and most widely used logic gates. Since any logic circuit can be constructed using only NAND and NOR gates, they are often referred to as the *universal building blocks*.

3-9.1 NAND Gates

The NAND, or not AND, gate is an AND gate followed by a NOT circuit. The operation of the NAND gate is described by one of De Morgan's theorems, which states that

$$\overline{A \cdot B} = \overline{A} + \overline{B} \tag{3-11}$$

The term on the left of the "equals" sign in Eq. 3-11 describes the ANDing operation followed by an inverting operation. The inverter is built into the same gate chip as the AND gate, as reflected by the NAND-gate symbol shown in Fig. 3-10 and the NAND-gate truth table in Table 3-6. Since the right-hand term in Eq. 3-11 is equal to the left side, a circuit that will perform the right-hand operation is also sometimes referred to as a NAND gate. Such a circuit consists of an OR gate preceded by two inverters, as shown in Fig. 3-11b. However, the most widely used and easily recognized NAND gate symbol is the symbol shown in Fig. 3-10a.

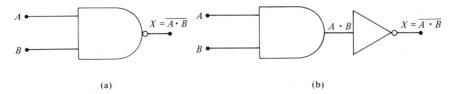

(a)   (b)

FIGURE 3-10   (a) Nand-gate symbol; (b) not AND equivalent of a NAND gate, or equivalent NAND-gate circuit using AND logic

TABLE 3-6. / NAND-Gate
Truth Table

A	B	$A \cdot B$	$\overline{A \cdot B}$
0	0	0	1
0	1	0	1
1	0	0	1
1	1	1	0

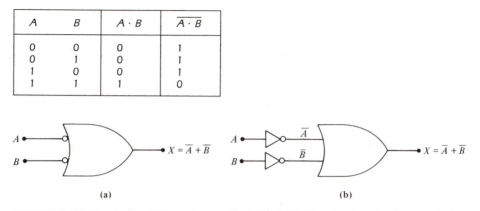

(a)   (b)

FIGURE 3-11 (a) NAND-gate symbol; (b) NAND gate by De Morgan's law implementation, or equivalent NAND-gate circuit using OR logic

3-9.2   NOR Gates

The NOR, or not OR, gate is an OR gate followed by a NOT circuit. The operation of the NOR gate is described by De Morgan's second theorem, which states that

$$\overline{A + B} = \overline{A} \cdot \overline{B} \tag{3-12}$$

The term on the left side of the "equals" sign in Eq. 3-12 describes the ORing operation of the NOR gate and the complementing bar describes the NOT operation. The inverter that performs the NOT operation is built into the same chip as the OR gate, as reflected by the NOR-gate symbol shown in Fig. 3-12 and the NOR-gate truth table shown in Table 3-7. Since the right-hand term in Eq. 3-12

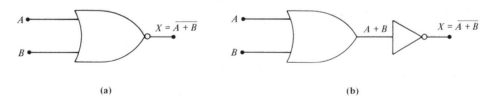

(a)   (b)

FIGURE 3-12 (a) NOR-gate symbol; (b) not OR equivalent of a NOR gate, or equivalent NOR-gate circuit using OR logic

TABLE 3-7. / NOR-Gate
Truth Table

A	B	$A + B$	$\overline{A + B}$
0	0	0	1
0	1	1	0
1	0	1	0
1	1	1	0

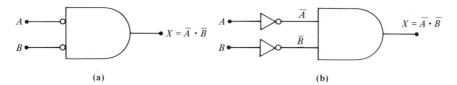

FIGURE 3-13 (a) NOR-gate symbol; (b) not AND equivalent of a NOR gate, or equivalent NOR-gate circuit using AND logic

is equal to the left side, a circuit that will perform the right-hand operation is also sometimes referred to as a NOR gate. Such a circuit consists of an AND gate preceded by two inverters, as shown in Fig. 3-13b. However, the most widely used and easily recognized NOR-gate symbol is shown in Fig. 3-12a.

The fact that any logic circuit can be constructed using only NAND gates is rather attractive to manufacturers of logic gates. This permits the manufacturer to design and manufacture a large number of only one kind of gate rather than a lesser number of several kinds of logic gates, thus reducing costs. Moreover, NAND gates are more versatile than the basic AND, OR, and NOT circuits and can easily be implemented with transistor circuitry.

The way in which NAND gates can be used to function as other basic logic circuits is shown in Fig. 3-14. NOR gates can also be used to implement any other logic function, as shown in Fig. 3-15; however, NAND gates are much more widely used for this purpose.

3-10
BOOLEAN EXPRESSIONS AND LOGIC DIAGRAMS

Boolean algebra expressions are frequently written to describe mathematically the behavior of a logic circuit such as the circuit shown in Fig. 3-16. Using a truth

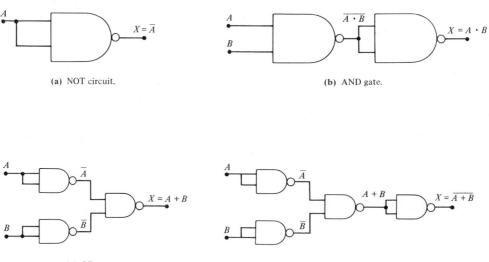

FIGURE 3-14 Logic circuits implemented with NAND gates: (a) NOT circuit; (b) AND gate; (c) OR gate; (d) NOR gate

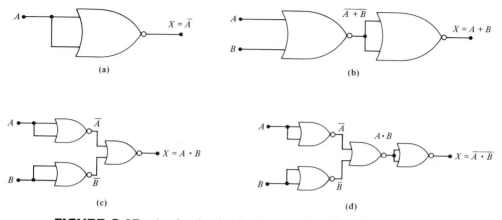

FIGURE 3-15   Logic circuits implemented with NOR gates: (a) NOT circuit; (b) OR gate; (c) AND gate; (d) NAND gate

FIGURE 3-16   Logic circuit for the Boolean expression $X = AB + \overline{C}$

table and the Boolean expression, one can determine which combinations of input signals cause the output signal to go "high" (logic 1).

EXAMPLE 3-4 ▬▬▬▬▬▬▬▬▬▬▬▬▬▬▬▬▬▬▬▬▬▬▬▬▬▬▬▬▬▬

Write the Boolean expression that describes mathematically the behavior of logic circuit shown in Fig. 3-17. Use a truth table to determine what input conditions produce a logic 1 output.

SOLUTION

The Boolean expression is written by starting at each input, tracing each signal through the circuit, writing the appropriate expression at the output of intermediate gate, and combining these terms to form the correct expression at the final output of the circuit. Intermediate expressions as well as the final expression are shown in Fig. 3-18. Table 3-8 is the truth table for the circuit.

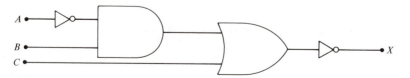

FIGURE 3-17   Circuit for Example 3-4

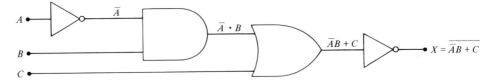

FIGURE 3-18   Circuit showing solution for Example 3-4

TABLE 3-8. / Truth Table for the Circuit in Fig. 3-18

A B C	$\overline{\overline{AB} + C}$ = X	
0 0 0	$\overline{\overline{00} + 0}$	1
0 0 1	$\overline{\overline{00} + 1}$	0
0 1 0	$\overline{\overline{01} + 0}$	0
0 1 1	$\overline{\overline{01} + 1}$	0
1 0 0	$\overline{\overline{10} + 0}$	1
1 0 1	$\overline{\overline{10} + 1}$	0
1 1 0	$\overline{\overline{11} + 0}$	1
1 1 1	$\overline{\overline{11} + 1}$	0

Boolean algebra expressions are also frequently written to describe the desired behavior of a digital system. Once the expression is written, we can construct the digital circuit that is required.   ❏

EXAMPLE 3-5 ━━━

Construct a logic circuit that will provide the following Boolean expression at its output:

$$X = A\overline{B} + AB\overline{C}$$

SOLUTION

The expression is the sum of two products. The products are obtained with AND gates, and the sum is obtained with an OR gate. Inputs B and C must also be inverted. The circuit required is shown in Fig. 3-19.   ❏

3-11
MINIMIZATION BY USE OF BOOLEAN ALGEBRA

In the early conceptual stages of the design of digital logic systems, more circuitry is frequently envisioned than is actually necessary. Boolean algebra allows us to simplify complex digital logic circuits by writing a Boolean expression for a

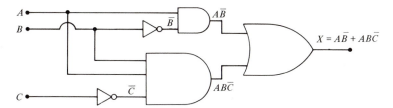

FIGURE 3-19 Circuit for Example 3-5

circuit, then simplifying the expression by applying the laws and theorems of Boolean algebra, and finally, simplifying the circuit to correspond to the simplified mathematical expression. This process, called *minimization,* provides real benefits to the circuit designer through a reduction in the number of logic gates required. Since fewer gates are thus required, costs are reduced and reliability is improved. An increase in speed may also be achieved.

In general, minimization is achieved by performing the following steps:

1. For equations with parentheses, expand using the same rules that apply to ordinary algebra.
2. Look for terms that can be simplified by direct substitution of simpler terms using Boolean algebra theorems.
3. Factor common terms as in ordinary algebra.
4. Apply Boolean algebra theorems wherever possible.
5. Repeat the entire procedure, if necessary, until total minimization is accomplished.

EXAMPLE 3-6 ━━━━━━━━━━━━━━━━━━━━━━━━━━━━━━━━━━━━━━

Given the Boolean expression

$$X = AB + ABC + A\bar{B}\bar{C} + A\bar{C}$$

(a) Draw the logic diagram for the expression.
(b) Minimize the expression.
(c) Draw the logic diagram for the reduced expression.

SOLUTION

(a) The logic diagram for the original expression is shown in Fig. 3-20.
(b) Steps in minimization process:

$$X = AB + ABC + A\bar{B}\bar{C} + A\bar{C}$$

$X = AB(1 + C) + A\bar{C}(1 + \bar{B})$ (factor out common variables)

$X = AB \cdot 1 + A\bar{C} \cdot 1$ (by law 17 of Table 3-4)

$X = AB + A\bar{C}$ (by law 18 of Table 3-4)

$X = A(B + \bar{C})$ (factor out common variable)

(c) The logic circuit corresponding to the minimized expression is shown in Fig. 3-21. ◻

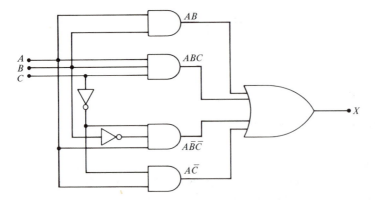

FIGURE 3-20   Logic diagram for Example 3-6

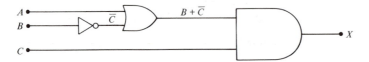

FIGURE 3-21   Minimized logic diagram for Example 3-6

EXAMPLE 3-7

Given the Boolean expression

$$X = \overline{AB + \overline{C}} + A\overline{C} + B$$

(a) Draw the logic diagram for the expression.
(b) Minimize the expression.
(c) Draw the logic diagram for the reduced expression.

SOLUTION

(a) The logic diagram for the original expression is shown in Fig. 3-22.
(b) Steps in the minimization process:

$X = \overline{AB + \overline{\overline{C}}} + A\overline{C} + B$

$\overline{AB + \overline{C}} = \overline{\overline{\overline{A} + \overline{B}}} \cdot \overline{\overline{\overline{C}}} = (\overline{A} + \overline{B})C$ (by De Morgan's theorem)

$X = (\overline{A} + \overline{B})C + A\overline{C} + B$

$X = \overline{A}C + \overline{B}C + A\overline{C} + B$

$X = B + \overline{B}C + \overline{A}C + A\overline{C}$

$X = B + C + \overline{C}A + \overline{A}C$ (by theorem 20 of Table 3-4, $B + \overline{B}C = B + C$)

$X = B + C + A + \overline{A}C$ (by theorem 20 of Table 3-4, $C + \overline{C}A = C + A$)

$X = B + C + A + C$ (by theorem 20 of Table 3-4, $A + \overline{A}C = A + C$)

$X = A + B + C$ (by law 9 of Table 3-4, $C + C = C$)

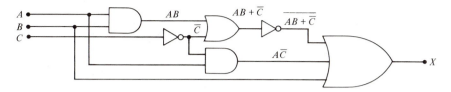

FIGURE 3-22 Logic diagram for Example 3-7

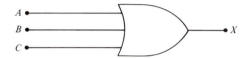

FIGURE 3-23 Minimized logic diagram for Example 3-7

(c) The logic circuit corresponding to the minimized expression is shown in Fig. 3-23. ❑

EXAMPLE 3-8

Given the Boolean expression

$$X = AB\overline{C} + \overline{A}BC + A\overline{B}\overline{C} + \overline{A}C$$

(a) Minimize the expression.
(b) Using only NAND gates, draw the logic diagram for the reduced expression.

SOLUTION

(a) Steps in the minimization process:

$$X = AB\overline{C} + \overline{A}BC + A\overline{B}\overline{C} + \overline{A}C$$
$$X = A\overline{C}(B + \overline{B}) + \overline{A}(\overline{C} + CB)$$
$$X = A\overline{C}(1) + \overline{A}(\overline{C} + B) \quad \text{(by law 11 and theorem 20 of Table 3-4)}$$
$$X = A\overline{C} + \overline{A}\overline{C} + \overline{A}B$$
$$X = \overline{C}(A + \overline{A}) + \overline{A}B$$
$$X = \overline{C} + \overline{A}B \quad \text{(by law 11 of Table 3-4)}$$

(b) The logic circuit that corresponds to the minimized expression is shown in Fig. 3-24, and Fig. 3-25 shows an equivalent circuit using only NAND

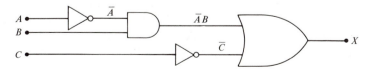

FIGURE 3-24 Minimized logic diagram for Example 3-8

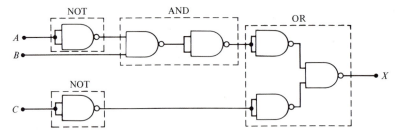

FIGURE 3-25   Circuit equivalent to Fig. 3-24 using only NAND gates

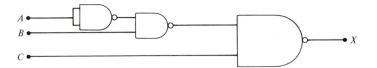

FIGURE 3-26   Circuit equivalent to Fig. 3-25 with redundant gates eliminated

gates. Figure 3-26 shows the required circuit after eliminating the redundant gate. ❏

3-12
WRITING BOOLEAN EQUATIONS FROM TRUTH TABLES

Sometimes Boolean equations can be written directly from problem statements or design specifications; however, it is generally advisable initially to develop a truth table to make sure that no critical combination of variables is overlooked. The following example illustrates the procedure.

EXAMPLE 3-9 ━━

Develop a logic circuit that has a *high* output when at least two out of three of the inputs *A, B,* and *C* are low by

(a) Developing a truth table.
(b) Writing the Boolean expression from the truth table.
(c) Minimizing the expression if possible.
(d) Implementing the logic circuit.

SOLUTION

Since there are three variables, the truth table will have eight entries. After the truth table is developed, the combinations of inputs of interest (all combinations where two inputs are low) can readily be identified and the corresponding Boolean term written. Each Boolean term is the product of three variables and is called a *minterm*. The truth table and minterms of interest are shown in Table 3-9.

TABLE 3-9. / Truth Table and Minterms for Example 3-9

A	B	C	Critical Minterms
0	0	0	$\overline{A}\ \overline{B}\ \overline{C}$
0	0	1	$\overline{A}\ \overline{B}\ C$
0	1	0	$\overline{A}\ B\ \overline{C}$
0	1	1	
1	0	0	$A\ \overline{B}\ \overline{C}$
1	0	1	
1	1	0	
1	1	1	

The Boolean expression desired is the sum of the minterms, written as

$$X = \overline{A}\overline{B}\overline{C} + \overline{A}\overline{B}C + \overline{A}B\overline{C} + A\overline{B}\overline{C}$$

The expression is minimized as follows:

$$X = \overline{A}\overline{B}\overline{C} + \overline{A}\overline{B}C + \overline{A}B\overline{C} + A\overline{B}\overline{C}$$

$$X = \overline{A}\overline{B}(\overline{C} + C) + \overline{A}B\overline{C} + A\overline{B}\overline{C}$$

$$X = \overline{A}\overline{B} \cdot 1 + \overline{A}B\overline{C} + A\overline{B}\overline{C}$$

$$X = \overline{B}(\overline{A} + A\overline{C}) + \overline{A}B\overline{C}$$

$$X = \overline{B}(\overline{A} + \overline{C}) + \overline{A}B\overline{C}$$

$$X = \overline{B}\overline{A} + \overline{B}\overline{C} + \overline{A}B\overline{C}$$

$$X = \overline{A}\overline{B} + \overline{C}(\overline{B} + B\overline{A})$$

$$X = \overline{A}\overline{B} + \overline{C}(\overline{B} + \overline{A})$$

$$X = \overline{A}\overline{B} + \overline{A}\overline{C} + \overline{B}\overline{C}$$

The logic circuit that corresponds to the minimized expression is shown in Fig. 3-27. ❏

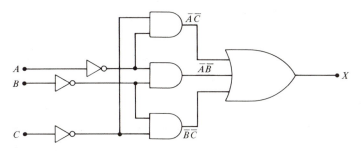

FIGURE 3-27 Minimized logic diagram for Example 3-9

3-13
APPLICATIONS

Boolean algebra finds many applications in logic design of digital circuits for consumer and industrial systems. Numerous applications can be found in alarm systems for protection of equipment or security of personnel, in sensing circuits for automobiles, airplanes, trains, and missiles, and in comparator circuits, vending machines, and elevator circuits, as well as in many other places. The following are only two of many examples of applications that could be cited.

EXAMPLE 3-10

A logic circuit is needed to monitor three elevators. Any time two of the three elevators are not at ground level, as indicated by a logic 0, a logic 1 at the output of the circuit is to bring a fourth elevator into service. Design a logic circuit that will provide the desired signal using symbols A, B, and C to represent the three elevators.

SOLUTION

A truth table showing all possible combinations for A, B, and C is shown in Table 3-10. The value of X is 1 for all combinations that should generate an output. The minterm for each critical case is written directly from the truth table. These terms are then used to write a sum-of-products Boolean expression for the problem statement. Our Boolean expression is

$$X = \overline{A}\,\overline{B}\,\overline{C} + \overline{A}\,\overline{B}C + \overline{A}B\overline{C} + A\overline{B}\,\overline{C}$$

This is minimized by the use of Boolean algebra theorems as follows:

$$X = \overline{A}\,\overline{B}(\overline{C} + C) + \overline{A}B\overline{C} + A\overline{B}\,\overline{C}$$
$$X = \overline{A}\,\overline{B} \cdot 1 + \overline{A}B\overline{C} + A\overline{B}\,\overline{C}$$
$$X = \overline{B}(\overline{A} + A\overline{C}) + \overline{A}B\overline{C}$$
$$X = \overline{B}(\overline{A} + \overline{C}) + \overline{A}B\overline{C}$$
$$X = \overline{B}\,\overline{A} + \overline{B}\,\overline{C} + \overline{A}B\overline{C}$$
$$X = \overline{A}(\overline{B} + B\overline{C}) + \overline{B}\,\overline{C}$$
$$X = \overline{A}(\overline{B} + \overline{C}) + \overline{B}\,\overline{C}$$
$$X = \overline{A}\,\overline{B} + \overline{A}\,\overline{C} + \overline{B}\,\overline{C}$$

The logic circuit that corresponds to the original expression is shown in Fig. 3-28.   ❏

EXAMPLE 3-11

A logic circuit is desired for use in an automobile. The circuit is to provide a signal if the ignition switch is *on* and a door is not closed, or if the ignition switch

TABLE 3-10. / Truth Table and Minterms for Example 3-10

A	B	C	X	Minterms
0	0	0	1	$\overline{A}\overline{B}\overline{C}$
0	0	1	1	$\overline{A}\overline{B}C$
0	1	0	1	$\overline{A}B\overline{C}$
0	1	1	0	
1	0	0	1	$A\overline{B}\overline{C}$
1	0	1	0	
1	1	0	0	
1	1	1	0	

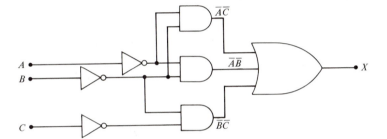

FIGURE 3-28 Minimized circuit for Example 3-10

is *on* and the seat belt is not fastened, or if the ignition switch is *on* and a door is not closed or the seat belt is not fastened. Design the required circuit.

SOLUTION

A truth table showing all possible combinations of variables is shown in Table 3-11. The following relationships apply:

$$S = 1 \qquad \text{switch } on$$

$$D = 1 \qquad \text{door closed}$$

$$B = 1 \qquad \text{seat belt fastened}$$

TABLE 3-11. / Truth Table and Minterms for Example 3-11

S	D	B	Critical Minterms
0	0	0	
0	0	1	
0	1	0	
0	1	1	
1	0	0	$S\overline{D}\overline{B}$
1	0	1	$S\overline{D}B$
1	1	0	$SD\overline{B}$
1	1	1	

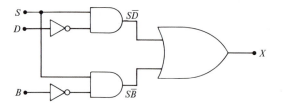

FIGURE 3-29 Minimized circuit for Example 3-11

The statement "The ignition switch is *on* and the door is not closed" means that when $S = 1$ and $D = 0$, a critical condition exists regardless of the value of B. The statement "The ignition switch is *on* and the seat belt is not fastened" means that when $S = 1$ and $B = 0$, a critical condition exists regardless of the value of D. The statement "The ignition switch is *on* and the door is not closed or the seat belt is not fastened" means that when $S = 1$ and either $D = 0$ or $B = 0$, a critical condition exists. These three sets of conditions provide the critical minterms listed in Table 3-11.

The Boolean expression is the sum of the critical minterms, written as

$$X = S\overline{D}\,\overline{B} + S\overline{D}B + SD\overline{B}$$
$$X = S\overline{D}(\overline{B} + B) + SD\overline{B}$$
$$X = S\overline{D} + SD\overline{B}$$
$$X = S(\overline{D} + D\overline{B})$$
$$X = S(\overline{D} + \overline{B})$$
$$X = S\overline{D} + S\overline{B}$$

The circuit that is necessary to implement the minimized expression is shown in Fig. 3-29. ❏

3-14 SUMMARY

Boolean algebra, developed by George Boole in the mid-nineteenth century, is a system of logical algebra that is useful in analyzing and describing digital circuits. Applying the laws and theorems of Boolean algebra allows us to minimize a given expression. This reduces the circuitry required, thus reducing cost and increasing reliability. The most useful and powerful Boolean algebra theorems are De Morgan's theorems, which are useful in describing NAND and NOR gates. De Morgan's theorems are also very useful in converting Boolean expressions from a sum-of-products form to a product-of-sums form, or vice versa. It is generally preferable to write Boolean expressions in the sum-of-products form, where the terms that are added are called minterms. The NAND and NOR gates are often referred to as universal building blocks, since they can be used to fabricate any of the other logic circuits.

PROBLEMS

1. List the theorem that applies to each of the following functions.

(a) $AB\overline{C} + A\overline{C}D = A\overline{C}(B + D)$
(b) $\overline{XYZ} = \overline{X} + \overline{Y} + \overline{Z}$
(c) $(X + Y)(X + YZ) = X + YZ$

2. Prove the following by use of a truth table:

$$A + AB = A$$

3. Prove the following by perfect induction:

$$A + \overline{A}B = A + B$$

4. Prove De Morgan's theorems by perfect induction.

5. Minimize the following expressions by use of Boolean algebra laws and theorems.

(a) $X = AB + AC + ABC$
(b) $X = ABC + \overline{A}B + AB\overline{C}$
(c) $X = AB + A(\overline{B} + C) + AB\overline{C}$

6. Minimize the following expressions.

(a) $X = W\overline{Z}(W + Y) + WY(\overline{Z} + \overline{W})$
(b) $X = (W\overline{Y} + W\overline{Z})(YZ + Y\overline{Z})(WYZ)$
(c) $X = \overline{A}BC + A\overline{B}C + ABC + AB\overline{C} + \overline{AB}C$

7. Minimize the following expression:

$$X = ABC + \overline{AB}C + AB\overline{C} + \overline{A}B\overline{C} + \overline{ABC} + A\overline{BC} + \overline{A}\overline{B}C$$

8. Minimize the following expressions.

(a) $X = \overline{(A + B)(\overline{C})}$
(b) $X = \overline{\overline{AB} + C}$
(c) $X = \overline{A\overline{B}C + \overline{BC}}$

9. Minimize the following expressions.

(a) $X = (A\overline{B}C + \overline{ABC})C$
(b) $X = (A + B + C)(\overline{ABC})$
(c) $X = (A\overline{B} + \overline{ABC})(A + \overline{B}C)$

10. Write the Boolean expression for the following logic circuit. Minimize the expression and draw the logic circuit that is necessary to implement the expression.

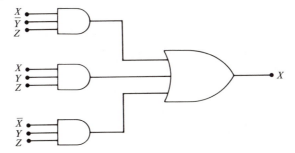

11. Write the Boolean expression for the following logic circuit. Minimize the expression and draw the logic circuit that is necessary to implement the expression.

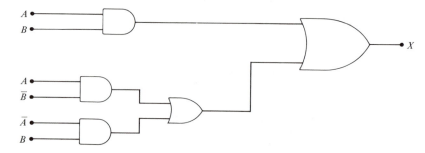

12. Write the Boolean expression for the following logic circuit. Minimize the expression and draw the logic circuit that is necessary to implement the minimized expression.

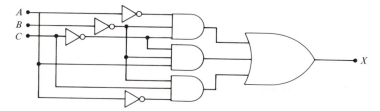

13. Write the Boolean expression at the output of the following logic circuit. Minimize the expression and draw the logic circuit that corresponds to the minimized expression.

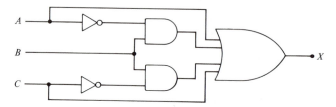

14. Draw a logic circuit using only NAND gates for which output expression is $X = AC + BC$.

15. Draw a logic circuit using only NOR gates for which the output expression is $X = A\overline{C} + \overline{B}C$.

16. Prove the following by use of a truth table:

$$\overline{A}B\overline{C} + \overline{A}BC + \overline{A}\overline{B}C = \overline{A}B + \overline{A}C$$

17. Three members of a family, Sue, Jim, and Bill, have inherited a large estate, which is to remain intact and be operated by majority-rule decisions. Design a circuit that will have a high output when a majority of the inputs are high.

18. Draw the logic circuit for a vending machine for coffee that costs 25 cents. The machine will accept quarters, dimes, and nickels and give change only for dimes.

19. An alarm system is needed to monitor three boilers in an industrial plant. Three sensors monitor water level, temperature, and pressure. The system should provide a high signal that can be connected to a buzzer if the water level is low and the temperature is too high, or if the water level is low and the pressure is too high, or if the water level is low and the temperature and pressure are too high, or if the water level is normal and the temperature or pressure is too high. Use the following relationships:

Water	low $= 0$	normal $= 1$
Temperature:	normal $= 0$	high $= 1$
Pressure:	normal $= 0$	high $= 1$

20. Design a monitoring system to be used during the preflight checkout of an airplane. The circuit is to have a logic 1 output if the engine temperature is normal, the engine rpm is normal, and the oil pressure is low; or if the engine temperature is high, the oil pressure is normal, and the engine rpm is normal; or the engine temperature is high, the oil pressure is low, and engine rpm is normal; or the engine temperature is high, the oil pressure is low, and the engine rpm is high. Use the following relationships:

Engine temperature:	normal $= 0$	high $= 1$
Oil pressure:	low $= 0$	normal $= 1$
Engine rpm:	normal $= 0$	high $= 1$

REFERENCES

Bukstein, Edward, *Practice Problems in Number Systems, Logic, and Boolean Algebra,* 2nd ed. Indianapolis, Ind.: Howard W. Sams & Company, Inc., 1977.

Mendelson, E., *Boolean Algebra and Switching Circuits*. Schaum's Outline Series. New York: McGraw-Hill Book Company, 1970.

Ward, Brice, *Boolean Algebra*. Indianapolis, Ind.: Howard W. Sams & Company, Inc., 1971.

Design of Combinational Logic Circuits

INSTRUCTIONAL OBJECTIVES

In this chapter we discuss fundamental concepts associated with the design of basic combinational logic circuits. After completing the chapter, you should be able to

1. List the four general areas of logic design.
2. Define the term *combinational logic*.
3. List the four tools used in analyzing Boolean functions.
4. List three nonalgebraic means of simplifying Boolean expressions.
5. Develop a truth table from an oral or written problem statement.
6. Prepare Karnaugh maps for two-, three-, four-, or five-variable Boolean expressions.
7. Transfer data from a truth table to a Karnaugh map.
8. Obtain a minimized expression from a Karnaugh map.
9. Implement a logic circuit from a minimized Boolean expression.
10. Perform basic logic design.
11. Define or describe the following terms:
    (a) Don't-care states.
    (b) Essential prime implicants.
    (c) Cell.
    (d) Enclosure.
    (e) Timing diagram.
    (f) Venn diagram.
    (g) Synthesis.

SELF-EVALUATION QUESTIONS

The following questions deal with the material presented in this chapter. Read the questions prior to studying the material and, as you read the chapter, watch for the answers to the questions. After completing the chapter, return to this section and evaluate your comprehension of the material by answering the questions again.

1. What is combinational logic?
2. What are the four general areas of logic design?

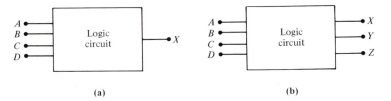

FIGURE 4-1   General models of combinational logic circuits

3. How many cells are in a Karnaugh map to be used to minimize a Boolean expression containing four variables?
4. In what form is a Boolean expression read from a Karnaugh map?
5. What is the Quine–McCluskey method used for?
6. What is a don't-care state?
7. What are prime implicants?
8. What are essential prime implicants?
9. What is one problem that may be encountered when implementing a logic circuit?

4-1
INTRODUCTION

Logic gates and various basic combinations of gates were presented in Chapters 2 and 3. However, the primary interests in those chapters were the individual gates themselves and the development of the analytical tools and skills required to analyze and describe the behavior of individual gates and basic logic circuits. We are now in a position to deal with more complex logic circuits.

In this chapter we introduce basic design considerations and techniques for combinational logic circuits. A *combinational logic circuit* is generally defined as a collection of individual logic gates which are connected such that a specified output is generated for a certain combination of input variables when there is *no* memory associated with the logic circuit. The circuitry is so named because it combines the input variables in such a way as to provide immediately the output signal or signals desired. Combinational logic circuits may have several inputs and a single output, as shown in Fig. 4-1a, or several inputs and two or more outputs, as shown in Fig. 4-1b.

There are four general areas of logic design: analysis, synthesis, minimization, and implementation. Each of these areas will be dealt with; however, major emphasis will be placed on a graphical tool for minimization called a Karnaugh map. This very powerful and useful tool is a favorite of engineers and technicians involved in the analysis and design of logic circuits.

4-2
ANALYSIS

There are four primary tools that are used in analyzing Boolean functions. Each tool provides a certain dimension to the analysis; therefore, each tool has its own sphere of usefulness. The four analytical tools are

1. Algebraic expressions.
2. Truth tables.
3. Logic diagrams.
4. Timing diagrams.

Algebraic expressions, which have already been discussed at some length, are frequently the starting point for analysis, since a Boolean equation can be written directly from a problem statement or from a logic diagram. The dimension to analysis offered by algebraic expressions is the variety of equivalent forms. One algebraic form may be the most concise form for expressing the function, while another form permits construction with the least delay between the input and output. Equivalent forms of algebraic expressions also refer to sum-of-products and product-of-sums forms. These are generally best dealt with by algebraic means.

Truth tables, which have been discussed previously, provide a tabular form of the Boolean function. Truth tables give us an overall view of the combinations of 1's and 0's for the function that may facilitate design by recognition of patterns. The tabular form is also convenient for the initial specification of a function, since its very construction assures us that, when complete, a function is completely specified for all possible values and combinations of the variables. Once an algebraic expression has been reduced to truth-table form, it is completely analyzed; therefore, the dimension to analysis provided by a truth table is its uniqueness.

Logic diagrams provide us with a great deal of very useful information about logic circuits that is not conveyed by the other analytical tools. For example, the number of levels of gating is clearly shown, which provides an indication of the amount of delay that a signal will encounter in passing through the circuit. Logic diagrams also convey information about the number of times input signals are used at various points in the diagram. The primary dimension offered by logic diagrams is the fact that they provide very explicit information.

Timing diagrams provide a very practical and indispensable dimension in the analysis of complex switching networks. The purpose of timing diagrams is to incorporate the dimension of time into Boolean functions. Although basic timing diagrams have already been used in early chapters and will be used in this chapter, their primary applications occur with sequential logic circuits, so we postpone further discussion until later.

4-3
SYNTHESIS

Synthesis is the reverse of analysis. Starting with a truth table, we must write (synthesize) a corresponding Boolean algebra equation. As a general statement, the following rules can usually be applied toward synthesis of a Boolean function.

1. The problem statement or Boolean function should be translated into a completed truth table.

2. A sum-of-product expression is obtained as follows: For each row of the truth table for which the term is to equal 1, the Boolean term is the product of variables that are equal to 1 and the complement of variables that are equal to 0. The sum of these products is the desired Boolean equation.

3. A product-of-sums expression is obtained as follows: For each row of the

truth table for which the term is to equal 0, the Boolean term is the sum of the variables that are equal to 0 plus the complement of the variables that are equal to 1. The product of these sums is the desired Boolean equation.

The use of these rules is demonstrated in the following example.

EXAMPLE 4-1

Synthesize (a) a sum-of-products equation and (b) a product-of-sums equation for the truth table in Table 4-1.

TABLE 4-1. / Truth Table for Example 4-1

Row	A	B	C	Function Value
0	0	0	0	0
1	0	0	1	1
2	0	1	0	1
3	0	1	1	0
4	1	0	0	1
5	1	0	1	0
6	1	1	0	1
7	1	1	1	0

SOLUTION

(a) Sum-of-products equation from rule 2:

$$X = \overline{A}\overline{B}C + \overline{A}B\overline{C} + A\overline{B}\overline{C} + AB\overline{C} \tag{4-1}$$

The four terms correspond to the variables necessary to make rows 1, 2, 4, and 6 equal to the function value of 1.

(b) Product-of-sums equation from rule 3:

$$Y = (A + B + C)(A + \overline{B} + \overline{C})(\overline{A} + B + \overline{C})(\overline{A} + \overline{B} + \overline{C}) \tag{4-2}$$

Each product term in Eq. 4-1, which is the sum-of-products form, is called a *minterm,* and each sum term in Eq. 4-2, which is the product-of-sums form, is called a *maxterm.* There is a reciprocal relationship between the sum-of-products and the product-of-sums expressions; that is, $X = \overline{Y}$ or $\overline{X} = Y$. This can be seen by observing the function values from which the two expressions were written. ❏

4-4
PROBLEM STATEMENT TO TRUTH TABLE

Often the most difficult part of logic design is developing a truth table from a vague or ambiguous English-language statement of the problem. With the problem statement written in the English language, some situation is described in which people or events in the outside world are to cause a logic device or system to respond in a particular way to a certain set of circumstances or events. However,

**TABLE 4-2. / Table of Connective
Translations from English to Logic**

English	Logic Translation
A or B or both	$A + B$
Either A or B but not both	$A\bar{B} + \bar{A}B$
A and B	$A \cdot B$
Not A	$\bar{A}$
Neither A nor B	$\bar{A} \cdot \bar{B}$
A as well as B	$A \cdot B$
A and not B	$A \cdot \bar{B}$
Not both A and B	$\overline{A \cdot B}$

written or verbal problem statements sometimes fail to account for all possible events that may occur. Sometimes only important events are clearly described, while less important, or unlikely, events are not considered or are overlooked.

If the problem statement can be completely and accurately transcribed from a written or verbal format into a truth table, then both important and trivial events are given equal status and dealt with equally. Going from a written or verbal problem statement to a truth table can often be a real challenge. For a person heavily involved in logic design, studying propositional calculus, which provides translation from many English statements to logic statements, would prove beneficial. Table 4-2 lists a few connective translations from propositional calculus that are useful in translating English statements to logic statements. If we start a logic design problem with a skeleton truth table listing all combinations of input variables and carefully translate the problem statement to logic expression, undue difficulty should not be encountered. The following example illustrates the procedure.

EXAMPLE 4-2

An alarm system to be used in conjunction with an automated bottling system is needed in a milk bottling plant. A conveyor belt carries empty bottles that are to be filled with milk. The alarm should sound if any of the following conditions occur:

(a) The milk tank is empty and bottles are on the conveyor belt.
(b) There are no bottles on the conveyor belt and there is milk in the tank.
(c) There is milk in the tank and bottles on the conveyor belt and electric power is lost.
(d) There is no milk in the tank and no bottles on the conveyor belt and electric power is lost.

Write the Boolean expression for the alarm system.

SOLUTION

We approach the problem by developing a skeleton truth table (Table 4-3) for the variables involved, according to the following assignments:

$$B = \text{bottles on the conveyor belt}$$

$$M = \text{milk in the tank}$$

$$P = \text{electric power in "on"}$$

**TABLE 4-3. / Skeleton Truth Table
for Example 4-1**

Combination	B	M	P	Output
0	0	0	0	
1	0	0	1	
2	0	1	0	
3	0	1	1	
4	1	0	0	
5	1	0	1	
6	1	1	0	
7	1	1	1	

Condition (a) in the problem describes combinations 4 and 5; therefore, the output should be *high* for these combinations. Condition (b) describes combinations 2 and 3; therefore, the output should be *high* for these combinations also. Condition (c) describes combination 6 and condition (d) describes combination 0; therefore, the output should be *high* for these combinations as well. The completed truth table and the minterms that will cause the output to be *high* for each combination of concern are shown in Table 4-4.

TABLE 4-4. / Completed Truth Table for Example 4-2

Combination	B	M	P	Output	Minterms
0	0	0	0	1	$\overline{B}\overline{M}\overline{P}$
1	0	0	1		
2	0	1	0	1	$\overline{B}M\overline{P}$
3	0	1	1	1	$\overline{B}MP$
4	1	0	0	1	$B\overline{M}\overline{P}$
5	1	0	1	1	$B\overline{M}P$
6	1	1	0	1	$BM\overline{P}$
7	1	1	1		

The desired Boolean expression is the sum of the minterms in the completed truth table and is written as

$$X = \overline{B}\overline{M}\overline{P} + \overline{B}M\overline{P} + \overline{B}MP + B\overline{M}\overline{P} + B\overline{M}P + BM\overline{P} \qquad (4\text{-}3)$$

Many times equations such as Eq. 4-3 can be simplified; however, this step will be postponed for the moment. ❑

4-5
GRAPHICAL METHODS OF SIMPLIFICATION

In Chapter 3 we discussed minimization of logic expressions by using laws and theorems of Boolean algebra. To minimize logic expressions in a consistently satisfactory manner requires considerable ingenuity, insight, and experience. The next few sections of this chapter discuss graphical methods for simplifying, or minimizing, logic expressions. A graphical representation of a logic expression usually makes the simplifying process much easier to visualize; therefore, the techniques presented here have found wide acceptance in logic analysis or design. However, it should be emphasized that graphical solutions do not always provide a minimized expression; a combination of algebraic and graphical solutions is sometimes required.

Several nonalgebraic methods for simplification that use charts, graphs, and maps have been, and are being, used. The graphical methods of John Venn, called Venn diagrams, are sometimes used to demonstrate visually basic laws and theorems of Boolean algebra. Venn diagrams are discussed briefly in the following section, primarily as a means of introducing Karnaugh maps, which are discussed at length.

4-6
VENN DIAGRAMS

Venn diagrams, named after John Venn, a mathematician and logician, are geometric diagrams that are used to represent logic statements by closed regions. The closed regions are generally represented by rectangles or circles.

To illustrate the use of Venn diagrams, consider the rectangle shown in Fig. 4-2a, which represents every person in the world. Within the rectangle in Fig. 4-2b is a circle that represents all Americans. For every person inside the circle, the statement "That person is an American" is true; however, outside the circle, the statement "That person is an American" is false. The circle therefore defines for whom the statement "That person is an American" is true.

In Fig. 4-3, a second circle, which represents all women, has been added within the rectangle. This circle overlaps the circle that represents all Americans. The common area represents all American women. The common area that is shaded defines the area for which the statement "That person is an American woman" is true. Also, for every person inside the shaded area, the statement "That person

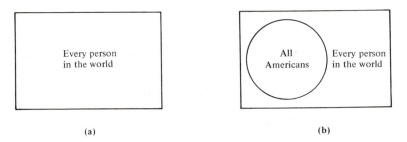

(a)   (b)

FIGURE 4-2 Concept of Venn diagrams

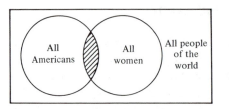

FIGURE 4-3 Venn diagram representation of an AND gate

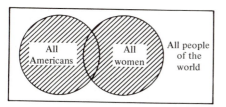

FIGURE 4-4   *Venn diagram representation of an OR gate*

is an American AND that person is a woman'' is true. This is a pictorial representation of the AND gate, which was introduced in Chapter 2.

In Fig. 4-4, both circles are completely shaded. The shaded area defines the area for which the statement ''That person is an American OR that person is a woman'' is true. This is a pictorial representation of the OR gate, which was introduced in Chapter 2.

4-7
MINIMIZATION BY USE OF MAPPING TECHNIQUES

The third general area of logic design is minimization. Minimization of Boolean expressions containing up to three variables can generally be accomplished quite readily by using the laws and theorems of Boolean algebra; however, using these laws and theorems becomes unwieldy if there are more than three variables in the expression. A more suitable technique for minimization of expressions containing four variables is that of mapping. In fact, mapping techniques can also be used for minimization of expressions containing two or three variables—but here by choice rather than necessity. The mapping technique discussed in the next few sections is called the *Karnaugh map* and is simply a modification of the Venn diagram. The use of Karnaugh maps to minimize expressions containing up to five variables will be discussed; however, one can make a good case for using a technique called the Quine–McCluskey method for minimizing expressions with five variables.

A Karnaugh map is a graphical form of a truth table and consists of a square or rectangular array of adjacent cells or blocks. The number of cells in a particular map depends on the number of variables in the Boolean expression to be minimized. The number of cells for a particular map is determined from the expression

$$N = 2^n$$

where N = number of cells required for the Karnaugh map
2 = base of the binary number system
n = number of variables in the Boolean expression

The configuration of the Karnaugh map for two-, three-, and four-variable expressions is shown in Fig. 4-5. The map required for a five-variable expression consists of two four-variable maps, with each being on a different plane of a three-dimensional figure, and will be introduced later in the chapter. In the Karnaugh map, each variable and its complement are assigned half of the cells in the map. The assigned cells consist of adjacent rows or columns. The sides of the map are labeled to show cell assignment, as shown for the two-variable map in Fig.

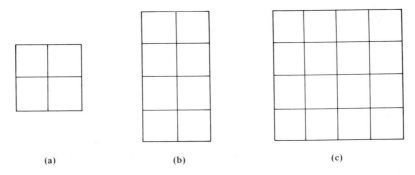

(a)   (b)   (c)

FIGURE 4-5   Karnaugh map configurations: (a) two-variable map; (b) three-variable map; (c) four-variable map

4-6. The two left-hand cells beneath A are assigned to A and the two right-hand cells are assigned to $\overline{A}$. Moving horizontally, the top two cells are assigned to variable B and the bottom two cells are assigned to $\overline{B}$. Each cell is assigned a unique address, which is specified by the row and column in which the cell resides. Since each variable and its complement are assigned half of the cells in the map, there obviously must be sharing of cells. Figure 4-7 shows which variables share each cell in Fig. 4-6. Although there is sharing of cells, no two terms, or complements of terms, are assigned all the same cells. The maps showing cell assignments for three- and four-variable expressions are shown in Fig. 4-8. The sides of the maps may be labeled in any convenient way as long as half of the adjacent cells are assigned to each variable and the other half to the complement of each variable.

Since the sides of a map are labeled arbitrarily, two cells of a map are considered to be adjacent as long as their respective addresses differ by no more than one variable. For example, $ABCD$ and $AB\overline{CD}$ are addresses of adjacent cells. Cells that are not adjacent graphically may be adjacent according to their address. This implies that maps should be viewed as being circular in both the horizontal and vertical planes. Diagonal cells are not adjacent, even though they share a common corner, because their addresses differ by more than one variable.

After learning how to draw a Karnaugh map with the correct number of cells and labeling the sides to assign cells to the variables, the next step is to plot the given expression. To be able to do this, the expression must be in the sum-of-products form. The expression $A\overline{B}C + ABC$ is the sum of two products, or two minterms, and is therefore in the correct form; however, the expression $(A + C)(B + C)$ is not in the correct form and cannot be plotted.

To plot an expression, we identify with a 1 each cell addressed by a minterm. The more variables the minterm consists of, the fewer the number of cells that

FIGURE 4-6   Cell assignment for two-variable Karnaugh map

	A	$\overline{A}$
B	AB	$\overline{A}B$
$\overline{B}$	$A\overline{B}$	$\overline{A}\,\overline{B}$

FIGURE 4-7   Cell sharing

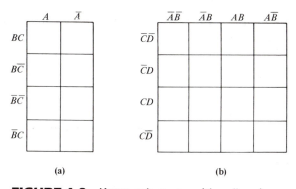

FIGURE 4-8 Karnaugh maps with cell assignments: (a) 3-variable; (b) 4-variable

are being addressed because the more variables the minterm contains, the more specific it becomes. For example, fewer cells are common to minterm ABC than to AB.

After placing a 1 in cells addressed by each minterm, adjacent cells containing a 1 are enclosed either singularly, in pairs, or in groups of 4, 8, or 16 (integral powers of 2). We enclose the largest number of adjacent cells containing a 1 as possible—as long as the enclosed 1's equal 2 raised to an integral power. The enclosed 1's are collectively called an *implicant* or a *group*. The following definitions will prove useful as you work with Karnaugh maps:

1. An *implicant* is one or more adjacent 1's in integral powers of 2.
2. A *prime implicant* is an implicant that cannot be enclosed by a larger implicant on a Karnaugh map. This is shown in Fig. 4-9.
3. An *essential prime implicant* is a prime implicant containing 1's that are not enclosed by any other prime implicant. This is shown in Fig. 4-10.

The desired minimized Boolean expression is obtained from the Karnaugh map by applying the following two steps:

1. All 1's must be included in at least one essential prime implicant. It is permissible, and desirable, to enclose a 1 more than once if it facilitates enlarging another enclosure.

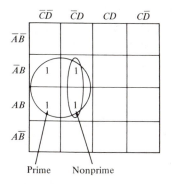

FIGURE 4-9 Comparison of prime implicants and nonprime implicants

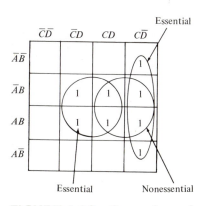

FIGURE 4-10 Comparison of essential prime implicants and nonessential prime implicants

2. Each essential prime implicant represents a minterm. The sum of the minterms that represent each essential prime implicant is the minimized Boolean expression in sum-of-products form corresponding to the given logic function.

The use of these rules is demonstrated in Example 4-3.

EXAMPLE 4-3

(a) Plot the Boolean expression

$$X = AB + A\overline{B} + BC$$

(b) Read out the minimized expression from the map.

SOLUTION

Since the expression contains three variables, we need a Karnaugh map containing cells equal to

$$N = 2^3 = 8$$

The map required is shown in Fig. 4-11. To plot the expression, we start by identifying all cells that are common to variables *A and B*. If we place an *A* in each cell assigned to *A* and a *B* in each cell assigned to *B*, as shown in Fig. 4-12, we see that the top two cells on the left are common to *A* and *B*. If we repeat this procedure for the terms $A\overline{B}$ and BC and place a 1 in cells common to the minterms, we have plotted the function as shown in Fig. 4-13. ❏

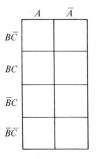

FIGURE 4-11 Karnaugh map for Example 4-3

FIGURE 4-12 Cell identification for Example 4-3

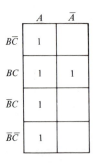

FIGURE 4-13 Addressed cells for Example 4-3 identified with 1's

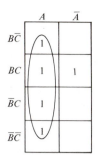

FIGURE 4-14 Enclosure for largest essential prime implicant

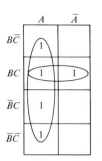

FIGURE 4-15 Essential prime implicants for Example 4-3

Once the Boolean expression is plotted, the cells containing a 1 are enclosed. The largest number of adjacent cells containing a 1 that can be enclosed is four, as shown in Fig. 4-14. The remaining cell that contains a 1 can be enclosed individually or as one of a pair by enclosing two adjacent cells, as shown in Fig. 4-15. Greater minimization is achieved by enclosing the pair of 1's as shown in Fig. 4-15. To "read out" the minimized Boolean expression, all 1's must be enclosed at least once. Greatest minimization is achieved by enclosing the largest number of adjacent 1's possible even if some have already been enclosed.

Read out the minimized expression by determining which variables are common to the cells containing enclosed 1's. In Fig. 4-15, the 1's making up the larger essential prime implicant are common only to A, and the 1's making up the smaller essential prime implicant are common to B and C; therefore, the minimized expression is

$$X = A + BC \qquad (4-4)$$

Since there are two essential prime implicants in Fig. 4-15, the minimized expression contains two minterms.

The rules for grouping the 1's and "reading out" the minimized Boolean expression are summarized as follows:

1. Enclose 1's in adjacent cells in maximum groups of 1, 2, 4, 8, or 16. The number of 1's enclosed should be as large as possible as long as it is an integral power of 2.

2. Draw as many maximum-sized enclosures (essential prime implicants) as possible until all 1's are enclosed at least once. There can and should be overlapping of enclosures to make each enclosure as large as possible.

3. On four-variable maps, 1's in cells on opposite sides of the map may be enclosed since the map is continuous, like a cylinder, in both the horizontal and vertical planes.

4. Each essential prime implicant represents the AND of the variables common to the circled 1's.

5. The minimized expression will contain the same number of minterms as there are essential prime implicants on the map.

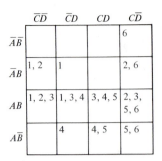

FIGURE 4-16　Identification of cells addressed by each term in the Boolean expression

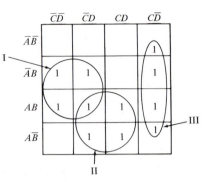

FIGURE 4-17　Essential prime implicants for Example 4-4

EXAMPLE 4-4

Minimize the following Boolean expression by use of the Karnaugh map:

$$X = B\overline{C} + B\overline{D} + AB + AD + AC + C\overline{D}$$

SOLUTION

The required four-variable Karnaugh map is shown in Fig. 4-16. For identification purposes, the following numbers will be used to indicate which cells on the Karnaugh map relate to each term in an expression:

$$
\begin{array}{ll}
1 = B\overline{C} & 4 = AD \\
2 = B\overline{D} & 5 = AC \\
3 = AB & 6 = C\overline{D}
\end{array}
$$

When the expression is plotted, the numbers above appear as shown. The different numbers are used in Fig. 4-16 only to indicate which cells are common to each term in the expression. Normally, a 1 is placed in the affected cells. If a 1 appears in a cell, there is no need for a second 1 in the same cell if the cell belongs to a second term in the expression. If we place a 1 in each cell in Fig. 4-16 that contains a number and enclose adjacent 1's, we obtain the map shown in Fig. 4-17. The cells with 1's included in enclosure I are common to B *and* $\overline{C}$. Common cells for enclosure II are A *and* D, and for enclosure III, cells C and $\overline{D}$ are common. The minimized expression is therefore

$$X = B\overline{C} + AD + C\overline{D}$$ ❏

EXAMPLE 4-5

Minimize the following Boolean expression by use of the Karnaugh map:

$$X = A\overline{C}D + \overline{A}B\overline{C}D + \overline{A}\overline{B}D + A\overline{B}CD$$

SOLUTION

Figure 4-18 shows the Karnaugh map for the expression. Since the map is considered to be a continuous cylinder, the pairs of 1's on both edges can be

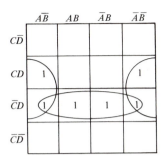

FIGURE 4-18   Essential prime implicants for Example 4-5

enclosed as an essential prime implicant, as shown. The two 1's in the center could be enclosed as a pair; however, by enclosing the group of four 1's as shown, greater minimization is achieved, since fewer variables share the four cells than share two. The minimized expression is

$$X = \overline{C}D + \overline{B}D$$ ❏

4-8
NONUNIQUE ESSENTIAL PRIME IMPLICANTS

Occasionally, we find a Karnaugh map for which more than one set of essential prime implicants exists. This situation is illustrated in Fig. 4-19. One can readily see that both maps yield Boolean expressions with three two-variable minterms. These equivalent expressions describe the same function; however, they differ in the way in which the variables are combined. Boolean functions that can be described by two or more equivalent minimized Boolean expressions, such as those shown in Fig. 4-19, are referred to as *nonunique*. On the other hand, functions that can be described by a single minimized Boolean expression are referred to as *unique*.

When a nonunique Boolean function is to be implemented with logic gates, it generally does not matter which of the possible minimized expressions is implemented. However, occasionally one expression may be preferred because certain variables or minterms are more accessible or may be available from an existing logic circuit.

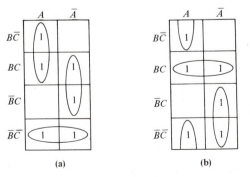

(a)　　　　　　　(b)

FIGURE 4-19   Boolean expressions from nonunique essential prime implicants: (a) $X_1 = AB + \overline{A}C + \overline{B}\,\overline{C}$; (b) $X_2 = A\overline{C} + BC + \overline{A}B$

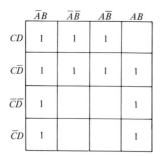

FIGURE 4-20   Karnaugh map for Example 4-6

EXAMPLE 4-6 ▬▬▬▬▬▬▬▬▬▬▬▬▬▬▬▬▬▬▬▬▬▬▬▬▬▬▬▬▬▬▬

Read out some of the possible Boolean expressions for the Karnaugh map shown in Fig. 4-20.

SOLUTION

As the problem statement implies, there are several possible combinations of implicants. Figure 4-21 shows several possible combinations of enclosures and the resulting Boolean expression. Although the expression for each map in Fig. 4-21 is valid, none are minimized, since each enclosure does not represent an

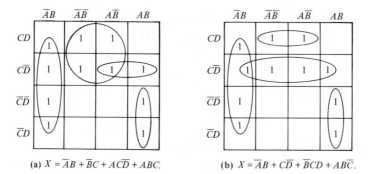

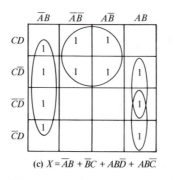

FIGURE 4-21   Various combinations of enclosures for Example 4-6: (a) $X = \overline{A}B$ $+ \overline{B}C + AC\overline{D} + ABC$; (b) $X = \overline{A}B + C\overline{D} + \overline{B}CD + AB\overline{C}$; (c) $X = \overline{A}B + \overline{B}C$ $+ AB\overline{D} + ABC$

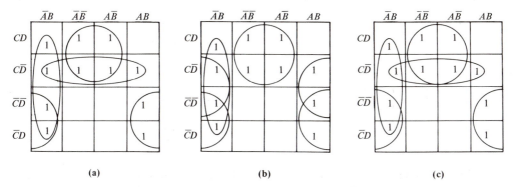

FIGURE 4-22 Possible combinations of nonunique essential prime implicants: (a) $X = \overline{A}B + \overline{B}C + \overline{C}D + B\overline{C}$; (b) $X = \overline{A}B + \overline{B}C + B\overline{D} + B\overline{C}$; (c) $X = \overline{A}B + BC + CD + BC$

essential prime implicant. Figure 4-22 shows three more possible combinations of implicants. Each set of implicants on each map in Fig. 4-22 are essential prime implicants; however, each map represents a nonunique combination. The enclosures used and the minimized Boolean expression are a matter of personal choice. As can be seen by comparing the Boolean expressions that correspond to the maps in Figs. 4-21 and 4-22, the minterms in the expressions contain fewer variables when essential prime implicants are used. ❏

4-9
DON'T-CARE STATES

In the design of combinational logic circuits we start with a written or verbal problem statement and translate this into a truth table. Using the truth table, we can list each combination of input variables that should cause a *high* output to exist. For some Boolean functions the output corresponding to certain combinations of input variables does not matter. This usually occurs because certain combinations of input variables cannot or should not exist. Also, there are times when we do not really care what value a function may take on. In both instances we call such a term a *don't-care state*. Don't-care states, which are denoted by various symbols, mean that we do not care whether the entry in a Karnaugh map corresponding to a certain combination of variables is a 1 or a 0. The symbols most frequently used to represent don't-care states are ϕ, $-$, d, or x. We shall use a d.

Don't-care states can be very important in the minimization process. Since we can assign either a 1 or 0 to a don't-care condition, we can choose whichever value will provide a larger enclosure and therefore a simpler, more economical circuit.

EXAMPLE 4-7 ▬▬▬▬▬▬▬▬▬▬▬▬▬▬▬▬▬▬▬▬▬▬▬▬

A logic circuit is to be constructed that will implement the Boolean expression

$$X = A\overline{B}C + \overline{A}B\overline{C} + \overline{A}\overline{B}\overline{C}$$

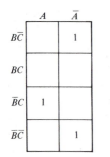

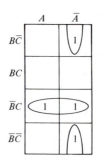

FIGURE 4-23 Karnaugh map for given expression for Example 4-7

FIGURE 4-24 Karnaugh map illustrating the use of don't-care states

Plot this expression on a Karnaugh map and reduce the expressions if the term $\overline{ABC}$ is a don't care.

SOLUTION

The expression is plotted in Fig. 4-23. Both the expression and the don't-care state are plotted in Fig. 4-24 and the essential prime implicants are enclosed. The expression corresponding to the map in Fig. 4-24 is

$$X = \overline{A}\,\overline{C} + \overline{B}C$$

Therefore, assigning a value of 1 to the don't-care state allowed us to reduce the original expression significantly.   ❏

4-10
MAPPING FIVE-VARIABLE BOOLEAN EXPRESSIONS

Five-variable Karnaugh maps contain 2^5 or 32 cells. As with the maps discussed previously, there are many ways to label the sides. The convention we shall use is shown in Fig. 4-25. The map can be thought of as two 16-cell square maps with one placed on top of the other. If this was done, the assignment of cells for variables A, B, C, and D are the same as for four-variable maps. All 16 cells in either the upper or lower plane are assigned to the fifth variable, E, while all 16

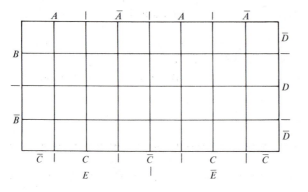

FIGURE 4-25 Five-variable Karnaugh map

cells in the other plane are assigned to $\overline{E}$. Although the map is three-dimensional, drawing the two planes side by side, as shown in Fig. 4-25, facilitates plotting Boolean expressions on the map.

Minimization of five-variable expressions is achieved as discussed previously for two-, three-, and four-variable maps. Cells whose address differs by only one variable are considered to be adjacent; therefore, cells that are at opposite edges of their 16-cell maps, or plane, are adjacent, as with four-variable maps.

EXAMPLE 4-8 ━━━━━━━━━━━━━━━━━━━━━━━━━━━━━━━━━━━

Plot the following expression on the five-variable Karnaugh map shown in Fig. 4-26.

$$X = \overline{A}C\overline{D}\overline{E} + BC\overline{D}\overline{E} + AB\overline{C}DE$$

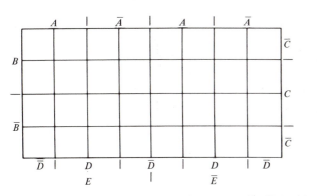

FIGURE 4-26   Five-variable Karnaugh for Example 4-8

SOLUTION

The first and second terms include an $\overline{E}$; therefore, both will be plotted on the right half of the map as though they were three-variable terms (excluding the $\overline{E}$). Both terms address two cells. The third term includes an E; therefore, it will be plotted on the left half of the map as a four-variable term (excluding the E). This term addresses only one cell, since it contains all five variables. The expression is plotted on the map in Fig. 4-27.   ❏

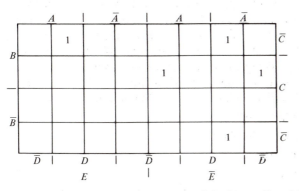

FIGURE 4-27   Karnaugh map showing cells addressed by the Boolean expression for Example 4-8

EXAMPLE 4-9

Use a five-variable Karnaugh map to minimize the Boolean expression

$$X = \overline{AB}\,\overline{CD}\,\overline{E} + \overline{A}C\overline{E} + \overline{A}B\overline{C}E + \overline{CD}E + \overline{A}B\overline{D}\overline{E} + \overline{A}BC\overline{D}E + A\overline{CD}\,\overline{E}$$

SOLUTION

The expression is shown plotted in Fig. 4-28. The minimized expression corresponding to the three essential prime implicants shown in Fig. 4-28 is

$$X = \overline{A}\,\overline{E} + \overline{CD} + \overline{A}BCD$$ ❏

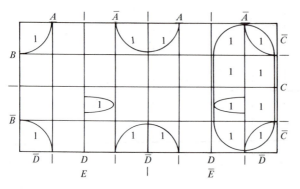

FIGURE 4-28   Karnaugh map for Example 4-9

4-11
CIRCUIT IMPLEMENTATION

The final step in the logic design process is that of *circuit implementation,* which means to interconnect basic logic gates to perform the minimized Boolean expression. Combinational logic circuits may be implemented with any of the basic logic gates or the NOT circuit; however, very often the logic circuit will have to be implemented from an existing supply of logic gates. When a company purchases large quantities of a particular logic gate, there is a price reduction; therefore, for economic reasons companies will usually purchase large quantities of a limited number of different off-the-shelf gate configurations. Although this may place a slight burden on the logic designer, it is certainly not a serious problem, since De Morgan's theorems allow us to transform an existing expression to a form suitable for the logic gates on hand. Usually, the following combinations of hardware are used in implementing logic circuits:

• AND gates, OR gates, and inverters.
• NAND gates and inverters.
• NOR gates and inverters.

Circuit implementation is accomplished in a relatively straightforward manner by working from the output of the circuit toward its input. Implementation is

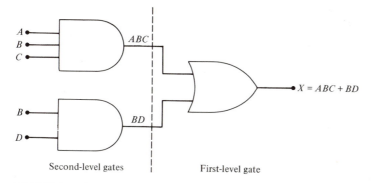

FIGURE 4-29   Example of circuit implementation of a Boolean expression

generally somewhat easier if the Boolean expression is in the sum-of-products form, such as

$$X = ABC + BD \qquad (4\text{-}5)$$

To implement a circuit that will produce this expression, the variables are first ANDed to produce the two minterms. These two terms are then ORed to produce the desired expression. If we start at the output and work toward the input, the first operation that we encounter is the OR operation, which is called a *first-level operation*. The gate that performed the first-level operation is called a *first-level gate*. Continuing toward the input, the AND operations are performed at the *second-level gates,* counting from the outputs, and are therefore called *second-level operations*. The circuit that is necessary to produce Eq. 4-5 is shown in Fig. 4-29 with levels identified. The following example illustrates the procedure.

EXAMPLE 4-10

Implement a logic circuit that will produce the Boolean expression

$$X = A\overline{B}(C\overline{D} + \overline{C}D)$$

SOLUTION

In analyzing the expression, we see that the terms $A\overline{B}$, $C\overline{D}$, and $\overline{C}D$ are ANDed. This ANDing operation is followed by an OR operation to produce the term $C\overline{D} + \overline{C}D$. The final operation is ANDing the terms $A\overline{B}$ and $C\overline{D} + \overline{C}D$. If we start at the output and work toward the input, the first-level gate is a two-input AND gate, as shown in Fig. 4-30. Second-level operations include ANDing of A and $\overline{B}$ and ORing $C\overline{D}$ and $\overline{C}D$, as shown in Fig. 4-31. Third-level operations involve the ANDing of the terms $C\overline{D}$ and $\overline{C}D$, as shown in Fig. 4-32. Fourth-

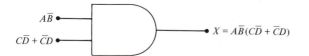

FIGURE 4-30   First-level gate for Example 4-10

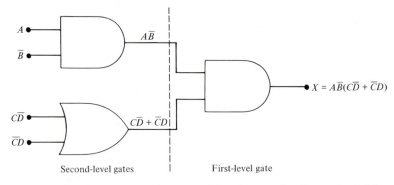

FIGURE 4-31   First- and second-level gates for Example 4-10

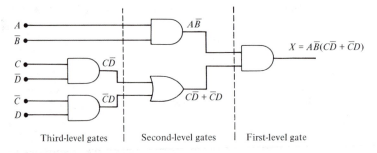

FIGURE 4-32   First-, second-, and third-level gates for Example 4-10

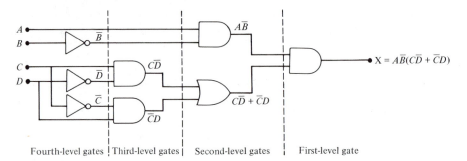

FIGURE 4-33   Complete circuit implemented for Example 4-10

level operations involve NOT circuits to obtain the necessary complemented variables, as shown in Fig. 4-33, which is the complete circuit necessary to produce the given Boolean expression. ❑

To facilitate transformation of Boolean expressions, De Morgan's theorems, which are very useful, are restated as follows:

$$\overline{A + B} = \overline{A} \ \overline{+} \ \overline{B} = \overline{A} \cdot \overline{B} \qquad \text{where } \overline{+} = \cdot$$
$$\overline{A \cdot B} = \overline{A} \ \overline{\cdot} \ \overline{B} = \overline{A} + \overline{B} \qquad \text{where } \overline{\cdot} = +$$

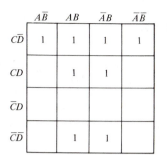

FIGURE 4-34   Karnaugh map for Example 4-11

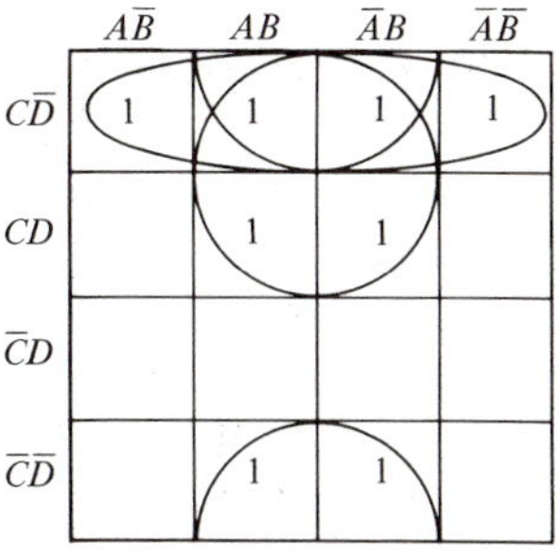

FIGURE 4-35   Karnaugh map showing prime implicants for Example 4-11

EXAMPLE 4-11 ▬▬▬▬▬▬▬▬▬▬▬▬▬▬▬▬▬▬▬▬▬▬▬▬▬

Figure 4-34 shows the map of a given problem statement. Obtain the minimized Boolean expression and implement the required circuit using only NAND gates.

SOLUTION

The prime implicants for Fig. 4-34 are shown in Fig. 4-35. The minimized expression is

$$X = BC + B\overline{D} + C\overline{D}$$

The NAND-gate-only circuit can be obtained by first using AND gates, OR gates, and inverters to implement the circuit and then converting to an all-NAND-gate circuit by using the circuits in Fig. 3-14. However, we can go directly to a NAND-gate-only circuit by first putting our expression in NAND form by reverse application of De Morgan's theorems as follows:

$$X = BC + B\overline{D} + C\overline{D}$$

$$X = \overline{\overline{BC + B\overline{D} + C\overline{D}}}$$

$$X = \overline{(\overline{BC}) \cdot (\overline{B\overline{D}}) \cdot (\overline{C\overline{D}})}$$

The transformation involves complementing each OR sign, each minterm, and the entire expression. The NAND-gate-only circuit that will implement the NAND form of the minimized expression is shown in Fig. 4-36.   ❑

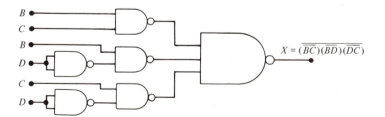

$$X = \overline{(\overline{BC})(\overline{BD})(\overline{DC})}$$

FIGURE 4-36   Logic circuit for Example 4-11 implemented with NAND gates only

4-12
MINIMIZATION OF MULTIPLE-OUTPUT CIRCUITS

Logic design problems often involve circuits in which common inputs generate multiple outputs. When dealing with such circuits, one can treat the outputs separately and perform a complete logic design procedure for each output. However, it is often possible to share intermediate terms that are common, thereby eliminating some duplication in the complete logic circuit. To appreciate the concept of *term sharing,* consider the following example.

EXAMPLE 4-12

The logic circuit illustrated in Fig. 4-37 is to have two outputs:

$$X = AB\overline{C} + \overline{A}C$$

$$Y = AB + \overline{B}C$$

Use the term sharing when designing a logic circuit that has the desired outputs.

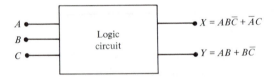

FIGURE 4-37 Multiple-output logic circuit in block diagram form

SOLUTION

The term that is common to both outputs is *AB*. Since both output expressions are the sum of two products, both expressions must appear at the output of an OR gate, as shown in Fig. 4-38. All the input expressions shown in Fig. 4-38 can be obtained with AND gates; however, *AB* is to be shared before it is ANDed with $\overline{C}$. This is shown in Fig. 4-39. The entire circuit with inputs *A, B,* and *C,* the shared terms *AB,* and the two outputs *X* and *Y* is shown in Fig. 4-40. ❑

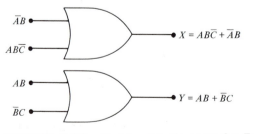

FIGURE 4-38 First-level logic gates for Example 4-12

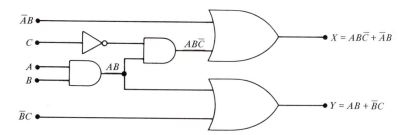

FIGURE 4-39 Partial logic circuit required to produce the Boolean expression of Example 4-12, illustrating term sharing

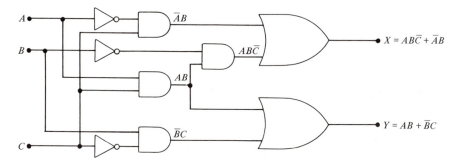

FIGURE 4-40 Complete logic circuit to produce the multiple outputs of Example 4-12

4-13
APPLICATIONS

Applying the principles and techniques that have been presented in this chapter to the basic design of combinational logic circuits presents an opportunity to deal with many interesting problems. The applications are limited only by one's imagination.

EXAMPLE 4-13 ━━━━━━━━━━━━━━━━━━━━━━━━━━━━━━━━━

During normal working hours two tellers, the supervisor, and the bank president have a key to a room holding safety-deposit boxes. After working hours the room can be opened only when the president's key and the supervisor's key are used at the same time. Any other combination of keys in use will sound an alarm. Develop a truth table and identify all combinations of variables that will sound the alarm. Plot these conditions on a Karnaugh map, and determine the minimized expression. Draw the logic circuit that corresponds to the minimized expression.

TABLE 4-5. / Truth Table for Example 4-13

T_1	T_2	S	P	X
0	0	0	0	0
0	0	0	1	1
0	0	1	0	1
0	0	1	1	0
0	1	0	0	1
0	1	0	1	1
0	1	1	0	1
0	1	1	1	1
1	0	0	0	1
1	0	0	1	1
1	0	1	0	1
1	0	1	1	1
1	1	0	0	1
1	1	0	1	1
1	1	1	0	1
1	1	1	1	1

SOLUTION

The truth table showing all combinations of variables that will activate the alarm is shown in Table 4-5. The Karnaugh map with all combinations of concern plotted is shown in Fig. 4-41. The minimized expression read from the map is

$$X = T_1 + T_2 + P\overline{S} + S\overline{P}$$

The logic circuit required to implement the expression is shown in Fig. 4-42.

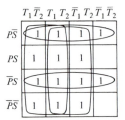

FIGURE 4-41   Karnaugh map for Example 4-13

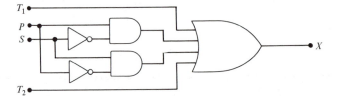

FIGURE 4-42   Logic circuit for Example 4-13

EXAMPLE 4-14 ━━━━━━━━━━━━━━━━━━━━━━━━━━━━━━━━━━━

Four loads are connected through relay contacts to a storage battery as shown in Fig. 4-43. Design a logic circuit that will provide a signal to initiate closure of relay contacts Re_5 to connect a second battery in parallel any time that three or more loads are drawing current.

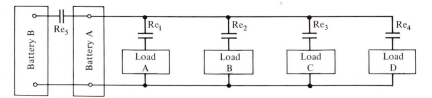

FIGURE 4-43 Electrical circuit schematic for Example 4-14

SOLUTION

The combinations of variables that will cause relay Re_5 to operate are shown in Table 4-6. The Karnaugh map with the combinations of variables that cause relay Re_5 to operate plotted is shown in Fig. 4-44. The minimized expression read from the map of Fig. 4-44 is

$$X = ABC + ABD + BCD + ACD$$

The logic circuit necessary to implement the minimized expression is shown in Fig. 4-45. ❑

TABLE 4-6. / Variables That Will Cause Re_5 to Operate

Loads				Output
A	B	C	D	X
0	0	0	0	
0	0	0	1	
0	0	1	0	
0	0	1	1	
0	1	0	0	
0	1	0	1	
0	1	1	0	
0	1	1	1	1
1	0	0	0	
1	0	0	1	
1	0	1	0	
1	0	1	1	1
1	1	0	0	
1	1	0	1	1
1	1	1	0	1
1	1	1	1	1

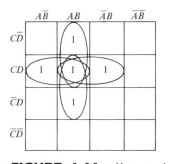

FIGURE 4-44 Karnaugh map for Example 4-14

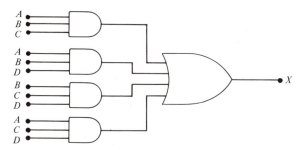

FIGURE 4-45 Minimized logic circuit for Example 4-14

EXAMPLE 4-15

A rich recluse is concerned that the maid or the butler or both are not completely trustworthy. The recluse wishes to have a security system built that will alert him if the maid or butler or both are in the library or the master bedroom without him or his trusted wife or both being present. Design a logic circuit that will provide the desired signal. Assume that the necessary input signals are available.

SOLUTION

The following symbols are used in developing Table 4-7 and identifying the essential prime implicants:

$R = 1$: recluse in library $\qquad$ $R = 0$: recluse in master bedroom

$W = 1$: wife in library $\qquad$ $W = 0$: wife in master bedroom

$M = 1$: maid in library $\qquad$ $M = 0$: maid in master bedroom

$B = 1$: butler in library $\qquad$ $B = 0$: butler in master bedroom

TABLE 4-7. / Truth Table for Example 4-15

R	W	M	B	X
0	0	0	0	
0	0	0	1	1
0	0	1	0	1
0	0	1	1	1
0	1	0	0	
0	1	0	1	
0	1	1	0	
0	1	1	1	
1	0	0	0	
1	0	0	1	
1	0	1	0	
1	0	1	1	
1	1	0	0	1
1	1	0	1	1
1	1	1	0	1
1	1	1	1	

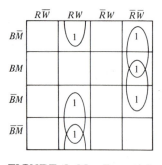

FIGURE 4-46 Essential prime implicants for Example 4-15

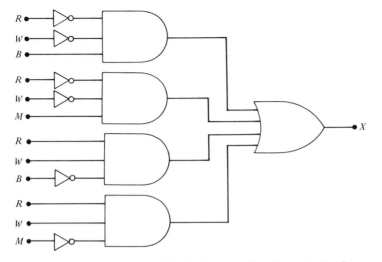

FIGURE 4-47 Required logic diagram for Example 4-15

The essential prime implicants are plotted in Fig. 4-46. The minimized Boolean expression read from the map is

$$X = \overline{R}\,\overline{W}B + \overline{R}\,\overline{W}M + RW\overline{B} + RW\overline{M}$$

The required logic circuit is shown in Fig. 4-47.   ❑

4-14
SUMMARY

A combinational logic circuit is a combination of logic elements connected in such a way that a specified output is generated for a certain combination of input variables when there is no memory element that is an integral part of the logic circuit. Combinational logic circuits may have a single output and several inputs, or they may have several outputs as well as several inputs.

The four general areas of logic design of combinational logic circuits are analysis, synthesis, minimization, and implementation. The primary tools used in the analysis of Boolean functions are algebraic expressions, truth tables, logic diagrams, and timing diagrams. Synthesis, which is the reverse of analysis, involves the development of a Boolean expression from a problem statement or from a truth table.

Karnaugh maps are the principal tool used to minimize Boolean expressions. Minimization by use of a Karnaugh map is achieved by plotting a given Boolean expression on the map, enclosing essential prime implicants, and reading out the minimized expression in the sum-of-products form.

Implementation refers to interconnecting basic logic gates in a manner necessary to perform the minimized Boolean expression. This frequently must be accomplished with existing logic gates by using De Morgan's theorems to transform the expression to the required form.

PROBLEMS

1. Write a sum-of-products and a product-of-sums equation for the following truth table.

A	B	C	X
0	0	0	0
0	0	1	0
0	1	0	1
0	1	1	1
1	0	0	0
1	0	1	1
1	1	0	0
1	1	1	1

2. Draw a Venn diagram representing all students, all students taking mathematics, all students taking electronics, and all students taking both.

3. Draw a Venn diagram representing all resistors, all composition resistors, and all wire-wound resistors.

4. Develop the truth table that corresponds to the following product-of-sums expression:

$$X = (A + B + \overline{C})(\overline{A} + B + C)(\overline{A} + \overline{B} + C)$$

5. Use the Karnaugh map to minimize the following expressions.
   (a) $X = AB + A\overline{B} + B\overline{C} + \overline{A}C$
   (b) $X = \overline{A}C + B\overline{C} + \overline{B}C + AC$

6. Use the Karnaugh map to minimize the following expressions.
   (a) $X = ABC + \overline{A}BC + A\overline{B}C + \overline{A}\overline{B}C$
   (b) $A = WX\overline{Y} + W\overline{X}Y + \overline{W}XY + W\overline{X}Y$

7. Use the Karnaugh map to minimize the following expressions.
   (a) $X = \overline{A}\overline{B}\overline{C} + ABC + A\overline{B}\overline{C} + \overline{A}B\overline{C} + \overline{A}\overline{B}C + AB\overline{C} + A\overline{B}C$
   (b) $A \cdot B + (A + B)(\overline{A} + C) + A(\overline{A} + C) = X$

8. Use the Karnaugh map to minimize the following expressions.
   (a) $X = \overline{B}\overline{C} + BD + \overline{C}D + A\overline{C} + AB + BC + \overline{A}C\overline{D} + \overline{A}BD$
   (b) $X = \overline{A}\overline{B}C\overline{D} + \overline{A}B\overline{D} + \overline{B}C\overline{D} + AB\overline{C}D + \overline{A}BD$

9. Convert the following expressions to sum-of-products form.
   (a) $\overline{\overline{A}B\ (C + \overline{D})}$
   (b) $(A + B\overline{C})(\overline{A}C + D)$

10. Use Boolean algebra theorems and the Karnaugh map to minimize the following expressions.
   (a) $X = \overline{\overline{A} \cdot \overline{B}} + ABC + \overline{A} \cdot \overline{C}$
   (b) $X = (A + \overline{B})(C + D) + A \cdot \overline{D} + A(B + \overline{C})$

11. Draw the truth table for the following circuits.

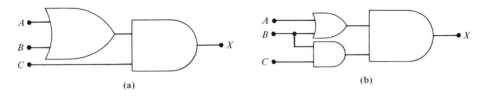

(a)   (b)

12. Use the Karnaugh map to minimize the following expressions.
(a) $X = CD\overline{E} + ABCDE + \overline{CD}\,\overline{E} + \overline{C}B + \overline{A}E + A\overline{CD}\overline{E} + ABC\overline{DE}$
(b) $X = AB\overline{C} + \overline{BC}DE + \overline{AD}\overline{E} + A\overline{C}E + \overline{ABD}\overline{E} + \overline{ABC}E + A\overline{CD}E$

13. Draw the logic diagram that is necessary to produce the outputs shown in the following figure.

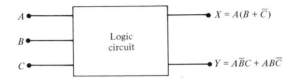

A

B — Logic circuit

C

$X = A(B + \overline{C})$

$Y = A\overline{B}C + AB\overline{C}$

14. Draw the logic circuit that is necessary to produce the expression

$$X = A\overline{B}C + AB\overline{C} + \overline{A}BC$$

15. Change the following expression to sum-of-products form and then draw the logic circuit that is necessary to product the expression in sum-of-products form.

$$X = (A + B\overline{C})(A\overline{B} + C)\,(C + \overline{B}D)$$

16. Draw a logic circuit using only NAND gates that will produce the Boolean expression

$$X = AB\overline{C} + A\overline{B}\overline{C}$$

17. Minimize the following expression by use of Boolean algebra theorems and by using a Karnaugh map. You should obtain the same, or equivalent, expressions for both techniques or you should be able to explain any differences.

$$X = AB + \overline{A}C + BC$$

18. An oil refinery has four bays from which tank trucks can be filled; however, no more than two trucks are to be filling at any one time. Design a logic circuit that will provide a signal if any three bays are in use at the same time.

19. A man has a beautiful daughter, Jerri. The man and his wife and daughter are going to spend the summer camping and fishing on a small island. Two of Jerri's suitors are going to accompany them. Their names are Walter and Charles. Jerri's father is very fond of fishing; however, he does not wish Jerri to be alone with Walter or Charles or both either at the beach or in the cabin. He intends to fish some distance offshore and does not want to divide his interest between fishing and concern for his daughter. Design a switching logic circuit for the man so that he can glance up from his fishing, close the switch for the people he sees on the beach, and observe a lighted red light if an unwanted condition exists: namely, if Jerri is alone with either or both boys without the mother being present.

20. An alarm system is needed for a holding tank that is part of an industrial process system. The alarm should be activated if low-level indicator L_2 is *low* and neither flow indicator F_A or F_B is *high*. The alarm should also be activated if high-level indicator L_1 is *high* at the same time that flow indicator F_A or F_B or both are *high*. Design the required logic circuit portion of the alarm system.

21. A large office building has an elevator system consisting of five elevators. Four of the elevators are normally available for use while the fifth is in reserve. Design a logic circuit that will provide an output signal any time three of the four elevators are in use. This signal will be used to bring the fifth elevator into standby operation. Signals are available from each of the four elevators, designated *A, B, C,* and *D.* The signal is *high* when the elevator is in use.

22. The motor pool at a college has a service station of limited pumping capacity. The station has five bays, three for cars and two for trucks. The system will handle the pumping demand for the following combinations:

(a) One car and two trucks.
(b) Two cars and one truck.
(c) Three cars.

Signals are available at each pumping station to identify a car as *C, A,* or *R* and a truck as *T* or *K.* Develop a truth table and list the minterms corresponding to when the system is being used at less than maximum capacity. Plot these combinations on a five-variable Karnaugh map and simplify if possible. Implement a logic circuit that will produce a high output when the system is being utilized at less than full capacity.

23. An alarm is needed to monitor the temperature, pressure, flow, viscosity, and liquid level of a certain solution in an industrial process plant. The alarm should be activated if the temperature or pressure is too *high* when the level is okay or if the temperature or pressure is too *high* when the viscosity is too *low* or if the temperature or pressure is too *high* when the flow rate or level is too low. Develop a truth table from the problem statement and write a Boolean expression from the truth table. Minimize the Boolean expression by use of a Karnaugh map and implement the required logic circuit. Use the following relationships to develop your truth table.

Parameter	Condition	Logic Level
Temperature	High	1
	Ok or low	0
Pressure	High	1
	Ok or low	0
Flow	OK	1
	Low	0
Viscosity	Ok	1
	Low	0
Level	Ok	1
	Low	0

REFERENCES

Culbertson, James T., *Mathematics and Logic for Digital Devices*. New York: Van Nostrand Reinhold Company, 1958.

Klenne, Stephen C., *Mathematical Logic*. New York: John Wiley & Sons, Inc., 1967.

Maley, G., and J. Earle, *The Logic Design of Transistor Digital Computers*. Englewood Cliffs, N.J.: Prentice-Hall, Inc., 1963.

Ward, Brice, *Boolean Algebra*. Indianapolis, Ind.: Howard W. Sams & Company, Inc., 1971.

IC Logic Families and Characteristics

INSTRUCTIONAL OBJECTIVES

In this very important chapter we describe today's most widely used digital logic families and discuss their characteristics. After completing the chapter, you should be able to

1. List logic families categorized as bipolar.
2. List logic families categorized as unipolar.
3. Define or describe the following terms:
    (a) SSI.
    (b) MSI.
    (c) LSI.
    (d) VLSI.
    (e) Unit load.
    (f) Fan-out.
    (g) Noise margin.
    (h) Propagation delay.
    (i) Wired logic.
    (j) Speed–power product.
4. Draw the electrical schematic for a TTL NAND gate and describe its operation.
5. List the members of the TTL logic family.
6. List the members of the MOS logic family.
7. Compare the most important characteristics of the various logic families.
8. List an advantage and a disadvantage of CMOS as compared with TTL.
9. Determine the fan-out for a gate that is driving gates of another series.
10. Describe a tri-state device and discuss its most important application.
11. Describe how interfacing is achieved between TTL and CMOS devices.
12. Describe the difference between series A and B in the CMOS 4000 family.
13. Describe the effect, if any, which the operating frequency of a CMOS device has on the total power dissipated by the device.

SELF-EVALUATION QUESTIONS

The following questions deal with the material presented in this chapter. Read the questions prior to studying the material and, as you read the chapter, watch for

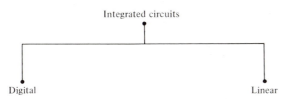

FIGURE 5-1 Classification of integrated circuits by function

the answers to the questions. After completing the chapter, return to this section and evaluate your comprehension of the material by answering the questions again.

1. What is the most widely used bipolar logic family?
2. How many gates must a digital circuit contain to be categorized as LSI?
3. Define the term *unit load*.
4. Which TTL series has the most desirable speed–power product?
5. List two applications where CMOS is widely used.
6. What circuit parameters affect the power dissipation of a CMOS device?
7. What logic family has the shortest propagation delay?
8. What is the primary application of tri-state devices?

5-1
INTRODUCTION

Integrated circuits (ICs) are miniature, low-cost electronic circuits whose components are fabricated on a single, continuous piece of semiconductor material and interconnected in such a way that they perform a high-level function. Such circuits, first introduced in 1958, have had a profound impact on the electronics industry; in fact, both *Business Week* and *Scientific American* have referred to the effect of integrated circuits as a "second industrial revolution."

Integrated circuits are generally categorized, according to the function that they perform, as either digital or linear ICs, as shown in Fig. 5-1. ICs can also be categorized according to the level of complexity of the IC as involving small-scale integration (SSI), medium-scale integration (MSI), large-scale integration (LSI), very large scale integration (VLSI), or ultra large scale integration (ULSI), on which work is being done today. The category in which an IC falls depends on the number of gates it contains, as shown in Table 5-1.

TABLE 5-1. / Categorization of ICs by Number of Gates

Category	Number of Gates
SSI	Fewer than 12
MSI	12 to 99
LSI	100 to 9999
VLSI	10,000 or more

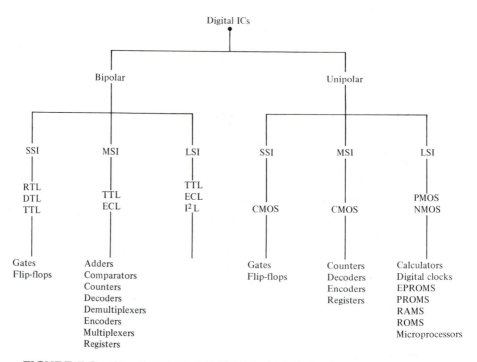

FIGURE 5-2   Classification of digital integrated circuits

Digital ICs can be further categorized, according to the manufacturing process, as either bipolar or unipolar (MOS). Figure 5-2 shows this categorization as well as the major logic families in conjunction with the levels of integration involved.

5-2
MANUFACTURING, PACKAGING, AND NUMBERING INTEGRATED CIRCUITS

Most ICs are fabricated on a single piece of semiconductor material and are referred to as *monolithic*. Monolithic ICs are categorized as either bipolar monolithic ICs or unipolar monolithic ICs. *Bipolar* ICs are those devices whose active components are current controlled. An example of a bipolar device is a bipolar junction transistor. *Unipolar* ICs are those devices whose active components are voltage controlled. An example of a unipolar device is a metal-oxide-semiconductor field-effect transistor (MOSFET).

The semiconductor wafer on which an integrated circuit is fabricated is extremely small, generally in the range of about 1 to 5 mm. After an IC is fabricated, it is packaged. Packaging serves several purposes, including protecting the chip from mechanical damage and chemical contamination as well as providing a completed unit large enough to handle and to which electrical connections can be made. The three most frequently used packages for ICs are axial lead (TO5), flat pack, and dual-in-line packages (DIPs), which are shown in Fig. 5-3. Of the three, dual-in-line packages are the most widely used. DIPs are manufactured in

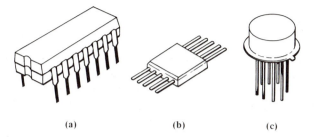

(a)　　　　　(b)　　　　　(c)

FIGURE 5-3 Most frequently used IC packages: (a) dual-in-line; (b) flat pack; (c) TO-5

a wide assortment of pin numbers ranging from 8 to 40. SSI packages usually have 8, 14, or 16 pins; MSI packages have 14, 16, or 24 pins; and LSI packages have 24, 28, or 40 pins. The least expensive package material is molded plastic, such as epoxy, resin, or silicone, which is used in virtually all SSI packages. Some MSI and LSI packages are made of ceramic because it has better thermal dissipation capabilities.

In order to check the electrical or mechanical characteristics of ICs, or to purchase them, it is necessary to know the part number, since there are many manufacturers of ICs in the United States as well as in other countries. Each manufacturer assigns a manufacturer's code number to the devices it makes. The complete identification system used by most manufacturers is an alphanumeric code. The first one to three symbols of the code are usually alphabetic characters that identify the manufacturer. This is followed by a group of four to seven alphanumeric characters that describe the device itself. When ordering ICs, the manufacturer's code and the device number should be suffixed with a single-letter code that specifies the type of package. Several manufacturers' codes, device numbers, and packaging codes are listed in Table 5-2.

TABLE 5-2. / Device Nomenclature for 7400 TTL Devices

Manufacturer	Prefix	Series		Packaging	
National Semiconductor	DM	Standard (no letter)		Epoxy-molded DIP	N
Texas Instruments	SN	Low power	L	Ceramic DIP	J
Motorola	MC	High speed	H	Glass/metal DIP	D
Intersil	IM	Clamped Schottky	S	Flat pack	W
Intel	P	Low-power Schottky	LS		
Fairchild	F				
Harris	H	**Device**		**Device Number**	
Signetics	N				
Advanced Micro Devices	AM	Quad two-input NAND gate		7400	
Monolithic Memories	MM	HEX inverter		7404	
Sprague	US	Quad two-input NOR gate		7402	
		Quad two-input AND gate		7408	
		Quad two-input OR gate		7432	
		Dual four-input NAND buffer		7440	
		BCD-to-decimal decoder		7442	
		Dual *D* flip-flop		7474	
		Quad Exclusive-OR gate		7486	

5-3
DIGITAL IC TERMINOLOGY

Although there are many manufacturers of digital ICs and several logic families, a fair amount of the terminology associated with digital ICs is somewhat standardized among manufacturers and between logic families. Some of the most common and most useful terms are discussed in the following paragraphs. These are parameters that logic designers consider when selecting logic devices or are used by people involved in analyzing or troubleshooting digital circuits.

5-3.1   Voltage and Current Levels

The symbols and definitions commonly used by digital IC manufacturers for logic levels and associated currents are listed in Table 5-3.

5-3.2   Fan-In and Fan-Out

The terms ''fan-in'' and ''fan-out'' are used to indicate the number of inputs or outputs that can be connected to a logic gate. The *fan-in*, or *input load factor,* of a gate is the number of input signals that can be connected to a gate without causing it to operate outside its intended operating range. Fan-in is expressed in terms of *standard inputs* or *unit loads* (ULs). A fan-in of 8 means that eight unit loads can safely be connected to the gate inputs. The eight unit loads correspond to eight separate input signals only if each separate input signal represents one unit load. To determine the fan-in (input current capabilities) for a gate, divide its HIGH-state unit load by the current value for one HIGH-state unit load, and its LOW-state current capabilities by the current value for one LOW-state unit load. The lower result will be the fan-in capabilities for the gate.

The *fan-out,* or *output load factor,* of a logic gate is the maximum number of inputs that can be driven by a logic gate. A fan-out of 10 means that 10 unit loads can be driven by the gate while still maintaining its output voltage within specifications for logic levels 0 and 1. To determine the fan-out for a gate, divide its HIGH-state current capabilities by the current value for one HIGH-state unit load, and its LOW-state current capabilities by the current value for 1 LOW-state unit load. The lower result will be the fan-out capabilities for the gate.

TABLE 5-3. / Voltage and Current Symbols and Definitions

Symbol	Definition
V_{IH}	HIGH-state input voltage, corresponding to logic 1 at an input.
V_{IL}	LOW-state input voltage, corresponding to logic 0 at an input.
V_{OH}	HIGH-state output voltage, corresponding to logic 1 at the output.
V_{OL}	LOW-state output voltage, corresponding to logic 0 at the output.
I_{IH}	HIGH-state input current: the current flowing into an input when the input voltage corresponds to logic 1.
I_{IL}	LOW-state input current: the current flowing from an input when the input voltage corresponds to logic 0.
I_{OH}	HIGH-state output current: the current flowing from the output when the output voltage corresponds to logic 1.
I_{OL}	LOW-state output current: the current flowing into an output when the output voltage corresponds to logic 0.

EXAMPLE 5-1 ━━

A unit load for some particular logic family is as follows:

$$1 \text{ UL} = \begin{cases} 50 \text{ } \mu\text{A} & \text{HIGH state} \\ 1 \text{ mA} & \text{LOW state} \end{cases}$$

Determine the fan-in and fan-out for a gate in this family that has the following parameters:

$$I_{OH} = 400 \text{ } \mu\text{A}$$

$$I_{OL} = 10 \text{ mA}$$

$$I_{IH} = 150 \text{ } \mu\text{A}$$

$$I_{IL} = 4 \text{ mA}$$

SOLUTION

The fan-in is computed as

$$\text{fan-in} = \frac{\text{HIGH-state curent}}{\text{HIGH-state unit load}}$$

$$= \frac{150 \text{ } \mu\text{A}}{50 \text{ } \mu\text{A}} = 3 \text{ UL}$$

$$= \frac{\text{LOW-state current}}{\text{LOW-state unit load}}$$

$$= \frac{4 \text{ mA}}{1 \text{ mA}} = 4 \text{ UL}$$

The lower of the two unit load values is the fan-in; therefore,

$$\text{fan-in} = 3$$

The fan-out is computed as

$$\text{fan-out} = \frac{\text{HIGH-state current}}{\text{HIGH-state unit load}}$$

$$= \frac{400 \text{ } \mu\text{A}}{40 \text{ } \mu\text{A}} = 8 \text{ UL}$$

$$= \frac{\text{LOW-state current}}{\text{LOW-state unit load}}$$

$$= \frac{10 \text{ mA}}{1 \text{ mA}} = 10 \text{ UL}$$

The lower of the two unit load values is the fan-out; therefore,

$$\text{fan-out} = 8 \text{ UL} \qquad \qquad ❑$$

5-3.3   Pulse Parameters

High-speed logic circuitry is generally pulse operated; therefore, terminology and parameters associated with pulse trains are frequently encountered. The pulse parameters of primary interest to us are

1. Rise and fall times.
2. Pulse width and amplitude.
3. Ringing.
4. Overshoot and undershoot.

Rise Time and Fall Time.   *Rise time* is defined as the time required for a pulse to increase from 10 percent of its maximum amplitude to 90 percent of its maximum amplitude, disregarding ringing, overshoot, or undershoot. The 10 and 90 percent levels are shown in Fig. 5-4. *Fall time* is the time required for a pulse to decrease from 90 percent to 10 percent of its maximum. Rise and fall times for pulses in digital circuitry are generally very short (in the nanosecond region); nonetheless, some finite time is required for the transition from minimum to maximum amplitude, or vice versa. Pulse amplitudes rise or fall exponentially due to capacitive effects.

Pulse Width.   Pulse width is the time duration of a pulse measured at 50 percent of the maximum amplitude, as shown in Fig. 5-4. This time duration is used in calculations involving the period or frequency of pulse trains.

Pulse Amplitude.   Pulse amplitude is the maximum steady-state pulse height. Pulse amplitude should be measured at a point on the waveform where overshoot or ringing is not a factor; therefore, such measurements are generally made toward the trailing edge of the waveform.

Overshoot and Undershoot.   When a voltage changes abruptly from a LOW state to a HIGH state, the sudden transition may cause the amplitude to rise beyond an eventual steady-state level. The added amplitude beyond the steady-state level is called *overshoot*. The same phenomenon occurs during the abrupt transition from

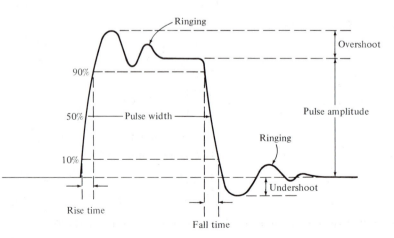

FIGURE 5-4   *Identification of pulse parameters for a nonideal pulse*

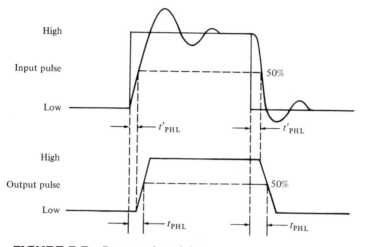

FIGURE 5-5 Propagation delay

the HIGH to LOW state and is called *undershoot* or *trailing-edge overshoot*. Both overshoot and undershoot, which are due to distributive capacitances and inductances, are shown in Fig. 5-4.

Ringing. As the energy associated with overshoot or undershoot is dissipated, the pulse amplitude decreases to its steady-state HIGH-state value or to zero amplitude. In the process of reaching a steady-state value, the amplitude experiences a damped oscillatory motion called *ringing,* shown in Fig. 5-4.

Propagation Delay. A pulse always experiences some delay in passing through a logic gate. The delay is called *propagation delay*. It is directly related to the rise and fall times shown in Fig. 5-4 and is defined as follows:

- t_{PLH}: delay time in going from logic 0 to logic 1.
- t_{PHL}: delay time in going from logic 1 to logic 0.

The relationship between rise and fall times and propagation delay is illustrated in Fig. 5-5. The input pulse shows a nonideal pulse superimposed over an ideal square wave. As a result of the rise time associated with the nonideal pulse there is a delay designated as t'_{PLH}. When the nonideal pulse reaches 50 percent of its maximum amplitude, the logic-gate output goes from a LOW state to a HIGH state, but due to the rise time of the output pulse there is additional delay. The total leading-edge delay, designated as t_{PLH} in Fig. 5-5, is called *turn-on delay*. The total delay at the trailing edge is called *turn-off delay*. In many logic gates, turn-on and turn-off delays are not equal and the larger of the two is the "worst-case" propagation delay.

Propagation delay is an important logic circuit parameter because the speed, or frequency, at which circuits can operate, is affected by it. High-speed logic circuits have short propagation delays, whereas low-speed circuits have larger delays.

5-3.4   Noise Immunity

Noise immunity is the ability of a logic gate to tolerate fluctuations of the voltage levels at its input and output terminals. Such fluctuations are called *noise* and are

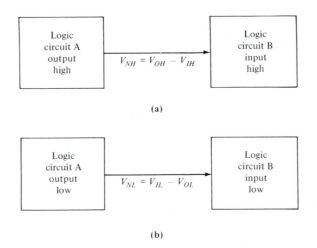

(a)

(b)

FIGURE 5-6   Noise margin

due to a variety of causes, including magnetic coupling of voltages from nearby circuits, radiated signals, and poorly filtered dc power supplies. The term *noise margin* permits us to describe quantitatively the immunity of a logic gate to noise, where noise margin is expressed as either

$$V_{NH} = V_{OH} - V_{IH}$$

or

$$V_{NL} = V_{IL} - V_{OL}$$

where

V_{NH} = HIGH-state noise margin
V_{NL} = LOW-state noise margin
V_{IL} = LOW-state input voltage
V_{IH} = HIGH-state input voltage
V_{OL} = LOW-state output voltage
V_{OH} = HIGH-state output voltage

The concept of noise margin is illustrated in Fig. 5-6. The HIGH-state output voltage (V_{OH}) of logic circuit A should be interpreted by logic circuit B as a logic 1. To help ensure that the voltage level is interpreted properly, there is a margin between the *minimum* value of V_{OH} and the *minimum* value of V_{IH}, as shown in Fig. 5-7. This margin was described earlier as the HIGH-state noise margin. Noise on the HIGH-state output pulse may cause this margin to be exceeded, as shown in Fig. 5-8, thus causing the gate to which the pulse is being applied to interpret

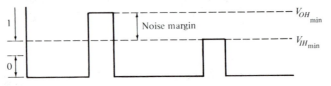

FIGURE 5-7   High-state noise margin

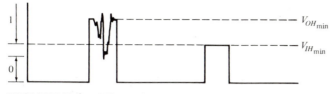

FIGURE 5-8   Effect of noise on digital signal

FIGURE 5-9   Low-state noise margin

the pulse as a logic 0 during the time interval in which the noise margin is exceeded.

In Fig. 5-6b, the LOW-state output of logic circuit A should be interpreted by logic circuit B as a logic 0. To help ensure that the voltage level is interpreted correctly, there is a margin between *maximum* value of V_{OL} and the *maximum* value of V_{IL}, as shown in Fig. 5-9. This margin was earlier termed the LOW-state noise margin. Noise on the LOW-state output pulse may cause this margin to be exceeded, thus causing the LOW-state pulse to be misinterpreted as a logic 1.

EXAMPLE 5-2 ▬▬▬▬▬▬▬▬▬▬▬▬▬▬▬▬▬▬▬▬▬▬▬▬▬▬

For a certain digital IC, calculate the noise margins for the input and output voltage parameters listed in Table 5-4.

SOLUTION

HIGH-state noise margin:

$$V_{NH} = V_{OH_{\min}} - V_{IH_{\min}}$$

$$= 2.8 \text{ V} - 2.0 \text{ V} = 0.8 \text{ V}$$

TABLE 5-4. / Noise Margins

	Minimum	Typical	Maximum
V_{OH}	2.8 V	3.6 V	
V_{OL}		0.2 V	0.4 V
V_{IH}	2.0 V		
V_{IL}			0.8 V

LOW-state noise margin:

$$V_{NL} = V_{IL_{\max}} - V_{OL_{\max}}$$

$$= 0.8 \text{ V} - 0.4 \text{ V} = 0.4 \text{ V}$$ ❏

The noise margins calculated in Example 5-2 are termed *dc noise margins,* which may seem somewhat inappropriate, since we are dealing with noise, which is generally considered to be an ac signal superimposed on a pulse train. However, when working with high-speed logic circuits, pulses 1 or 2 μs in duration may be treated as dc for logic circuits capable of responding to nanosecond-duration pulses.

5-3.5 Power Dissipation

Power dissipation is the amount of power, generally measured in milliwatts per gate, that an IC dissipates in the form of heat. Usually, power measurements are made with half of the circuits on the chip in the HIGH state and the other half in the LOW state.

Power dissipation is an important consideration for several reasons, including the fact that it increases the temperature of the ICs as well as the temperature of the entire electronic package. Since excessive heat is one of the primary causes of failure in electronic circuits, we can readily see that a reduction in power dissipation can reduce repair costs.

5-3.6 Current Source and Current Sink

The current at the output of a logic gate will flow either out of or into the gate, depending on its logic level. If a gate output is at logic 1, it provides current to the gate that it is driving, as shown in Fig. 5-10a; therefore, the driving gate is referred to as a *current source.*

If the output of a logic gate is at logic 0, current from the input of gates being driven low flows into the output of the driving gate as shown in Fig. 5-10b. The driving gate is therefore referred to as a *current sink.*

The amount of current that a logic gate can source or sink is generally expressed in unit loads and is an important consideration as we discuss the various logic families.

5-3.7 Speed–Power Relationship

An ideal logic gate dissipates no power and changes states instantaneously; therefore, the product of speed times power for an ideal gate is zero. Although a *speed–power product* of zero cannot be achieved with actual logic gates, a low-speed–power product is attainable and desirable. Logic gates are often compared

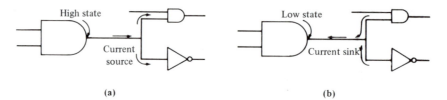

(a)   (b)

FIGURE 5-10   Logic gate acting as (a) current source; (b) current sink

on the basis of their speed–power product, where a lower product is generally more desirable.

5-4
IC DATA SHEETS

All manufacturers of integrated circuits print data sheets for each type of IC they manufacture. The purpose of the data sheets is to provide the user with information about the IC. Typical data sheets provide the following kinds of information:

- Part number.
- Pin assignments.
- Electrical and mechanical specifications.
- Logic function and logic diagram.
- Maximum ratings.

Data sheets for specific ICs in the various logic families to be discussed are presented in the following paragraphs.

5-5
EVOLUTION OF IC LOGIC FAMILIES

The remainder of this chapter deals with specific IC logic families that are in wide use in today's modern equipment. During the past quarter century digital principles have not changed appreciably, although digital technology has changed at an incredible rate.

Early logic gates, such as those discussed in Chapter 2, were implemented with diodes and resistors and were called *diode-logic circuits*. Such circuits were simple and easy to implement but had a serious shortcoming: The input voltage levels deteriorated with each succeeding diode due to the inherent voltage drop across a diode.

Diode logic gave way to *resistor–transistor logic* (RTL). The basic RTL gate was a NOR gate, as shown in Fig. 5-11. RTL also had several shortcomings,

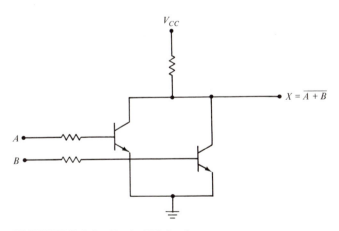

FIGURE 5-11 Basic RTL logic gate

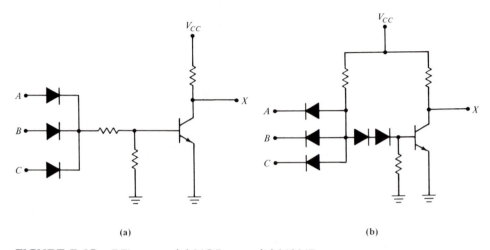

(a)   (b)

FIGURE 5-12   *DTL gates: (a) NOR gate; (b) NAND gate*

including limited fan-in, fan-out capabilities. In addition, the input resistance of RTL gates changed as inputs were added or removed and was quite low when several input signals were connected.

However, despite its limitations, RTL has the distinction of being the first logic family to be used in integrated circuits. Beyond that distinction from a historical vantage point, little more needs to be said with regard to RTL, since it is obsolete and finds few applications.

The next-generation logic family following RTL was *diode–transistor logic* (DTL). Two examples of DTL gates are shown in Fig. 5-12. DTL offered several advantages over the logic families that preceded it, including increased fan-in and fan-out capabilities and a higher-value, nearly constant, input resistance. Although rarely used in new design, DTL provided an important link in the evolution process, which brought us to where we are today in modern high-speed digital circuits. Most digital circuitry of recent design utilizes IC chips belonging to one or more of the following very popular families:

   1. TTL: transistor–transistor logic.
   2. ECL: emitter-coupled logic.
   3. IIL: integrated injection logic.
   4. MOS ICs: metal-oxide-semiconductor ICs.

As would be expected, each logic family has certain advantages and disadvantages compared with the other IC families. There are many important factors, such as speed, power dissipation, noise immunity, and cost, to consider when selecting an IC. To provide a basis for comparison, each of these logic families is discussed in some detail in the following paragraphs.

5-6
TRANSISTOR–TRANSISTOR LOGIC

The continuing quest for higher speed and reduced power dissipation in logic circuits lead to the development of *transistor–transistor logic* (TTL). Also called

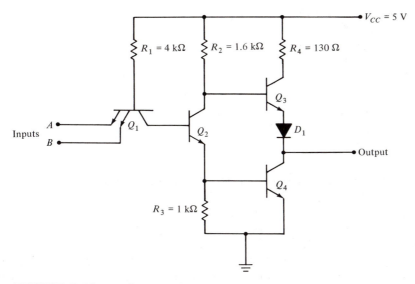

FIGURE 5-13 Basic TTL NAND gate

T^2L, TTL is the most popular and most widely used IC logic family. Its popularity is due primarily to its very low cost and the wide variety of SSI logic elements and MSI functional circuits that are available. Its ease of use, high performance characteristics, and interfacing capabilities also contribute to its popularity. Although TTL ICs have been available commercially for several years, they continue to be widely used in new design work. In fact, TTL has so dominated the industry that one of the major selling points for devices in other logic families is their compatibility with TTL devices.

The basic TTL gate is a NAND gate, as shown in Fig. 5-13. If we compare Fig. 5-12b with Fig. 5-13, we see some similarities. Two significant modifications are the replacement of the diodes at the inputs of the DTL NAND gate with the multiple-emitter transistor in the TTL gate and the addition of the push-pull output stage (generally called a totem-pole configuration) in the TTL gate. The introduction of the multiple-emitter transistor reduced costs and IC chip space while permitting TTL gates to retain the high-speed switching characteristics associated with DTL. The totem-pole configuration at the output provides excellent fan-out capabilities and high-speed switching.

5-7
THEORY OF OPERATION OF THE BASIC TTL NAND GATE

With regard to logic levels, the basic TTL NAND gate obeys the same logic rules as those discussed previously for NAND gates; these are summarized in Table 5-5. As can be seen in this truth table, if both inputs are at logic 0, the output is at logic 1. The circuit operates as follows. Logic 0 at either input causes transistor Q_1 to be biased ON and operating at saturation. With Q_1 operating at saturation, its collector voltage will be LOW; therefore, transistors Q_2 and Q_4 will be biased OFF. When Q_2 is OFF, its collector voltage is HIGH; therefore, Q_3 is biased ON and functions as an emitter follower. Current flows from the 5-V source through

TABLE 5-5. / NAND-Gate Truth Table

Inputs		Output
A	B	X
0	0	1
0	1	1
1	0	1
1	1	0

Q_3, through forward-biased diode D_1, and to the load; thus the gate is acting as a *current source*, as shown in Fig. 5-14.

Referring again to Table 5-5, when both inputs are at logic 1, we see that the output is at logic 0. With both inputs HIGH, the circuit operates as follows. Transistor Q_1 is biased OFF; therefore, its collector voltage is sufficiently positive to bias Q_2 into saturation. Current flow through the 1-kΩ resistor develops a voltage drop that biases Q_4 into saturation. The function of diode D_1 comes into play under these conditions. With transistors Q_2 and Q_4 biased ON, the collector voltage of Q_2 will be at approximately 1.2 V. This voltage would turn Q_3 ON if D_1 were not in the circuit. The combined voltage drop across the emitter-to-collector of Q_4 and across D_1 reverse-biases the base-to-emitter junction of Q_3, thus keeping it turned OFF. With Q_4 biased ON, current flows from the load through Q_4 to ground; thus the gate is acting as a *current sink,* as shown in Fig. 5-15. Typical values of current that TTL gates can source or sink are listed in Table 5-6.

5-8
SIGNIFICANCE OF THE TOTEM-POLE OUTPUT CIRCUIT

The truth table shown in Table 5-5 would be obtained if transistor Q_3 and diode D_1 were removed from the circuit of Fig. 5-13, R_4 connected directly to Q_4, and the operation of the circuit observed. However, power dissipated by the gate would increase appreciably due to increased current through R_4 and Q_4 when the gate

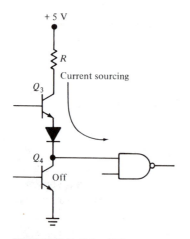

FIGURE 5-14   TTL current sourcing

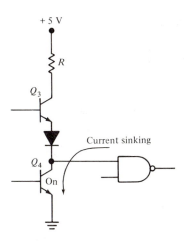

FIGURE 5-15 TTL current sinking

output is LOW. With Q_3 in the circuit, there is no current through R_4 when the gate output is LOW.

The totem-pole configuration also provides a desirable circuit characteristic when the gate output is HIGH. In this state, Q_3 is operating as an emitter follower. Such circuits have a very low output impedance. This low output impedance permits any capacitive load to be charged very rapidly due to the small RC time constant. This action, which is usually called *active pull-up*, permits TTL gates to have a very fast transition from the LOW state to the HIGH state.

5-9
THE TTL LOGIC FAMILY

In 1964 Texas Instruments introduced the TTL product line, which has since been expanded to encompass an entire family or series of logic devices. These are presently available from a large number of manufacturers, which use a rather standardized identifying numbering system designated as the 5400/7400 series. The TTL family includes a wide variety of logical and functional characteristics. To provide greater flexibility with regard to speed and power dissipation considerations, the following subfamilies have been developed:

- 7400 standard series.
- 74L00 low-power series.
- 74H00 high-speed series.
- 74S00 Schottky series.
- 74LS00 low-power Schottky series.

TABLE 5-6. / Summary of TTL Sink–Source Capabilities

Driving Gate		Driven Gate	
Output LOW	Output HIGH	Input LOW	Input HIGH
16-mA sink	400-μA Source	1.6 mA	40 μA

**TABLE 5-7. / Partial List of Logic Devices
in the Standard TTL Logic Series**

Manufacturer's Number	Description
7400	Quad two-input NAND gate
7401	Quad two-input NAND gate (open collector)
7402	Quad two-input NOR gate
7403	Quad two-input NOR gate (open collector)
7404	Hex inverter
7408	Quad two-input AND gate
7410	Triple three-input NAND gate
7420	Dual four-input NAND gate
7430	Eight-input NAND gate
7432	Quad two-input OR gate
7437	Quad two-input NAND buffer
7440	Dual four-input NAND buffer
7442	BCD-to-decimal converter
7486	Quad Exclusive-OR gate

Within each subfamily a wide range of IC logic chips is available. A list of some of the logic devices that are available in the standard 7400 series is provided in Table 5-7. Although these devices are available with either a 5400 number or a 7400 number, our primary interest will lie with the 7400 numbers. The only difference between the two is the wider operating temperature range and power supply voltage range of the 5400 series, since they are meant for military use.

5-10
CHARACTERISTICS OF STANDARD TTL

The standard TTL family is the most popular logic family ever developed. It has been the most readily available and widely used logic family for a number of years; however, it is presently receiving a serious challenge from the low-power Schottky series. The recommended operating conditions, principal electrical characteristics, and switching parameters are listed in Table 5-8.

5-10.1   Noise Margins

Inspection of the electrical characteristics in Table 5-8 shows a guaranteed maximum logic 0 output (V_{OL}) of 0.4 V and a guaranteed maximum logic 0 input (V_{IL}) of 0.8 V. Using these values, the LOW-state dc noise margin is computed as

$$V_{NL} = V_{IL} - V_{OL}$$

$$= 0.8 \text{ V} - 0.4 \text{ V} = 400 \text{ mV}$$

Similarly, the table shows a guaranteed minimum logic 1 output (V_{OH}) of 2.4 V and a guaranteed minimum logic 1 input (V_{IH}) of 2.0 V. Using these values, the HIGH-state dc noise margin is computed as

$$V_{NH} = V_{OH} - V_{IH}$$

$$= 2.4 \text{ V} - 2.0 \text{ V} = 400 \text{ mV}$$

TABLE 5-8. / Standard TTL AND-Gate Characteristics and Operating Conditions

Recommended Operating Conditions

Parameter	5400			7400			Units
	Min.	Typ.	Max.	Min.	Typ.	Max.	
Supply voltage (V_{CC})	4.5	5.0	5.5	4.75	5.0	5.25	V
Temperature range	−55	25	125	.0	25	70	°C
Fan-out			10			10	UL

Electrical Characteristics

Symbol			Min.	Typ.	Max.	Units
V_{IH}	Logic 1 input voltage	$V_{CC} = 4.75$ V	2.0			V
V_{IL}	Logic 0 input voltage	$V_{CC} = 4.75$ V			0.8	V
V_{OH}	Logic 1 output voltage	$V_{CC} = 4.75$ V, $V_{in} = 0.8$ V, I_o $= -400$ μA	2.4	3.6		V
V_{OL}	Logic 0 output voltage	$V_{CC} = 4.75$ V, $V_{in} = 2.0$ V, I_o $= 15$ mA		0.1	0.4	V
I_{IH}	Logic 1 input current	$V_{CC} = 5.25$ V, $V_{in} = 2.4$ V			4.0	μA
I_{IH}	Logic 1 input current	$V_{CC} = 5.25$ V, $V_{in} = 0.4$ V			1	mA
I_{IL}	Logic 1 input current	$V_{CC} = 5.25$ V, $V_{in} = 0.4$ V			−1.6	mA
I_{OS}	Output short-circuit current	$V_{CC} = 5.25$ V, $V_{in} = 0$ V, V_o $= 0$ V	−20		−55	mA
I_{CCH}	Supply current, logic 1 output	$V_{CC} = 5.25$ V, $V_{in} = 0$ V		1	1.8	mA
I_{CCL}	Supply current, logic 0 output	$V_{CC} = 5.25$ V, $V_{in} = 5.0$ V		3	5.1	mA

Switching Characteristics

Symbol			Min.	Typ.	Max.	Units
t_{PLH}	Turn-off propagation delay	$V_{CC} = 5.0$ V		11	22	ns
t_{PHL}	Turn-on propagation delay	$V_{CC} = 5.0$ V		7	15	ns

The calculations above are for *guaranteed worst-case* dc noise margins for the standard 7400 TTL series, which are 400 mV in both cases. *Actual* dc noise margins are somewhat higher, typical values being $V_{NL} = 1$ V and $V_{NH} = 1.6$ V.

5-10.2 Propagation Delay

In general, turn-off (t_{PLH}) and turn-on (t_{PHL}) propagation delays have different values, with both dependent on the loading conditions. The basic NAND gate of the standard TTL series has typical turn-on and turn-off propagation delays of 7 and 11 ns, respectively, as shown in Table 5-8. Using these values, we obtain an *average* propagation delay of 9 ns. Average propagation delay provides us with a value for comparing the relative speed of logic gates. For example, a circuit using logic gates with an average propagation delay of 10 ns is said to be faster than a circuit using logic gates with an average propagation delay of 15 ns.

5-10.3   Power Dissipation

The power dissipated by a logic gate can be measured when the output is either HIGH or LOW. It is necessary to obtain values for both states to determine how much power must be supplied by the power supply. The current drawn by a standard TTL NAND gate connected to a 5-V power supply is 3 mA when the gate output is LOW. This LOW-state supply current (I_{CCL}) represents a power dissipation given as

$$P_{D_L} = V_{CC}I_{CCL}$$

$$= (5 \text{ V})(3 \text{ mA}) = 15 \text{ mW}$$

When the gate output is HIGH, the power supply delivers 1 mA to it; therefore, the power dissipation is

$$P_{D_H} = V_{CC}I_{CCH}$$

$$= (5 \text{ V})(1 \text{ mA}) = 5 \text{ mW}$$

If we assume that the gate output is HIGH and LOW equal amounts of time, we can compute the *average power dissipation* as

$$P_{D_{AVE}} = \frac{P_{D_H} + P_{D_L}}{2}$$

$$= \frac{5 \text{ mW} + 15 \text{ mW}}{2} = 10 \text{ mW}$$

5-11
CHARACTERISTICS OF LOW-POWER TTL

Low-power TTL is designated as the 74L00 series. The circuit for low-power TTL is essentially the same as for standard TTL except that resistor values have been *increased*. The larger resistance values decrease power consumption by decreasing circuit current, but there is a corresponding decrease in speed because of the increased resistance values. A typical low-power series NAND gate dissipates 1 mW of power and has a propagation delay of 33 ns.

Low-power TTL gates have a fan-out of 10 other low-power TTL gates but will only drive one standard series TTL gate.

The 74L00 series is well suited for applications such as battery-powered calculators, where power dissipation is of greater concern than speed; however, CMOS logic devices are providing increasingly stiff competition for the low-power TTL series for such applications due to their low power consumption.

5-12
CHARACTERISTICS OF HIGH-SPEED TTL

High-speed TTL is designated as the 74H00 series. The circuit for high-speed TTL is essentially the same as for standard TTL except that resistor values have been *decreased* and transistor Q_4 has been replaced with a Darlington pair, as

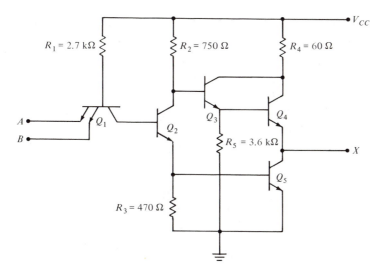

FIGURE 5-16 *High-speed TTL NAND gate*

shown in Fig. 5-16, to increase switching speed. Average propagation delay has been reduced from 9 ns for the standard TTL to 6 ns for high-speed TTL, but the increase in speed is achieved at the expense of power dissipation. High-speed TTL gates dissipate an average of 22 mW. The fan-out for high-speed TTL is 10; however, its fan-in is only about 1.3. Most high-speed logic circuits now use Schottky TTL devices.

5-13
CHARACTERISTICS OF SCHOTTKY CLAMPED TTL

Schottky clamped TTL, which is designed as the 74S00 series, has the highest speed of any member of the TTL family. Its propagation delay of 3 ns is achieved by connecting a Schottky barrier diode from the base to the collector of each circuit transistor as shown in Fig. 5-17a. This keeps the transistor from going into saturation when heavily forward biased. The schematic diagram for a Schottky clamped TTL AND gate is shown in Fig. 5-18.

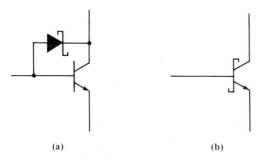

(a)   (b)

FIGURE 5-17 (a) Transistor with Schottky barrier diode clamp; (b) symbol for transistor with diode clamp

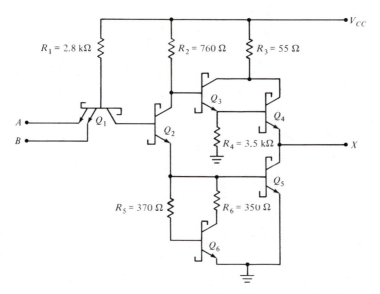

FIGURE 5-18   TTL NAND gate using Schottky barrier diode clamped transistors

5-14
CHARACTERISTICS OF LOW-POWER SCHOTTKY TTL

Low-power Schottky TTL is designated as the 74LS00 series. This very popular TTL series offers the same operating speed as standard TTL, but power dissipation is only 2 mW per gate compared to 10 mW for a standard TTL gate. Table 5-9 provides a comparison of the different members of the TTL family.

TABLE 5-9. / Comparison of Parameters for TTL Series

TTL Series	Propagation Delay (ns)	Power Dissipation (mW)	LOW-State Noise Margin (mV)	HIGH-State Noise Margin (mV)	Fan-out Driving (Same Series)
7400	9	10	400	400	10
74L00	33	1	400	400	10
74H00	6	22	400	400	10
74S00	3	19	300	700	10
74LS00	9	2	300	700	10

5-15
LOADING CONDITIONS FOR TTL

A logic gate connected to the output of another gate in a logic circuit is referred to as a *load*. One of the major considerations in designing digital systems is the maximum load that can be connected to the output of a gate. The term *fan-out*, which was introduced in Section 5-3, is used to provide a description of the drive capability of a gate. Fan-out is expressed quantitatively by use of the *standard input* or *unit load* (UL), where one unit load for any TTL device is defined as

$$1 \text{ UL} = \begin{cases} 40 \text{ } \mu A & \text{in the HIGH state} \\ 1.6 \text{ mA} & \text{in the LOW state} \end{cases}$$

The following example illustrates how unit loads are used with the various TTL series to determine fan-out or drive capabilities.

EXAMPLE 5-3 ━━━━━━━━━━━━━━━━━━━━━━━━━━━━━━━━━━

How many 7400 quad two-input NAND gate ICs can be successfully driven by the NAND gate IC shown in Fig. 5-19?

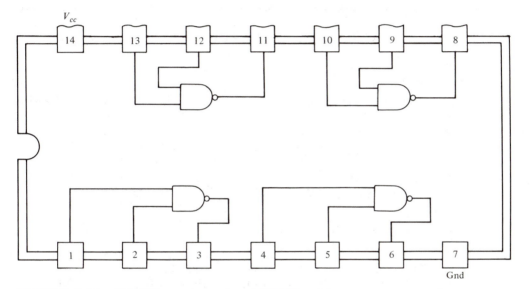

FIGURE 5-19   SN7400 quad two-input NAND gate

SOLUTION

The fan-out for each of the four NAND gates in the IC is 10; therefore, four gates can drive 40 gates that are contained within 10 quad two-input NAND gate ICs.   ❏

EXAMPLE 5-4 ━━━━━━━━━━━━━━━━━━━━━━━━━━━━━━━━━━

How many 74S00 NAND gates can be driven by a 7400 NAND gate?

SOLUTION

As can be seen in Appendix E, each 74S00 gate represents 1.25 UL. This represents a HIGH-state input current of

$$I_{IH} = (40 \text{ } \mu A)(1.25 \text{ UL}) = 50 \text{ } \mu A$$

and a low-state input current of

$$I_{IL} = (1.6 \text{ mA})(1.25 \text{ UL}) = 2 \text{ mA}$$

Therefore, the fan-out for a 7400 NAND gate driving a 74S00 NAND gate is

$$\text{fan-out (HIGH state)} = \frac{400 \text{ μA}}{50 \text{ μA}} = 8 \text{ gates}$$

$$\text{fan-out (LOW state)} = \frac{16 \text{ mA}}{2 \text{ mA}} = 8 \text{ gates}$$

In both the HIGH and LOW states the fan-out is 8 UL; therefore, we can say that a 7400 NAND gate will drive eight 74S00 NAND gates.   ❏

EXAMPLE 5-5

How many 7400 NAND gates can be driven by a 74LS00 NAND gate?

SOLUTION

Each 7400 NAND gate represents 1 UL, so fan-out for the 74LS00 NAND gate is

$$\text{fan-out (HIGH state)} = \frac{400 \text{ μA}}{40 \text{ μA}} = 10 \text{ UL}$$

$$\text{fan-out (LOW state)} = \frac{8 \text{ mA}}{1.6 \text{ mA}} = 5 \text{ UL}$$

The lower value is the fan-out; therefore, one 74LS00 gate can drive five 7400 NAND gates.   ❏

A summary of the fan-out capabilities for each member of the TTL family when driving gates from each other series in the family is shown in Table 5-10. The figures may vary slightly from manufacturer to manufacturer.

5-16
SPEED–POWER PRODUCT FOR THE TTL FAMILY

An ideal logic gate is one that has an infinitely fast switching speed and dissipates no power. The *speed–power product* of an ideal gate would, therefore, be zero. While a speed–power product of zero is not possible, a low value is both possible and desirable. This product provides us with a quantitative means of comparing the members of TTL family with regard to speed and power. This comparison is shown in tabular form in Table 5-11. The speed–power relationship is shown graphically in Fig. 5-20. As can readily be seen in both Table 5-11 and Fig. 5-20, the low-power Schottky series has a very low speed–power product compared to other members of the TTL family. This is one reason it has become very popular.

TABLE 5-10. / Summary of Fan-Out Capabilities for the TTL Series

TTL Driving Device	TTL Load Device				
	7400	74L00	74H00	74S00	74LS00
7400	10	50	8	8	20
74L00	2	10	1	1	10
74H00	12	50	10	10	25
74S00	12	100	10	10	50
74LS00	5	40	4	4	10

TABLE 5-11. / Speed–Power Relationship for TTL Series

Series	Speed–Power Product (pJ)	Propagation Delay (ns)	Power Dissipation (mW)
7400	90	9	10
74L00	33	33	1
74H00	132	6	22
74S00	57	3	19
74LS00	18	9	2

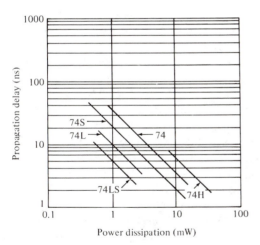

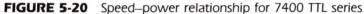

FIGURE 5-20   Speed–power relationship for 7400 TTL series

5-17
UNUSED INPUTS ON TTL DEVICES

Unused inputs on TTL gates behave as though a logic 1 is connected to them. With OR gates or NOR gates, this is a problem. With AND gates or NAND gates there is no problem with regard to the logic; however, it is usually undesirable to allow an input to ''float'' because the gate is more susceptible to noise.

Figure 5-21 shows two acceptable ways to connect any unused inputs. In Fig. 5-21a, the unused input is connected to an input that is being used. This technique

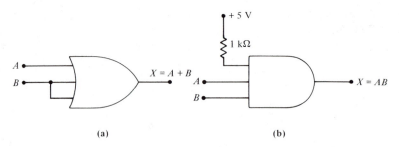

(a)   (b)

FIGURE 5-21   Acceptable ways to connect unused inputs on TTL gates

can be used with any type of TTL gate as long as the fan-out of a driving gate is not exceeded. In Fig. 5-21b, the unused input is connected to $+5$ V through a 1-kΩ resistor. This technique can be used with AND gates and NAND gates but cannot be used with OR gates and NOR gates. Their unused inputs should be grounded.

5-18
WIRED LOGIC

To connect the output of any two logic gates discussed to this point requires the use of an additional two-input logic gate, identified as gate C in Fig. 5-22. If our discussion were limited to connecting two logic gate outputs, wired logic would hardly warrant discussion. However, there are many applications, particularly in *bus-organized* digital systems, where 16 or more signals must be ANDed. Using TTL gates with totem-pole outputs, we would need an AND gate with as many input lines as we have signals to be ANDed. In situations where many signals are to be ANDed, wired logic, shown in Fig. 5-23, is a very attractive alternative although totem-pole TTL devices cannot be used. The reason for this can be seen by considering Fig. 5-24, which shows two NAND gates with their totem-pole outputs connected. If the output of gate A is HIGH, transistor Q_{3_A} will be ON and Q_{4_A} will be OFF. If the output of gate B is LOW, transistor Q_{3_B} will be OFF and Q_{4_B} will be ON. Under these conditions, Q_{4_B} will be required to sink current which may exceed its guaranteed 16-mA sink capacity by a factor of 3 or more for some TTL family members. To overcome this problem, an open-collector TTL series, which is very useful in wired logic, has been developed.

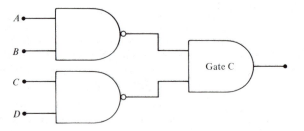

FIGURE 5-22   Logic gate C driven by totem-pole TTL NAND gates

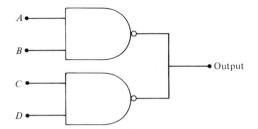

FIGURE 5-23   Concept of wired logic

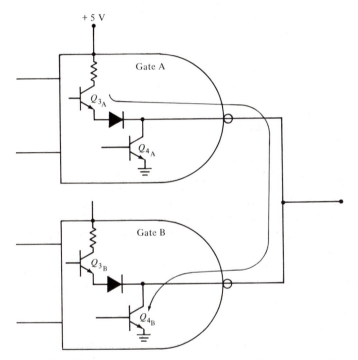

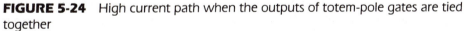

FIGURE 5-24   High current path when the outputs of totem-pole gates are tied together

5-19
TTL OPEN-COLLECTOR DEVICES

To accomplish wired logic, manufacturers of TTL devices have developed a series of open-collector logic devices. An open-collector TTL NAND gate is shown in Fig. 5-25. As can be seen, the collector of Q_3 is tied only to a pin on the IC package. An external resistor, called a *pull-up resistor*, must be connected to the collector via the IC pin. The other lead on the pull-up resistor is connected to V_{CC}. If the input signals to the gate are such as to cause base current to flow in transistor Q_3, it will be driven into saturation, thus pulling the output voltage nearly to ground potential (logic 0). If no base current flows, Q_3 acts like an open circuit.

Whether the logic that results from direct wiring of the outputs of TTL open-

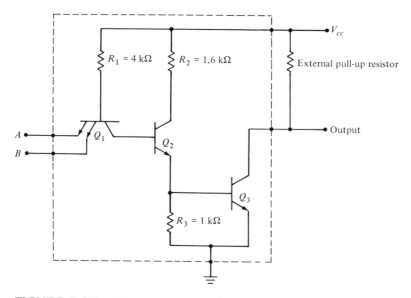

FIGURE 5-25   TTL open-collector NAND gate

collector NAND gates represents wire AND or wire OR is a matter of what the term is describing and the form in which the Boolean expression for the output is written. Consider the logic circuit shown in Fig. 5-26. The level of the output is influenced by both signals into the AND gate, $\overline{AB}$ and $\overline{CD}$. Clearly, the output is the AND of the two complemented terms, $\overline{AB}$ and $\overline{CD}$. If $\overline{AB}$ and $\overline{CD}$ are HIGH, the output is HIGH.

Wiring the open-collector NAND-gate outputs directly, as shown in Fig. 5-27, no longer permits separate gate outputs. However, the common output is obtained in the same manner as the output of the circuit in Fig. 5-26 and can therefore be described with the same Boolean expression, which is $X = \overline{AB} \cdot \overline{CD}$. As is readily apparent, the Boolean expression is for an AND function. However, the expression can easily be converted to OR-function form by use of De Morgan's theorem.

The circuit shown in Fig. 5-27 is called both *wired AND* and *wired OR* logic. Both descriptions are accurate depending on what terminals are viewed as input and output terminals. When described as wired AND, a dashed AND gate is frequently drawn around the output tie point, as shown in Fig. 5-28, for emphasis. The output is HIGH only when inputs $\overline{AB}$ and $\overline{CD}$ are HIGH; therefore, the term wired AND refers to the ANDing of signals $\overline{AB}$ and $\overline{CD}$.

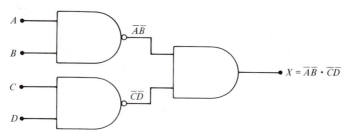

FIGURE 5-26   ANDing with logic gates

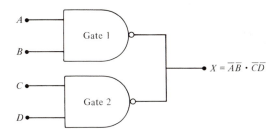

FIGURE 5-27   Wired logic circuit using open-collector NAND gates

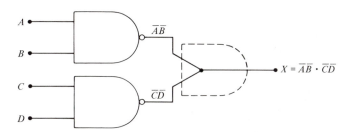

FIGURE 5-28   Concept of wired AND logic

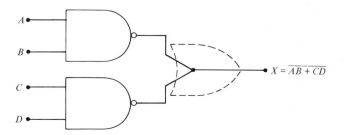

FIGURE 5-29   Concept of wired OR logic

Applying De Morgan's theorem to the output expression for Fig. 5-28 yields

$$\overline{AB} \cdot \overline{CD} = \overline{\overline{AB} + \overline{CD}} = \overline{AB + CD}$$

This alternative-form logic expression suggests ORing, which is represented symbolically as shown in Fig. 5-29. As can be seen from the output expression, the output is LOW when A and B are HIGH *or* when C and D are HIGH. Consequently, we can see that the term "wired OR" refers to the circuit rather than the tie point. Actually, the circuit behaves as a NOR gate; none the less, it is described as a wired OR logic gate with a "LOW equal to TRUE," or "active LOW," output.

EXAMPLE 5-6 ━━

Write the Boolean expression for the wired AND gate in Fig. 5-30.

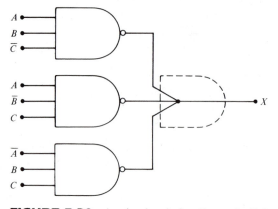

FIGURE 5-30   Logic circuit for Example 5-6

SOLUTION

The output expression is

$$X = \overline{AB\overline{C}} \cdot \overline{A\overline{B}C} \cdot \overline{\overline{A}BC}$$ ❏

EXAMPLE 5-7 ━━━━━━━━━━━━━━━━━━━━━━━━━━━━━━━━

Write the Boolean expression for the wired OR circuit in Fig. 5-31.

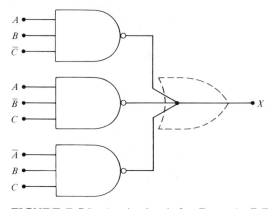

FIGURE 5-31   Logic circuit for Example 5-7

SOLUTION

The output expression is

$$X = \overline{AB\overline{C} + A\overline{B}C + \overline{A}BC}$$

Table 5-12 is a partial list of open-collector gates available in the standard TTL series. To function correctly, open-collector gate outputs must be tied to V_{CC} through a pull-up resistor. The value of the pull-up resistor depends on the number of gate outputs being wired ANDed and the number of loads being driven. Unless power dissipation is critical, use a 1-kΩ pull-up resistor for circuits with seven or

TABLE 5-12. / Partial List of Totem-Pole and Open-Collector TTL Gates

Open-Collector Device	Description	Equivalent Totem-Pole Device
7401	Quad two-input NAND gates	7400
7405	Hex inverter	7404
7409	Quad two-input AND gates	7408
7412	Triple three-input NAND gates	7410
7422	Dual four-input NAND gates	7420
7433	Quad two-input NOR gates	7402
7438	Quad two-input NAND buffers	7437

fewer source gates and a fan-out of up to 7. If either the number of source gates or the fan-out is greater than 7, some care must be taken in determining the value of the pull-up resistor.

5-20
TRI-STATE DEVICES

Totem-pole TTL devices have very high switching speeds compared to open-collector devices; however, they cannot be used in wired-logic applications. Tri-state devices were developed to incorporate into a single device the advantages of both totem-pole and open-collector devices. These devices have a totem-pole output for fast switching speeds but are designed so that their outputs can be tied directly together to provide a wired logic function.

All modern microcomputers and their peripherals use tri-state devices. Their use permits direct connection or disconnection of devices without affecting circuit operation. The three possible output states for tri-state devices are

- *HIGH state:* low impedance from output terminal to V_{CC}.
- *LOW state:* low impedance from output terminal to ground.
- *Disabled state:* high impedance from output terminal to V_{CC} and to ground.

The third state is achieved by adding an Enable line to a standard TTL gate such as the NAND gate shown in Fig. 5-32. When the Enable input is LOW, the gate functions as a conventional TTL NAND gate. The output will be either HIGH or LOW, depending on inputs A and B, and will change states at the same speed as a standard series TTL NAND gate. When the Enable input is HIGH, both totem-pole transistors, Q_3 and Q_4, are biased OFF, thus disabling the gate. In the disabled state, the impedance at the output, looking toward either ground or V_{CC}, is very high (nearly an open circuit). Tri-state devices may be designed so that they operate in a normal mode when the Enable line is LOW, as described above, or HIGH. If the device functions normally with the Enable line LOW, the device is referred to as an *active LOW* device. Conversely, an *active HIGH* device functions normally when the Enable line is HIGH. The voltage truth table for the tri-state NAND gate shown in Fig. 5-32 is shown in Table 5-13.

Tri-state devices are designed for use in systems where a large number of gate outputs are to be connected to a common line, or set of lines, called a *bus*. For example, microprocessors use buses to transmit all data, instructions, and controls

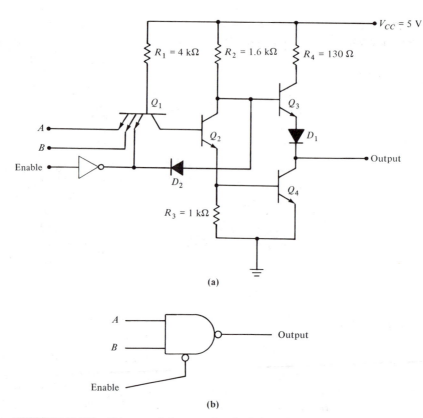

(a)

(b)

FIGURE 5-32   Tri-state (a) gate circuit; (b) logic symbol

TABLE 5-13. / Tri-State NAND-Gate Voltage Truth Table

Enable	A	B	Q_1	Q_2	Q_3	Q_4	Output
0 V	0 V	0 V	ON	OFF	ON	OFF	3.6 V
0 V	0 V	5 V	ON	OFF	ON	OFF	3.6 V
0 V	5 V	0 V	ON	OFF	ON	OFF	3.6 V
0 V	5 V	5 V	OFF	ON	OFF	ON	0 V
5 V	0 V	0 V	ON	OFF	OFF	OFF	High Z
5 V	0 V	5 V	ON	OFF	OFF	OFF	High Z
5 V	5 V	0 V	ON	OFF	OFF	OFF	High Z
5 V	5 V	5 V	ON	OFF	OFF	OFF	High Z

from one part of the microprocessor to another, as shown in Fig. 5-33. The system must be designed so that digital signals can be transmitted from any device connected to the bus to any other single device on the bus without interfering with the operation of any other device. Figure 5-34 shows four tri-state devices connected to a bus. Device 1 will operate in the normal two-state logic mode because its Enable line is LOW; therefore, the logic 1 at its output will be placed on the bus. The other three tri-state devices are disabled because their Enable lines are HIGH; therefore, these devices look essentially like an open circuit to the logic 1 at the output of device 1.

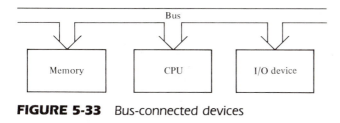

FIGURE 5-33 *Bus-connected devices*

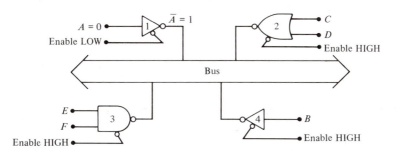

FIGURE 5-34 *Bus-connected tri-state devices*

Each tri-state device connected to the bus in Fig. 5-34 can output digital signals to the bus, but none can accept data from the bus. Digital signals are placed on the bus or received from the bus by using bidirectional tri-state devices called *bidirectional tri-state buffers* or *tri-state bus transceivers*. The principle of operation of a bidirectional tri-state buffer is shown in Fig. 5-35. If the Transmit/receive line is LOW, tri-state buffer 1 will operate in a normal two-state mode, while tri-state buffer 2 will present a high impedance to both the bus and the bus-connected device. Under these conditions, the bus-connected device can send data to the bus. When the Transmit/receive line is HIGH, data are sent from the bus to the bus-connected device via tri-state buffer 2. The logic diagram of a commercially produced bidirectional tri-state buffer is shown in Fig. 5-36.

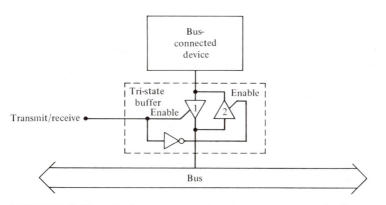

FIGURE 5-35 *Device connected to bus via tri-state buffer*

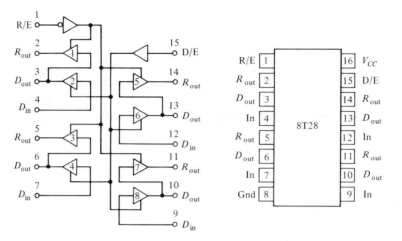

FIGURE 5-36   Logic diagram and pin assignment for 8T28 bidirectional tri-state buffer

5-21
MOS INTEGRATED CIRCUITS

Unipolar monolithic ICs are devices whose components are primarily *voltage controlled*. Such devices are categorized, according to the type of transistors in the ICs, as junction field-effect transistors (JFETs) or as metal-oxide-semiconductor field-effect transistors (MOSFETs). The schematic symbol for *n*- and *p*-channel MOSFETs is shown in Fig. 5-37. Junction field-effect transistors have found limited applications in ICs but are widely used as discrete transistors. On the other hand, MOS devices constructed with MOSFETs are widely used in digital integrated circuits and their use continues to increase as manufacturing technologies develop. The MOS family is made up of the following:

- *p*-Channel MOS (PMOS).
- *n*-Channel MOS (NMOS).
- Vertical MOS (VMOS).
- Complementary MOS (CMOS).

MOS devices offer some very attractive advantages over bipolar devices, such as TTL, including relatively simple and inexpensive fabrication, very low power consumption, and small size. MOSFETs occupy about one-fiftieth as much space

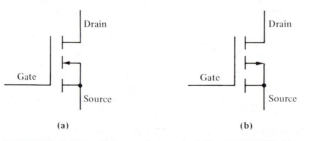

FIGURE 5-37   Schematic symbol for MOSFET: (a) *n*-channel; (b) *p*-channel

as bipolar transistors. Additionally, MOS ICs generally do not use IC resistor elements, which means that they can accommodate a much larger number of circuit elements on a single chip than can bipolar ICs. Consequently, MOS ICs are experiencing wide usage in the areas of large-scale integration (LSI) and very large scale integration (VLSI). In addition to a significant size reduction, MOS ICs provide increased reliability because the number of external connections is reduced.

MOS ICs operate at relatively slow speeds compared with bipolar ICs; however, there are many applications where speed is not a critical factor. In these applications MOS logic devices offer a very attractive alternative to bipolar logic devices.

5-22
CHARACTERISTICS OF MOS LOGIC DEVICES

The primary characteristics of interest for MOS logic devices are fan-out, propagation delay, noise margin, power dissipation, and supply voltage.

5-22.1 Fan-Out

MOSFET transistors have very high input resistance; therefore, one might reasonably expect that MOS logic devices would have very high fan-out capabilities. Fan-out is, in fact, quite high for signals in the audio-frequency range; however, at higher frequencies input capacitances cause a deterioration in switching speeds. The deterioration is proportional to the number of loads connected; nonetheless, MOS logic devices can operate well within specifications with a fan-out of 50.

5-22.2 Propagation Delay

The propagation delay for MOS logic devices is about 50 ns. This fairly long delay is due primarily to the relatively high output resistance of the driving gate and the capacitance at the inputs of the gates being driven.

5-22.3 Noise Margins

MOS logic devices have very good noise margins. The noise margin for all MOS devices is almost always greater than 1 V and is considerably higher for CMOS devices.

5-22.4 Power Dissipation

The power dissipated by MOS logic devices is very low, due primarily to the large resistances of the MOS devices themselves. The *average* power dissipation is on the order of 0.5 mW per gate. This low power consumption makes MOS devices very suitable for use in LSI and VLSI applications, where high-density packaging of gates and other logic devices occurs. MOS devices are also very suitable for use in battery-operated systems such as watches and calculators.

5-22.5 Supply Voltage

Unlike TTL, which will not tolerate a wide range of dc supply voltages, MOS logic devices can be safely operated over a supply voltage range of 3 to 15 V. Some CMOS devices are designed for a voltage range of 3 to 18 V.

5-23
THE MOS LOGIC FAMILY

As has already been stated, the members of the MOS logic family are PMOS, NMOS, VMOS, and CMOS. The NMOS and CMOS devices are presently widely used and their use will continue to grow. On the other hand, PMOS is rapidly becoming obsolete. VMOS is the newest member of the MOS family.

5-23.1 PMOS and NMOS

The basis PMOS and NMOS logic circuit, which is an inverter, is shown in Fig. 5-38. Transistor Q_2, in both the PMOS and NMOS inverters, functions as a load resistor for the active transistor Q_1. The basic inverter circuit can be expanded to build any of the basic logic gates that are available with TTL. For example, Fig. 5-39 shows a two-input NAND gate fabricated with NMOS transistors.

PMOS and NMOS are widely used in LSI and VLSI but find limited application in SSI and MSI circuits. NMOS technology is newer and faster than PMOS and is being used in place of PMOS in new designs.

5-23.2 VMOS

Conventional MOSFETS are three-terminal devices: the source, the gate, and the drain. This structure is such that current must flow in a narrow channel next to the gate. Maximum current and maximum power ratings are very low. With VMOS (vertical MOS) there are two source terminals, a V-shaped gate and a drain terminal, as shown in Fig. 5-40. As can be seen, the n^+ substrate acts as the drain. With the V-shaped gate, the current path is much wider; therefore, the current density is increased. This provides us with a power transistor in the MOS

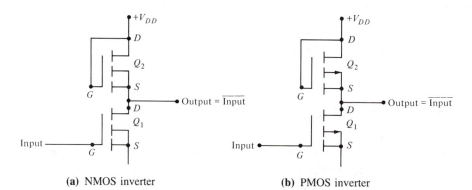

(a) NMOS inverter   (b) PMOS inverter

FIGURE 5-38 Basic NMOS and PMOS logic circuit: (a) NMOS inverter; (b) PMOS inverter

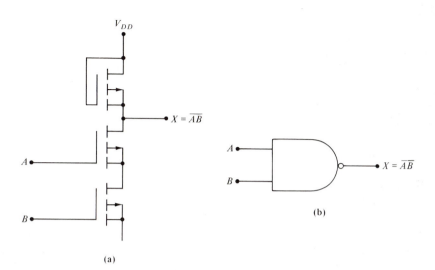

FIGURE 5-39 (a) NMOS NAND gate; (b) gate symbol

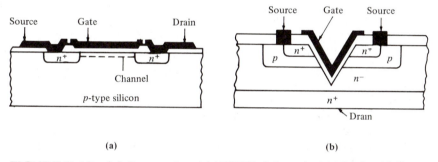

FIGURE 5-40 (a) Conventional MOSFET; (b) vertical MOS (VMOS)

family. Some VMOS devices are rated as high as 12 A of current and 80 W of power. These levels of current and power ratings, together with the very high input impedance characteristic of all MOS devices, make VMOS suitable as an interface between logic circuits and high current loads.

5-23.3 CMOS

CMOS devices are made by combining n-channel and p-channel MOS transistors into a single device. The most basic CMOS device is the inverter, shown in Fig. 5-41. When input signal A is LOW, the p-channel MOSFET (upper device) is ON and the n-channel MOSFET (lower device) is OFF. When the n-channel MOSFET is OFF, the output signal is HIGH. On the other hand, if input signal A is HIGH, the p-channel device is OFF, the n-channel device is ON, and the output is LOW. In either case, the current through the MOSFET devices in series is very small and is equal to the *leakage current* of the OFF device. Since the current is very low, the total power dissipation is extremely low, typically about 10 nW.

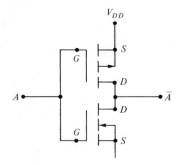

FIGURE 5-41   CMOS inverter

CMOS NAND and NOR Gates.   The schematic diagram for the CMOS NAND gate is shown in Fig. 5-42. The NAND gate is fabricated by modifying the CMOS inverter of Fig. 5-41. The modification involves adding a parallel *p*-channel MOSFET and a series *n*-channel MOSFET.

The gate operates as follows. If input *A* is at 0 V, the right *p*-channel MOSFET is turned ON and the upper *n*-channel MOSFET is turned OFF. If input *B* is at 0 V, the left-hand *p*-channel MOSFET and the lower *n*-channel MOSFET behave the same way. If input *A* or *B* is at a voltage level corresponding to logic 1, the corresponding *p*-channel MOSFET is turned OFF and the *n*-channel MOSFET is turned ON. Therefore, we can see that the only time the output will be LOW is when both *n*-channel MOSFETS are ON, which occurs when both input signals are HIGH.

The CMOS NOR gate shown in Fig. 5-43 is also obtained by modifying the CMOS inverter. The modification involves adding a series *p*-channel MOSFET and a parallel *n*-channel MOSFET. As with the NAND gate, a logic 0 level voltage turns the corresponding *p*-channel MOSFET ON, and a logic 1 level turns the corresponding *n*-channel MOSFET ON.

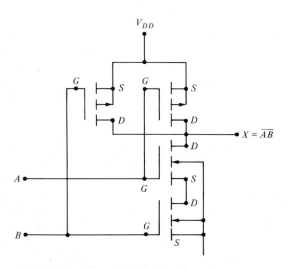

FIGURE 5-42   CMOS NAND gate

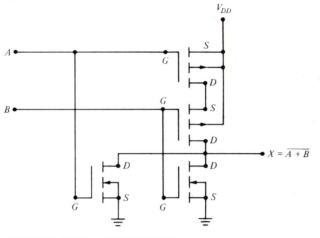

FIGURE 5-43   CMOS NOR gate

5-24
CMOS FAMILIES

There are three families of CMOS logic devices:

 • 4000 series.

 • 74C00 series.

 • Silicon-on-sapphire series.

The 4000 series, which was introduced in 1968, is presently the most popular series, but the recently introduced 74C00 series is destined to be a "star." From a performance viewpoint, silicon-on-sapphire devices compete well; however, at present their cost is prohibitive for widespread general use.

5-24.1   CMOS 4000 Logic Devices

The CMOS 4000 logic family consists of two series:

 • Series A or UB (unbuffered) series.

 • Series B.

Series B is a fairly recently modified version of the original 4000 family. To distinguish between the two, the original series has been designated as series A. Series B was developed primarily in recognition of the need for standardization between manufacturers of CMOS devices; however, buffered outputs that improve sink and source current capabilities have also been incorporated, as shown in Fig. 5-44. The buffers also improve the switching characteristics, as can be seen by observing the transfer curve of Fig. 5-45. In addition to the increased drive current capabilities, the maximum supply voltage limit has been increased from 15 V for series A devices to 18 V for series B.

A wide range of logic devices are available in the CMOS 4000 family in either the A or B series. Table 5-14 is a partial list of the devices that are available.

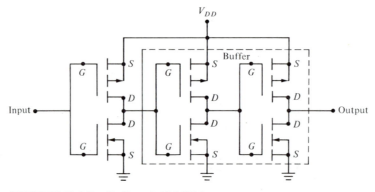

FIGURE 5-44   Buffered CMOS inverter

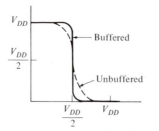

FIGURE 5-45   Transfer curve

TABLE 5-14. / Partial List of CMOS 4000B Logic Gates

Number	Description
4001B	Quad two-input NOR gate
4002B	Dual four-input NOR gate
4011B	Quad two-input NAND gate
4012B	Dual four-input NAND gate
4023B	Triple three-input NAND gate
4025B	Triple three-input NOR gate
4050B	Hex buffer
4069B	Hex inverter
4071B	Quad two-input OR gate
4072B	Dual four-input OR gate
4073B	Triple three-input AND gate
4075B	Triple three-input OR gate
4077B	Quad exclusive-NOR gate

5-24.2   Characteristics of CMOS 4000B Logic Devices

Some of the outstanding characteristics of CMOS logic devices are

1. *Low power dissipation.* The quiescent, or dc, power dissipation of CMOS logic devices is on the order of 10 μW. Power dissipation when conducting

maximum current is about 0.5 to 2 mW per gate, compared to 15 mW per gate for standard TTL.

2. *Wide range of supply voltages.* The 4000B series devices operate properly with power supply voltages from approximately 3 to 18 V. Because of the wide range of operating voltages, power supply regulation is not particularly critical.

3. *Good noise immunity.* The 4000B series has the same noise margin in both the LOW and HIGH states. Most manufacturers guarantee a noise margin of 30 percent of V_{DD}; however, in typical applications the noise margin is generally about 45 percent of V_{DD}. Thus, if $V_{DD} = 15$ V, the HIGH and LOW noise margins are guaranteed to be 4.5 V but are more likely to be greater than 6.5 V.

4. *High fan-out capability.* Since CMOS devices have a very high input resistance, one might think that fan-out would be extremely high; however, fan-out is limited by the load capacitance connected to the gate, which is approximately 7.5 pF per gate. In practical applications, this limits fan-out to approximately 50.

5-24.3 Selected Electrical Specifications for CMOS 4000B Devices

Table 5-15 lists some of the more important electrical characteristics for CMOS 4000B devices.

TABLE 5-15. / Electrical Characteristics for CMOS 4000B Devices

Symbol	Parameter	Conditions			Limits at 25°C			Units
		V_o (V)	V_{in} (V)	V_{DD} (V)	Min.	Typ.	Max.	
I_{DD}	Quiescent		0.5	5		0.01	0.25	µA
	current		0.10	10		0.01	0.50	
			0.15	15		0.01	1	
I_{OL}	Logic 0	0.4	0.5	5	0.51	1		mA
	output	0.5	0.10	10	1.3	2.6		
	current	1.5	0.15	15	3.4	6.8		
I_{OH}	Logic 1	4.6	0.5	5	-0.51	-1		mA
	output	2.5	0.5	5	-1.6	-3.2		
	current	9.5	0.10	10	-1.3	-2.6		
		13.5	0.15	15	-3.4	-6.8		
V_{OL}	Logic 0		0.5	5		0	0.05	V
	output		0.10	10		0	0.05	
	voltage		0.15	15		0	0.05	
V_{OH}	Logic 1		0.5	5	4.95	5		V
	output		0.10	10	9.95	10		
	voltage		0.15	15	14.95	15		
V_{IL}	Logic 0	0.5, 4.5		5			1.5	V
	input	1.9		10			3	
	voltage	1.5, 3.5		15			4	
V_{IH}	Logic 1	0.5		5	3.5			V
	input	1		10	7			
	voltage	1.5		15	11			
I_{in}	Input		0.18	18		$\pm 10^{-5}$	± 0.1	µA
	current							

5-25
POWER DISSIPATION VERSUS FREQUENCY FOR CMOS DEVICES

The total power dissipated by CMOS devices is the sum of the quiescent, or dc power, and the ac power, expressed quantitatively as

$$P_T = P_{dc} + P_{ac}$$

The dc power is given as

$$P_{dc} = V_{DD} \times I_{leakage}$$

and the ac power is expressed as

$$P_{ac} = CNfV_{DD}^2$$

where C = input capacitance per driven gate (7.5 pF maximum)
        N = number of gates being driven
        f = operating frequency
        V_{DD} = supply voltage

As can be seen from the expression for the ac power, the total power is affected by both the frequency and load capacitance; therefore, at higher frequencies CMOS loses some of its competitive edge with regard to power dissipation, compared with TTL.

EXAMPLE 5-8

A CMOS NAND gate that dissipates 25 μW of dc power is driving five CMOS gates at a frequency of 100 kHz. What is the total power dissipation if $V_{DD} = 15$ V?

SOLUTION

Each CMOS has a 7.5-pF input capacitance; therefore,

$$P_{ac} = (7.5 \times 10^{-12})(5)(1 \times 10^5)(15^2)$$

$$= 8437.5 \times 10^{-7} = 843.75 \ \mu W$$

$$P_T = P_{dc} + P_{ac}$$

$$= 25 \ \mu W + 843.75 \ \mu W = 868.75 \ \mu W \qquad \square$$

5-26
UNUSED INPUTS ON CMOS DEVICES

Unused inputs on CMOS chips cannot be left disconnected, due to the possibility of incorrect operation and/or excessive power dissipation by the CMOS chip due to noise pickup. Unused inputs can be connected to ground or V_{DD}, or to one of the other inputs that is being used.

5-27
HIGH-SPEED CMOS

Considerable work has been, and is being, done with 74C00 series high-speed CMOS. As a result, a fairly complete range of high-speed CMOS devices that are pin compatible and will compete directly with the LS TTL series is available.

The use of silicon-gate processing technology, as opposed to metal-gate technology used in the standard CMOS, permits the high-speed CMOS devices to combine the switching speeds and operating frequencies of TTL with the low-power-consumption advantages of CMOS. The devices also enjoy high noise immunity and a substantial range of operating voltages which are characteristic of CMOS devices. Table 5-16 lists some of the characteristics of the various CMOS 74C00 series.

TABLE 5-16. / Characteristics of High-Speed CMOS Devices

Type of Logic	Supply Voltage Range (V)	Quiescent Current (μA)	Speed (ns)
74C00	3–15	10	100
74SC00	3–6.5	30	20–30
74HC00	2–6	25	8–14
74HCT00	2–6	30	9–15
74HCU00	2–6	25	16

5-28
CMOS SILICON-ON-SAPPHIRE DEVICES

The only major problem associated with CMOS devices since their inception has been slow operating speeds due to input capacitance. One way to reduce this capacitance significantly is to use silicon-on-sapphire processing techniques. This process, which is still in its infancy, shows promise of increasing operating speeds of CMOS devices by four or five times their present speed. However, mastery of the process has been elusive to date, so devices using SOS processing techniques are relatively expensive.

5-29
CMOS TRANSMISSION GATES

There is one device within the CMOS family for which there is no equivalent in any of the other logic families. This unique device is the transmission gate. Its schematic diagram is shown in Fig. 5-46. The transmission gate is a bilateral device that acts essentially as a single-pole, single-throw switch. When the control

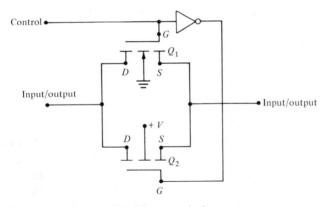

FIGURE 5-46   CMOS transmission gate

input is HIGH, the parallel p and n MOSFETS are both ON and the device functions as a closed switch. On the other hand, when the control input is LOW, the device behaves as an open switch. The transmission gate, which passes signals equally well in either direction, finds applications in both analog and digital circuits.

5-30
INTERFACING CMOS AND TTL DEVICES

Optimum performance of digital systems is often achieved by using more than one logic family. By doing this, the logic family that exhibits the best character- istics for each part of the total system can be used. The logic families that are combined most frequently are CMOS and TTL. The CMOS device is used in those parts of the circuit where speed is not critical, thereby reducing power dissipation, while TTL is used in areas where speed is an important consideration. As high-speed CMOS that are pin compatible with TTL devices continue to de- velop, CMOS-TTL interfacing will be of less concern; however, at this time it is still a topic of some concern.

5-30.1   TTL Driving CMOS

Figure 5-47 shows a TTL gate driving a CMOS gate with the supply voltage for both devices at 5 V. When the output of the TTL gate is LOW, it easily drives the CMOS gate, since $V_{OL(max)} = 0.4$ V for the TTL gate, which is well below the $V_{IL(max)} = 1.5$ V for the CMOS gate. When the output of the TTL gate is HIGH, the situation is not quite as straightforward but is easily resolved. The voltage corresponding to a logic 1 output for a TTL gate has a minimum value of 2.4 V with a typical value of 3.6 V. The minimum voltage at the input of a CMOS gate that is interpreted as a logic 1 is 3.5 V. This means that the noise margin is reduced to an undesirable level of 0.1 V when $V_{OH} = 3.6$ V for the TTL gate. This situation is remedied by using an external *pull-up* resistor R_p as shown in Fig. 5-47.

If the CMOS gate is connected to a supply voltage greater than 5 V, a special interface circuit must be used with the gates. The circuit, which is called a trans-

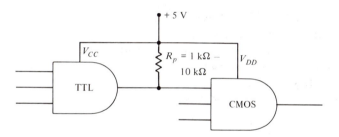

FIGURE 5-47 TTL-CMOS 5-V interface

lator, accepts lower-voltage inputs from the TTL gate and converts them to higher-voltage outputs, which are required by the CMOS gate.

5-30.2 CMOS Driving TTL

Figure 5-48 shows a CMOS gate driving a TTL gate with the supply voltage for both devices at 5 V. When the output of the CMOS gate is HIGH, it easily drives the TTL gate, since $V_{OH} = 5$ V. This is recognized as a logic 1 by the TTL gate. However, a problem may occur when the output of the CMOS gate is LOW because it cannot reliably sink the 1.6 mA (I_{IL}) of the TTL gate. There are two solutions to this problem. The first solution is to use 74L00 or 74LS00 series TTL gate, since CMOS can easily sink the LOW-state input current I_{IL} for either of these series. The second solution is to use an interface circuit called a CMOS *buffer* between the conventional CMOS gate and the TTL load.

5-31
PRECAUTIONS FOR HANDLING MOS DEVICES

The high input resistance of MOS devices makes them susceptible to damage from static charges. Most of the newer MOS devices have protective diodes on the inputs, so the problem is not as serious as it once was. However, precautions should still be observed when handling any MOS device. Some of the precautions that should be observed are

1. When MOS devices are removed from the conductive material in which they are shipped, do not touch the pins.
2. Make sure that you have no static charge on your body when handling MOS devices.

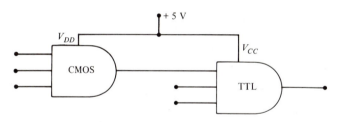

FIGURE 5-48 CMOS-TTL 5-V interface

3. MOS devices should be placed with their pins down on a grounded surface when removed from protective material or from a circuit.
4. MOS devices should not be stored in plastic or polystyrene trays.
5. All tools, test equipment, and metal work surfaces should be grounded when working with MOS devices.
6. Remove input drive before turning off V_{CC}.

5-32
EMITTER-COUPLED-LOGIC DEVICES

When evaluating any logic series, speed and propagation delay are usually important considerations. If these are prime considerations, emitter-coupled-logic (ECL) devices will be strong contenders. Emitter-coupled logic, like TTL, is a bipolar technology. Unlike TTL, the transistors in ECL circuits never operate at saturation, which accounts for their very high switching speed.

5-32.1 Basic ECL Circuit

The basic ECL circuit is a differential amplifier as shown in Fig. 5-49. The relationship between logic levels and input voltages for ECL are

Logic Level	V_{in}
0	-1.6 V
1	-0.8 V

Therefore, the operation of the ECL circuit shown in Fig. 5-49 will be analyzed with each of these input voltage levels applied. When $V_{in} = -1.6$ V, transistor Q_1 will be biased OFF, since the voltage at its emitter is at approximately -2 V ($V_{BB} - V_{BE} = -2$ V). The current through resistor R_E is

$$I_E = \frac{V_{EE} - 2 \text{ V}}{1 \text{ k}\Omega} = 3 \text{ mA}$$

This current flows through forward-biased transistor Q_2. When $V_{in} = -0.8$ V, the base of Q_1 is more positive, by 0.5 V, than the reference voltage at the base

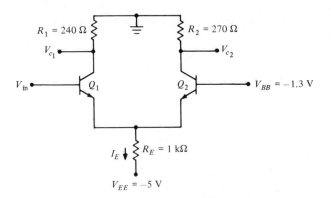

FIGURE 5-49 Basic ECL circuit

of Q_2. Thus Q_2 is biased OFF and the emitter voltage of both transistors is

$$V_E = V_{in} + V_{BE}$$

$$= -0.8 \text{ V} - 0.7 \text{ V} = -1.5 \text{ V}$$

The current through resistor R_E is therefore

$$I_E = \frac{V_{EE} - 1.5 \text{ V}}{1 \text{ k}\Omega} = 3.5 \text{ mA}$$

We can see that the current I_E remains at a relatively stable value regardless of which transistor is ON. Because of this current-switching mode of operation, this type of logic device is sometimes referred to as *current-mode logic* (CML).

When transistor Q_2 is ON due to the logic 0 (-1.6 V) at the input, its collector voltage is

$$V_{C_2} = 3 \text{ mA} \times 270 \text{ }\Omega$$

$$= -0.8 \text{ V}$$

Since transistor Q_1 is OFF when Q_2 is ON, its collector voltage is 0 V. When transistor Q_1 is ON due to a logic 1 (-0.8 V) at the input, its collector voltage is

$$V_{C_1} = 3.5 \text{ mA} \times 240 \text{ }\Omega$$

$$= -0.8 \text{ V}$$

These voltage levels are tabulated in Table 5-17. With regard to the voltages, two points should be made. First, the output voltage levels are not directly related to the input logic levels. Second, the collector voltages V_{C_1} and V_{C_2} are *complements* of each other.

5-32.2 Basic ECL Gate

The basic ECL circuit can easily be modified to function as a logic gate by connecting a transistor in parallel with Q_1 and by connecting the collectors of Q_1 and Q_2 to emitter-follower stages as shown in Fig. 5-50. The circuit shown

TABLE 5-17. / Operating States for the Basic ECL Circuit of Fig. 5-49

V_{in}	V_{C_1}	V_{C_2}
-1.6 V (logic 0)	0 V	-0.8 V
-8.0 V (logic 1)	-0.8 V	0 V

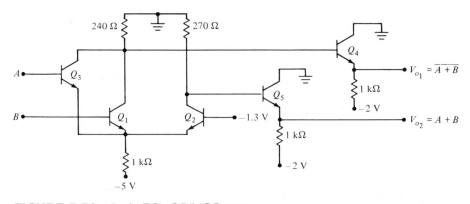

FIGURE 5-50 Basic ECL OR/NOR gate

functions as an OR/NOR gate. The addition of transistor Q_3 provides us with a second input terminal, thus giving us a circuit that can be described as a gate. The addition of Q_4 and Q_5 serves two very important functions, which are to subtract 0.8 V from the voltage levels of V_{C_1} and V_{C_2} to shift the output levels to the correct ECL logic levels of -0.8 V and -1.6 V and to provide a low output impedance for large fan-out capabilities. The ECL OR/NOR gate symbol, together with voltage levels and the corresponding truth table, are shown in Fig. 5-51.

5-32.3 Characteristics of ECL Devices

The following characteristics are those of most importance for ECL devices:

1. Very high switching rate, since the transistors in ECL circuits never operate at saturation. ECL circuits that operate at 500 MHz with 60 mW of power dissipation or 250 MHz with 25 mW of power dissipation are available.

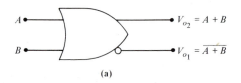

(a)

Voltage Levels

A	B	V_{o_1}	V_{o_2}
-1.6	-1.6	-0.8	-1.6
-1.6	-0.8	-1.6	-0.8
-0.8	-1.6	-1.6	-0.8
-0.8	-0.8	-1.6	-0.8

(b)

Truth Table

A	B	V_{o_1}	V_{o_2}
0	0	1	0
0	1	0	1
1	0	0	1
1	1	0	1

(c)

FIGURE 5-51 (a) ECL OR/NOR-gate symbol; (b) voltage levels; (c) truth table

2. Extremely short propagation delays. ECL devices offer the shortest propagation delays of any logic family. The fastest ECL series available has a propagation delay of approximately 1 ns.

3. Complementary outputs, which cause a function and its complement to appear simultaneously at the device output without using external inverters. This reduces chip count by eliminating the need for associated inverters, reduces system power required, and eliminates propagation delay associated with external inverters.

4. High input and low output impedance, due to the emitter-follower outputs, permits large fan-out, which is typically about 25.

5. Relatively constant power supply current drain, which simplifies power supply design, thus reducing costs. The constant current also reduces noise.

6. Power dissipation of 25 mW for a basic ECL gate. This compares favorably with the Schottky TTL series.

7. Noise margin of about 200 to 250 mV. This is less than for TTL and may cause ECL to operate unreliably if used in a high-noise environment.

5-33
INTEGRATED-INJECTION-LOGIC DEVICES

The newest development in bipolar logic technology is *integrated-injection logic* (I^2L). This family of bipolar devices was developed to compete with MOS devices in LSI and VLSI applications.

The basic I^2L circuit employs a *pnp* transistor that drives a multiple-collector *npn* transistor as shown in Fig. 5-52. The *pnp* transistor, Q_1, functions as a constant-current source. The current through Q_1 is determined by the value of V_{CC} or by the value of the resistor R_{ext}, which is externally connected to the IC chip. Adjusting either the voltage source or the resistor value affects both power dissipation and propagation delay. The product of power dissipation and propagation delay gives us the *speed–power product,* expressed in units of joules, for an I^2L gate. The value of the voltage or resistor should be such as to permit a current flow ranging from 1 nA at low speeds to 1 mA at high speeds.

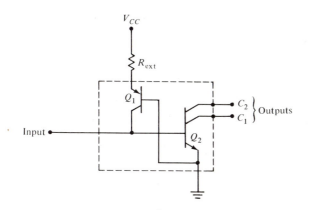

FIGURE 5-52 Basic I^2L circuit

EXAMPLE 5-9 ━━━━━━━━━━━━━━━━━━━━━━━━━━━━━━━━━━━━━━━

When operating at a current of 0.5 mA from a 5-V power supply, the propagation delay of an I²L gate is 12 ns. When operating at a current of 0.5 μA from a 5-V supply, the same gate has a propagation delay of 10 μs. Determine the speed–power product for both operating conditions.

SOLUTION

For the high-current, high-speed condition we obtain a speed–power product as follows:

$$P_D = 5 \text{ V} \times 0.5 \text{ mA} = 2.5 \text{ mW}$$

$$\text{speed–power product} = 12 \text{ ns} \times 2.5 \text{ mW} = 30 \text{ pJ}$$

For the low-current, low-speed condition we obtain a speed–power product as follows:

$$P_D = 5 \text{V} \times 0.5 \text{ μA} = 2.5 \text{ μW}$$

$$\text{speed–power product} = 10 \text{ μs} \times 2.5 \text{ μW} = 2.5 \text{ pJ}$$

It can be seen from the example that the speed–power product remains fairly constant over a wide range of operating conditions.   ❑

 The basic I²L circuit of Fig. 5-53a can be represented by the equivalent circuit model shown in Fig. 5-53b, where Q_1, which is sometimes called a *current-injection transistor,* has been replaced with the constant-current source.
 Since I²L circuits have no resistors that are internal to the chip, very high component densities, which may equal or exceed MOS, can be achieved. Additionally, when I²L circuits are operated at slow speeds (typically 100 ns), they dissipate less power than any logic family available.
 The basic I²L circuit is easily modified to form logic gates as shown in Fig. 5-53. The circuit shown operates as follows. If both inputs are grounded (logic 0), Q_1 and Q_2 will be OFF and Q_3 will be ON, as shown in Fig. 5-53a; therefore, output X will be LOW (logic 0). If either input is HIGH (logic 1), the transistor connected to that input will be ON and Q_3 will be OFF, as shown in Fig. 5-53b; therefore, output X will be HIGH (logic 1). By observing Table 5-18, which

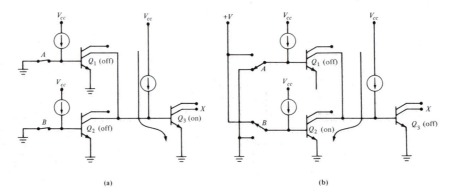

(a)   (b)

FIGURE 5-53   Basic I²L gate

TABLE 5-18. / Table of Voltage and Truth Table

| Input Voltages | | Output Voltage | | Input | | Output |
A	B	X		A	B	X
0 V	0 V	0 V		0	0	0
0 V	+ V	+ V		0	1	1
+ V	0 V	+ V		1	0	1
+ V	+ V	+ V		1	1	1

shows a table of voltages and a truth table for the circuit, it can readily be seen that the circuit functions as an OR gate. The unused collector terminals on Q_1 and Q_2 in Fig. 5-53b may be used to form other logic functions, such as NOR gates.

I^2L appears to have tremendous potential as a logic family. The technique is extremely versatile in that both digital and analog circuits can be built on the same chip. An impressive array of applications where I^2L devices are being used includes watches, clocks, video games, digital tuning, and color controls for television sets, complex calculators chips, digital voltmeters, and electronic organs.

5-34
GALLIUM-ARSENIDE ICs

Recent advances in processing techniques of semiconductor materials promise to move gallium-arsenide (GaAs) ICs from developmental laboratories into practical commercial use. Because of their very high operating speeds and relatively low power consumption, these devices will probably enable the development of systems that are impractical or even impossible using silicon semiconductor technology.

The performance characteristics associated with gallium-arsenide ICs become especially important in the design of very fast, very densely packaged systems. GaAs provides higher electron mobility and lower parasitic capacitances than silicon semiconductors. This permits fabrication of much smaller transistors than is possible with silicon and provides speed–power performance characteristics far superior to those of silicon.

As IC manufacturers gain experience in producing GaAs devices, they will no doubt use this technology extensively in the manufacture of many LSI chips, including random-access memory ICs and gate arrays, which are discussed in Chapter 15.

5-35
SUMMARY

The logic families that are most widely used at the present time are TTL, MOS, ECL, and I^2L. There are several series of TTL, with each series offering the user certain advantages with regard to speed, power dissipation, and propagation delay. The TTL family also includes several special devices, including open-collector and tri-state devices.

The most important member of the MOS family is CMOS. Until recently it was used primarily in applications where power dissipation was more important than speed. However, the recently developed high-speed CMOS series will broaden significantly the applications where CMOS is an attractive choice.

ECL is used where speed and propagation delay are of utmost importance. The basic ECL circuit always includes a logic function and its complement, thus eliminating the need for additional inverter chips.

The newest member of the bipolar logic family is I^2L, which was developed to compete with MOS devices in LSI and VLSI applications. This very versatile logic family dissipates less power than any other logic family when operated at slow speeds and competes very favorably at high speeds. I^2L, which appears destined for greatness, will have a significant impact in the LSI and VLSI fields as well as in other areas of electronics.

PROBLEMS

1. For the device described by the nomenclature AM74H32J list the following:
(a) Manufacturer
(b) Series
(c) Device
(d) Type of package

2. Which TTL series dissipates the least power?

3. Which TTL series has the longest propagation delay?

4. Which TTL series can operate at the highest frequency?

5. Which TTL series has the most desirable speed–power product?

6. What is the basic difference between bipolar and unipolar devices?

7. The characteristics for three different logic devices are listed below.
(a) Which device can operate at the highest frequency?
(b) Which device has the best HIGH-state noise margin?
(c) Which device has the best LOW-state noise margin?
(d) Which device has the poorest HIGH-state noise margin?
(e) Which device has the poorest LOW-state noise margin?
(f) Which device draws the most current from the power supply?
(g) Which device draws the least current from the power supply?

Characteristic	Device A	Device B	Device C
V_{supply}	5 V	6 V	6 V
V_{IL}	0.7 V	0.9 V	0.6 V
V_{IH}	1.6 V	1.8 V	1.7 V
V_{OL}	0.3 V	0.4 V	0.3 V
V_{OH}	2.5 V	2.2 V	2.4 V
P_D	12 mW	16 mW	14 mW
t_{PLH}	16 ns	10 ns	12 ns
t_{PHL}	14 ns	9 ns	8 ns

8. How many 74H00 gates can be driven by a 74LS00 gate?

9. What is the rise and fall time for the following pulse?

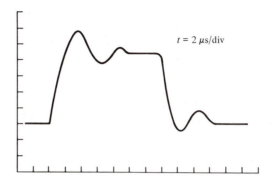

$t = 2\ \mu s/div$

10. What are the upper and lower limits of V_{CC} for TTL devices?

11. What is the maximum input voltage that TTL will recognize as a logic 0?

12. What is the minimum input voltage that TTL will recognize as a logic 1?

13. A standard TTL gate is sinking current from eight unit loads. How much current is this?

14. A standard TTL gate is sinking current from 12 74LS00 gates. How many unit loads does this represent?

15. If all the gates in a 7400 IC are connected in series, what total propagation delay will a pulse applied to the gates experience?

16. What are two advantages of wired logic?

17. Complete the truth table for the following circuit.

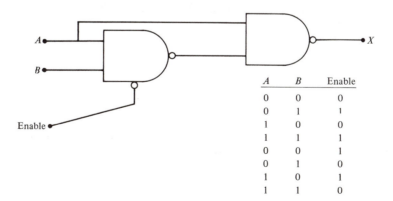

A	B	Enable
0	0	0
0	1	1
1	0	0
1	1	1
0	0	1
0	1	0
1	0	1
1	1	0

18. List two advantages of CMOS over TTL.

19. List two advantages of ECL over TTL.

20. What is the primary factor that limits fan-out with CMOS devices?

21. What is the approximate voltage at the points identified in the following NAND gate for the input conditions shown?

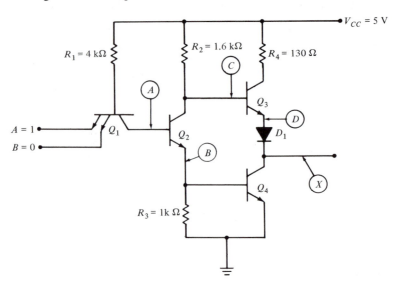

22. What is the major physical difference between the series A and B CMOS devices?

23. A CMOS NAND gate connected to a 12-V power supply draws a dc current of 2.5 mA when operated at a frequency of 120 kHz. What total power is dissipated by the gate if it is driving 12 CMOS gates?

24. What unique logic device operates as a bilateral switch?

25. What is an alternative name for ECL?

26. A certain standard TTL gate is sinking 10 unit loads. What is the maximum output resistance that the gate can have in its LOW state?

27. Write the Boolean equation that is valid for the output of each of the following TTL gates.

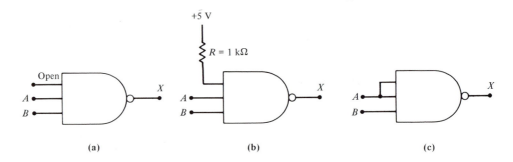

28. Which logic family operates least reliability in high-noise environments?

29. Determine the logic level of X for the input conditions shown in the following circuit.

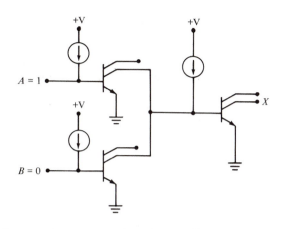

30. Determine the speed–power product for an I²L gate that has a propagation delay of 2 ns when drawing 0.2 mA from a 5-V power supply.

31. What is the propagation delay of an I²L gate whose speed–power product is 24 pJ when the gate is drawing 1.2 μA from a 4.8-V power supply?

REFERENCES

Aitman, Lawrence, ed., *Large Scale Integration.* New York: McGraw-Hill Book Company, 1976.

Anderson, Steve, "The Smart Machine Revolution," *Business Week,* July 5, 1976, pp. 39–44.

Cushman, Robert H., "High-Speed CMOS Replacements for LS TTL Promise Continued Viability of Jellybean Logic," *EDN,* March 17, 1983, pp. 64–74.

ECL Product Family. Sunnyvale, Calif.: Signetics Corp., 1982.

Greenfield, Joseph D., *Practical Digital Design Using ICs,* 2nd ed. New York: John Wiley & Sons, Inc., 1983.

Hnatek, Eugene R., *User's Guidebook to Digital CMOS Integrated Circuits.* New York: McGraw-Hill Book Company, 1981.

Noyce, R. N., "Microelectronics Technology," *Scientific American,* September 1977, pp. 63–69.

Tocci, Ronald J., *Digital Systems: Principles and Applications,* rev. ed. Englewood Cliffs, N.J.: Prentice-Hall, Inc., 1980.

The TTL Data Book. Dallas, Tex.: Texas Instruments, Inc., 1976.

Wojslaw, Charles F., *Integrated Circuits: Theory and Applications.* Reston, Va.: Reston Publishing Co., Inc., 1978.

Number Systems and Arithmetic Operations

INSTRUCTIONAL OBJECTIVES

In this chapter we discuss the number systems that are of primary concern when working with digital systems designed for computational purposes. The major emphasis is on the binary number system, with lesser emphasis given to the octal and hexadecimal systems. Techniques for converting from one number system to another as well as arithmetic operations within each system are discussed. After completing the chapter, you should be able to:

1. Convert decimal numbers to their binary, octal, or hexadecimal equivalent, or vice versa.
2. Convert binary numbers to their octal or hexadecimal equivalent, or vice versa.
3. Convert octal numbers to their hexadecimal equivalent, or vice versa.
4. Add, subtract, multiply, and divide binary numbers.
5. Add and subtract octal and hexadecimal numbers.
6. Determine the 1's and 2's complement of a binary number.
7. Define the terms *base, radix,* and *bit.*
8. Describe how the weight of a position within a number is determined.
9. Describe the term *end-around carry.*
10. Describe why complementary addition to perform subtraction is often done.

SELF-EVALUATION QUESTIONS

The following questions deal with the material presented in this chapter. Read the questions prior to studying the material and, as you read the chapter, watch for answers to the questions. After completing the chapter, return to this section and evaluate your comprehension of the material by answering the questions again.

1. Why is the binary number system ideal for use in digital computing systems?
2. What condensed term is used to describe a binary digit?
3. What does the term *radix* mean?
4. How is the weight of a position in a binary number determined?
5. How is the weight of a position in an octal or hexadecimal number determined?
6. How many symbols are used to represent numbers in the octal and hexadecimal number systems?
7. What is the advantage of complementary addition to perform subtraction over straight subtraction?
8. How is the 1's complement of a binary number obtained?

6-1
THE DECIMAL NUMBER SYSTEM

The decimal number system has 10 symbols, called Arabic numerals, which are used for counting. Since there are 10 separate symbols, the number system is also called the *base* 10 or *radix* 10 system. Using the 10 symbols 0, 1, 2, 3, . . . , 9 individually permits us to represent increasing numbers of individual units. To represent more than nine units, we must either develop additional symbols or use those we have in combination, which we obviously do. When used in combination, the value of the symbol depends on its position in the combination of symbols. We refer to this as *positional notation* and refer to the position as having a *weight* designated as units, tens, hundreds, thousands, and so on.

The *units* symbol occupies the first position to the left of the decimal point. The designation of the position is the units position because the base of the number system raised to a power equal to the number of positions away from the decimal points equals one "unit"; that is, $10^0 = 1$. This is expressed mathematically for each of the first five positions as

$$10^4 \quad 10^3 \quad 10^2 \quad 10^1 \quad 10^0$$

ten thousands   thousands   hundreds   tens   units

The value of the combination of symbols, 234 is determined by adding the *weight* of each position as

$$2 \times 10^2 + 3 \times 10^1 + 4 \times 10^0$$

which can be written as

$$2 \times 100 + 3 \times 10 + 4 \times 1$$

or

$$200 + 30 + 4 = 234$$

The positions to the right of the decimal point carry a positional notation and corresponding weight as well. The exponents to the right of the decimal point are negative and increase in integer steps starting with -1. This is expressed mathematically for each of the first three positions as

$$.10^{-1} \quad 10^{-2} \quad 10^{-3}$$

tenths   hundredths   thousandths

These concepts apply to each of the number systems to be discussed in this chapter.

6-2
THE BINARY NUMBER SYSTEM

The binary number system is a number system of *base* or *radix* equal to 2, which means that there are two symbols used to represent numbers: 0 and 1.

A seventeenth-century German mathematician, Gottfried Wilhelm von Leibniz, was a strong advocate of the binary number system. If it seems somewhat strange that one of the leading mathematicians of his day would advocate such a simple number system, we might also note that he was a philosopher and something of a mystic. Leibniz felt that there was beauty in the analogy between the zero representing a great void and the one representing the Diety. There was little additional support at the time for the binary number system; however, it has become extremely important in the computer age.

The symbols of the binary number system are used to represent numbers in the same way as in the decimal system. Each symbol is used individually; then the symbols are used in combination. Since there are only two symbols, we can only represent two numbers, 0 and 1, with individual symbols. We then combine the 1 with the 0 and with itself to obtain two additional numbers. These four numbers are the first four numbers of the binary number system and are

$$
\begin{matrix} 0 \\ 1 \end{matrix} \left\{ \text{each symbol used individually} \right.
$$

$$
\begin{matrix} 10 \\ 11 \end{matrix} \left\{ \begin{matrix} \text{symbol 1 moved to the second column and} \\ \text{combined with each individual symbol} \end{matrix} \right.
$$

In order to use the symbols in combination to form larger numbers, the second column in the first two numbers should be filled with zeros, which serve as a placeholder. Thus our first four numbers can be written as

$$
\begin{matrix} 0 & 0 \\ 0 & 1 \\ 1 & 0 \\ 1 & 1 \end{matrix}
$$

To form larger numbers, a 1 is placed in the third column. This is to be combined with all possible combinations of 1's and 0's in the first two columns. This is the same as carrying a 1 to the next-higher column and repeating all numbers previously formed in the lower-order columns. This is illustrated in Table 6-1.

**TABLE 6-1. / Binary
and Decimal Number
Correspondence**

Binary Number	Decimal Number
0 0	0
0 1	1
1 0	2
1 1	3
1 0 0	4
1 0 1	5
1 1 0	6
1 1 1	7

The same type of positional weighted system is used with binary numbers as in the decimal system. In the decimal system, the weight of a position is determined by raising the base of the number system (10) by a power equal to the number of positions away from the decimal point. The same technique is used to determine the weight of a position in the binary system. The base (2) is raised to a power equal to the number of positions away from the *binary point*. The weight and designation of the several positions are as follows:

Power equal to position

Base $\quad 2^5 \quad 2^4 \quad 2^3 \quad 2^2 \quad 2^1 \quad 2^0. \quad 2^{-1} \quad 2^{-2} \quad 2^{-3}$

Weight of position

thirty-twos	sixteenths	eights	fours	twos	units	one-halves	one-fourths	one-eighths

When the symbols 0 and 1 are used to represent binary numbers, each symbol is called a *binary digit* or, for short, a *bit*. Thus the binary number 1010 is a four-digit binary number or a 4-bit binary number.

6-2.1 Binary-to-Decimal Conversion

Since we are "programmed" to count in the decimal number system, it is only natural that we think in terms of the decimal equivalent value when we see a binary number. The conversion process is straightforward and is done as follows. Multiply the binary digit (1 or 0) in each position by the weight of the position and add the results. The following example illustrates the conversion of binary numbers to their decimal equivalent.

EXAMPLE 6-1 ━━━━━━━━━━━━━━━━━━━━━━━━━━━━━━━

Convert the following binary numbers to their decimal equivalent.

(a) 101   (b) 1001

SOLUTION

(a) $101 = (1 \times 2^2) + (0 \times 2^1) + (1 \times 2^0)$
$\qquad\quad = (1 \times 4) + (0 \times 2) + (1 \times 1)$
$\qquad\quad = 4 + 0 + 1 = 5$

(b) $1001 = (1 \times 2^3) + (0 \times 2^2) + (0 \times 2^1) + (1 \times 2^0)$
$\qquad\qquad = (1 \times 8) + (0 \times 4) + (0 \times 2) + (1 \times 1)$
$\qquad\qquad = 8 + 0 + 0 + 1 = 9$ ❏

Fractional binary numbers are converted to their decimal equivalent by using the process illustrated in the following example.

EXAMPLE 6-2 ━━

Convert the following binary numbers to their decimal equivalent.

(a) 0.011   (b) 0.101

SOLUTION

$$
\begin{aligned}
\text{(a)} \quad 0.011 &= (0 \times 2^{-1}) + (1 \times 2^{-2}) + (1 \times 2^{-3}) \\
&= (0 \times (\tfrac{1}{2})^1) + (1 \times (\tfrac{1}{2})^2) + (1 \times (\tfrac{1}{2})^3) \\
&= (0 \times 0.5) + (1 \times 0.25) + (1 \times 0.125) \\
&= 0 + 0.25 + 1.125 = 0.375
\end{aligned}
$$

$$
\begin{aligned}
\text{(b)} \quad 0.101 &= (1 \times 2^{-1}) + (0 \times 2^{-2}) + (1 \times 2^{-3}) \\
&= (1 \times 0.5) + (0 \times 0.25) + (1 \times 0.125) \\
&= 0.5 + 0 + 0.125 = 0.625
\end{aligned}
$$
❏

EXAMPLE 6-3 ━━

Convert the binary number 110.010 to its decimal equivalent.

SOLUTION

$$
\begin{aligned}
110.010 &= (1 \times 2^2) + (1 \times 2^1) + (0 \times 2^0) + (1 \times 2^{-1}) \\
&\quad + (1 \times 2^{-2}) + (0 \times 2^{-3}) \\
&= (1 \times 4) + (1 \times 2) + (0 \times 1) + (0 \times 0.5) \\
&\quad + (1 \times 0.25) + (0 \times 0.125) \\
&= 4 + 2 + 0 + 0.25 = 6.25
\end{aligned}
$$
❏

6-2.2 Decimal-to-Binary Conversion

It is frequently necessary to convert decimal numbers to equivalent binary numbers. The two most frequently used methods for making the conversion are the

- Repeated division-by-2 or multiplication-by-2 method.
- Sum-of-weights method.

Each method is discussed below.

Repeated Division-by-2 or Multiplication-by-2 Method.   To convert a decimal whole number to an equivalent number in a new base, the decimal number is repeatedly divided by the new base. For the case of interest here, the new base is 2, hence the repeated division by 2. Repeated division by 2 means that the original number is divided by 2, the resulting quotient is divided by 2, and each resulting quotient thereafter is divided by 2 until the quotient is 0. The remainder resulting from each division forms the binary number. The first remainder to be produced is the digit nearest the binary point regardless of whether the number being divided is greater or less than 1. The following examples illustrate the technique.

EXAMPLE 6-4 ━━━━━━━━━━━━━━━━━━━━━━━━━━━━━━━━━━━━

Convert the decimal number 18 to binary.

SOLUTION

$$
\begin{array}{cll}
2\,\underline{|\,18} & \text{Remainders} \\
2\,\underline{|\,9} & 0 & \text{(bit nearest the binary point)} \\
2\,\underline{|\,4} & 1 \\
2\,\underline{|\,2} & 0 \\
2\,\underline{|\,1} & 0 \\
\quad\ \ 0 & 1 & \text{(most significant bit)}
\end{array}
$$

Therefore, $18 = 10010$. ❏

When converting decimal fractions to binary, multiply repeatedly by 2 any fractional part. The equivalent binary number is formed from the 1 or 0 in the units position. The following example illustrates the procedure.

EXAMPLE 6-5 ━━━━━━━━━━━━━━━━━━━━━━━━━━━━━━━━━━━━

Convert the decimal number 0.625 to binary.

SOLUTION

$$
\begin{array}{lll}
0.625 \times 2 = 1.250 & 1 & \text{(bit nearest binary point)} \\
0.250 \times 2 = 0.50 & 0 \\
0.50 \times 2 = 1.00 & 1
\end{array}
$$

Therefore, $0.625 = 0.101$. ❏

Any further multiplication by 2 in Example 6-5 will equal 0; therefore, the multiplication can be terminated. However, this is not always so. Often it will be necessary to terminate the multiplication when an acceptable degree of accuracy is obtained. The binary number obtained will then be an approximation. ❏

EXAMPLE 6-6 ▬▬▬▬▬▬▬▬▬▬▬▬▬▬▬▬▬▬▬▬▬▬▬

Convert the binary number 0.6 to binary.

SOLUTION

$$0.6 \times 2 = 1.2 \qquad 1$$
$$0.2 \times 2 = 0.4 \qquad 0$$
$$0.4 \times 2 = 0.8 \qquad 0$$
$$0.8 \times 2 = 1.6 \qquad 1$$
$$0.6 \times 2 = 1.2 \qquad 1$$

Therefore, $0.6 = 0.10011$.   ❏

Sum-of-Weights Method.   The second method that is frequently used to convert decimal numbers to their equivalent in binary is the sum-of-weights method. This involves expressing the decimal number as the sum of binary weight values whose sum equals the decimal number. For example, the decimal number 10 can be expressed as the sum of binary weights as

$$10 = 8 + 2 = 2^3 + 2^1$$

By placing a 1 in the binary weight positions that make up the decimal number and an 0 in the remaining positions, we obtain the equivalent binary number. Completing the example for the conversion of decimal 10 to binary yields $10 = 2^3 + 2^1 = 1010$.

EXAMPLE 6-7 ▬▬▬▬▬▬▬▬▬▬▬▬▬▬▬▬▬▬▬▬▬▬▬

Convert the following decimal numbers to their binary equivalent using the sum-of-weights method.

(a) 15   (b) 59

SOLUTION

(a) $15 = 8 + 4 + 2 + 1$
    $= 2^3 + 2^2 + 2^1 + 2^0$
    $= 1111$

(b) $59 = 32 + 16 + 8 + 2 + 1$
    $= 2^5 + 2^4 + 2^3 + 2^1 + 2^0$
    $= 111011$ ❏

6-3
BINARY ARITHMETIC

Binary arithmetic includes the basic arithmetic operations of addition, subtraction, multiplication, and division. The following sections present the rules that apply to these operations when they are performed on binary numbers.

6-3.1 Binary Addition

Binary addition is performed in the same way as addition in the decimal system and is, in fact, much easier to master. Binary addition obeys the following four basic rules:

$$\begin{array}{cccc} 0 & 0 & 1 & 1 \\ +\,0 & +\,1 & +\,0 & +\,1 \\ \hline 0 & 1 & 1 & 10 \end{array}$$

The results of the last rule may seem somewhat strange, but remember that these are binary numbers. Put into words, the last rule states that

binary one + binary one = binary two = binary ''one zero''

When working with numbers where more than one base is involved, it is sometimes advisable to use a subscript to indicate the base. For example, writing the fourth rule as

$$\begin{array}{c} 1_2 \\ +\,1_2 \\ \hline 10_2 \end{array}$$

should remove any confusion.

When adding more than single-digit binary numbers, carry into higher-order columns as is done when adding decimal numbers. For example, 11 and 10 are added as follows:

$$\begin{array}{c} 11 \\ +\,10 \\ \hline 101 \end{array}$$

In the units (2^0) column 1 plus 0 equals 1. In the twos (2^1) column, 1 plus 1 equals 0 with a carry of 1 into the fours (2^2) column.

As another example, add $1 + 1 + 1$. This is done by adding two 1's and then adding the remaining 1 to this sum as shown below.

$$\begin{array}{c} 1 \\ +\,1 \\ \hline 10 \\ +\,1 \\ \hline 11 \end{array}$$

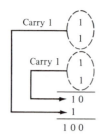

FIGURE 6-1

If a long column of 1's are to be added, group them in pairs and carry a 1 into the next higher-order column for each pair. The sum for the column is 0 if there is an even number of 1's and 1 if there is an odd number of 1's. For example, add $1 + 1 + 1 + 1$.

There is an even number of 1's; therefore, the sum for the column is 0. Two 1's are carried, one for each pair, into the next higher-order column and added as shown in Fig. 6-1.

Binary addition is quite straightforward if the basic rules presented are followed, even if the numbers contain several digits, as in the following examples.

EXAMPLE 6-8

Add 111 and 101.

SOLUTION

$$
\begin{array}{r}
111 \\
+\,101 \\
\hline
1100
\end{array}
$$ ❏

EXAMPLE 6-9

Add 1010, 1001, and 1101.

SOLUTION

$$
\begin{array}{r}
1010 \\
1001 \\
+\,1101 \\
\hline
100000
\end{array}
$$ ❏

6-3.2 Binary Subtraction

As with addition, there are few basic rules for subtracting binary numbers, which are as follows:

$$
\begin{array}{cccc}
0 & 1 & 1 & 10 \\
-0 & -1 & -0 & -1 \\
\hline
0 & 0 & 1 & 1
\end{array}
$$

When doing subtracting, it is sometimes necessary to borrow from the next higher-order column. The only time it will be necessary to borrow is when we try to subtract a 1 from a 0. In this case, a 1 is borrowed from the next higher-order column, which leaves a 0 in that column and creates a 10 in the column being subtracted. The following examples illustrate binary subtraction.

EXAMPLE 6-10

Perform the following subtractions.

(a) $11 - 01$ (b) $11 - 10$

SOLUTION

$$
\begin{array}{llll}
\text{(a)} & \begin{array}{r} 11 \\ -01 \\ \hline 10 \end{array} & \text{(b)} & \begin{array}{r} 11 \\ -10 \\ \hline 01 \end{array}
\end{array}
$$

❏

It was not necessary to borrow in Example 6-10. Now consider an example in which it is necessary to borrow.

EXAMPLE 6-11

Perform the following subtraction: $100 - 011$.

SOLUTION

$$
\begin{array}{r}
100 \\
-011 \\
\hline
001
\end{array}
$$

❏

This example involves two borrows, which are handled as follows. Since a 1 is to be subtracted from a 0 in the units column, a borrow is required from the next higher-order column. However, it also contains a 0; therefore, the twos column must borrow the 1 in the fours column. This leaves a 0 in the fours column and places a 10 in the twos column. Borrowing a 1 from 10 leaves a 1 in the twos column and places a 10 in the units column. This is shown in Fig. 6-2 together with the result.

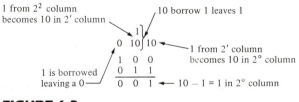

FIGURE 6-2

When subtracting a larger number from a smaller number, the results will be negative. To perform this subtraction, one must subtract the smaller number from the larger and prefix the results with the sign of the larger number, as illustrated in the following example.

EXAMPLE 6-12 ━━━

Perform the following subtraction: $101 - 111$.

SOLUTION

Subtract the smaller number from the larger and prefix the results with the sign of the larger number.

$$\begin{array}{r} 111 \\ - \underline{101} \\ 010 \end{array}$$ ❏

6-3.3   Binary Subtraction By 1's-Complement Addition

The method of subtraction that we have just examined is used in some machines used for computation; however, there are other methods of performing subtraction that require less circuitry in the machine. One of these methods is referred to as *1's-complement addition*. To use this method, one must first obtain the 1's complement of the smaller number. This is easily obtained by subtracting each digit in the binary number from the largest digit in the number set. For binary numbers, the largest digit in its number set is 1; therefore, subtract each digit from 1. The 1's complement of binary numbers can also be obtained by replacing 1's with 0's and 0's with 1's, but this should not be viewed as a basic rule for obtaining the complement of a number.

EXAMPLE 6-13 ━━━

Obtain the 1's complement of the following binary numbers.

(a) 0011   (b) 1010   (c) 1100

SOLUTION

The 1's complement for each number is obtained by subtracting each digit from 1.

$$\begin{array}{lll} \text{(a)} \quad \begin{array}{r} 1111 \\ - \underline{0011} \\ 1100 \end{array} & \text{(b)} \quad \begin{array}{r} 1111 \\ - \underline{1010} \\ 0101 \end{array} & \text{(c)} \quad \begin{array}{r} 1111 \\ - \underline{1100} \\ 0011 \end{array} \begin{array}{l} \text{given numbers} \\ \text{1's complements} \end{array} \end{array}$$

Subtraction by 1's-complement addition involves adding the 1's complement of the smaller number to the larger number. The most significant digit (the carry) is removed and added to the remaining digits in the result. This is called *end-around carry*. The following example illustrates the procedure. ❏

EXAMPLE 6-14 ▬▬▬▬▬▬▬▬▬▬▬▬▬▬▬▬▬▬▬▬▬▬▬▬▬

Perform the following subtraction by ordinary binary subtraction and by 1's-complement addition: $10101 - 10010$.

SOLUTION

Binary Subtraction   1's-Complement Addition

```
    10101              10101
  - 10010            + 01101   (1's complement of 10010)
  -------            -------
       11            100010
                     +    1    (end-around carry)
                     -------
                         11
```

As shown, only the subtrahend is converted to the 1's-complement form. ❏

When subtracting a larger number from a smaller using the 1's-complement method, the following steps are involved:

1. Obtain the 1's complement of the larger number.
2. Add the 1's complement of the larger number to the smaller number.
3. The most significant bit should always be 0; therefore, the end-around carry is always 0. The final answer is negative and is the 1's complement of the result of the addition.

EXAMPLE 6-15 ▬▬▬▬▬▬▬▬▬▬▬▬▬▬▬▬▬▬▬▬▬▬▬▬▬

Perform the following subtraction by ordinary binary subtraction and by 1's-complement addition:
$1011 - 1101$.

SOLUTION

Binary Subtraction   1's-Complement Addition

```
    1101              0010   (1's complement of 1101)
  - 1011            + 1011
  ------            ------
  - 0010            01101
                    +   0   (end-around carry is 0)
                    ------
                     1101   (take 1's complement of result)
                  -  0010   (final result, which is negative)
```
❏

When performing subtraction, the subtrahend may contain fewer digits than the augend. This presents no problem in ordinary subtraction, but when subtraction is performed using 1's-complement addition, the subtrahend must contain the same number of digits as the augend. If it contains fewer digits, zeros must be added on the left until the number of digits in the augend and subtrahend are equal. The following steps illustrate the procedure.

Binary subtraction   Zero added to subtrahend   1's-Complement form

$$10001 \atop - 1010$$ $$10001 \atop - 01010$$ $$10001 \atop + 10101$$

If the leftmost zero had not been added to the subtrahend, a totally different 1's complement would have been obtained.   ❏

6-3.4 Binary Subtraction Using 2's-Complement Addition

Another method of performing subtraction is by 2's-complement addition. When subtracting a smaller number from a larger, the following steps are involved:

1. Obtain the 2's complement of the smaller number. The 2's complement is the 1's complement plus 1.
2. Add the 2's complement of the smaller number to the larger number.
3. Discard the most significant bit (the carry), which will always be 1 if the result is positive.

The following example illustrates the method.

EXAMPLE 6-16 ▬▬▬▬▬▬▬▬▬▬▬▬▬▬▬▬▬▬▬▬▬▬▬▬▬▬▬▬▬▬

Perform the following subtraction by ordinary binary subtraction and by 2's-complement addition: $1101 - 1010$.

SOLUTION

Binary Subtraction   2's-Complement Addition

$$1101 \atop - 1010 \atop \overline{0011}$$ $$1101 \atop + 0110 \quad \text{(2's complement of 1010)} \atop \cancel{1}0011 \quad \text{(discard most significant bit)}$$ ❏

When subtracting a larger number from a smaller using 2's-complement addition, the following steps are involved:

1. Obtain the 2's complement of the larger number.
2. Add the 2's complement of the larger number to the smaller number.
3. If the most significant bit (the carry) is 0, as it will always be when subtracting a larger number from a smaller, the final answer is negative and is the 2's complement of the result of the addition.

The following example illustrates the method.

EXAMPLE 6-17 ▬▬▬▬▬▬▬▬▬▬▬▬▬▬▬▬▬▬▬▬▬▬▬▬▬▬▬▬▬▬

Perform the following subtraction by ordinary binary subtraction and by 2's-complement addition: $10010 - 11000$.

SOLUTION

Binary Subtraction   2's-Complement Addition

```
    11000              01000   (2's complement of 11000)
  − 10010            + 10010
  − 00110             011010   carry is 0; final answer is 2's complement
                               of sum, not including carry
                      00101    (1's complement)
                      +  1     (plus 1 to obtain 2's complement)
                    − 00110    final answer                        ❏
```

6-4
BINARY MULTIPLICATION

Binary multiplication follows the same rules as multiplication in the decimal system but is much easier because the multiplication tables are not nearly as extensive. In fact, the multiplication tables for the binary system have only four entries:

$$0 \times 0 = 0$$

$$0 \times 1 = 0$$

$$1 \times 0 = 0$$

$$1 \times 1 = 1$$

The multiplication process is divided into two steps. First, multiplication of 1's and 0's is performed beginning with the least significant bit in the multiplier to form partial products. Second, as in decimal multiplication, a summation is made following the rules of binary addition. The following examples illustrate the procedure.

EXAMPLE 6-18 ━━━━━━━━━━━━━━━━━━━━━━━━━━━━━━━━━━━━

Multiply the following binary numbers.

(a) 101×11 (b) 1101×10 (c) 1010×101 (d) 1011×1010

SOLUTION

```
(a)     101      (b)       1101
      ×  11              ×   10
        101              0000
       101               1101
       1111              11010
```

```
(c)    1010      (d)       1011
     ×  101              × 1010
       1010              0000
      0000               1011
     1010                0000
     110010              1011
                         1101110
```
❏

Multiplication of fractional numbers is performed in the same way as with fractional numbers in the decimal system, as illustrated in the following example.

EXAMPLE 6-19 ━━━

Perform the binary multiplication: 0.01×11.

SOLUTION

$$
\begin{array}{r}
0.01 \\
\times\ \ 11 \\
\hline
01 \\
01\ \ \\
\hline
0.11
\end{array}
$$
❑

6-5
BINARY DIVISION

Division in the binary number system employs the same procedure as division in the decimal system, as will be seen in the following examples.

EXAMPLE 6-20 ━━━

Perform the following binary division.

(a) $110 \div 11$ (b) $1100 \div 11$

SOLUTION

$$
\text{(a)}\quad 11\ \overline{)\,110\,}^{\,10} \qquad \text{(b)}\quad 11\ \overline{)\,1100\,}^{\,100}
$$

(a)
```
        10
  11 )110
     11
     ──
     00
     00
     ──
```
(b)
```
        100
  11 )1100
     11
     ──
     00
     00
     ──
     00
     00
     ──
```
❑

Binary division problems with remainders are also treated the same as in the decimal system, as the following examples show.

EXAMPLE 6-21 ━━━

Perform the following binary division.

(a) $1111 \div 110$ (b) $1100 \div 101$

SOLUTION

$$
\begin{array}{r}
10.1 \\
\text{(a)}\quad 110\ \overline{)1111.0} \\
\underline{110} \\
11 \\
\underline{00} \\
110 \\
\underline{110} \\
000
\end{array}
\qquad
\begin{array}{r}
10.011 \\
\text{(b)}\quad 101\ \overline{)1100.000} \\
\underline{101} \\
10 \\
\underline{00} \\
100 \\
\underline{000} \\
1000 \\
\underline{101} \\
110 \\
\underline{101} \\
1\quad \text{remainder}
\end{array}
\qquad \square
$$

6-6
THE OCTAL NUMBER SYSTEM

The octal number system is used extensively in digital work because of the ease of converting numbers from octal to binary, or vice versa. The octal system has a base, or radix, of 8, which means that there are eight symbols which are used to form octal numbers. Any eight symbols could be used; however, by convention, the first eight symbols of the decimal number system are used. Therefore, the single-digit numbers of the octal number system are

$$0, 1, 2, 3, 4, 5, 6, 7$$

To count beyond 7, a 1 is carried to the next higher-order column and combined with each of the other symbols, as in the decimal system. The weight of the different positions for the octal system is the base raised to the appropriate power as shown below.

Weight	8^3	8^2	8^1	8^0	.	8^{-1}	8^{-2}
Positional notation (decimal value)	five hundred twelves	sixty-fours	eights	units		one-eighths	one sixty-fours

Octal numbers look just like decimal numbers except that the symbols 8 and 9 are not used. Table 6-2 shows octal numbers 0 through 47 and their decimal equivalent. To distinguish between octal numbers and decimal numbers, we *must* subscript the octal numbers with their base. For example, $23_8 = 19_{10}$.

TABLE 6-2. / Octal Numbers and Their Decimal Equivalent

Octal	Decimal	Octal	Decimal	Octal	Decimal	Octal	Decimal	Octal	Decimal
0	0	10	8	20	16	30	24	40	32
1	1	11	9	21	17	31	25	41	33
2	2	12	10	22	18	32	26	42	34
3	3	13	11	23	19	33	27	43	35
4	4	14	12	24	20	34	28	44	36
5	5	15	13	25	21	35	29	45	37
6	6	16	14	26	22	36	30	46	38
7	7	17	15	27	23	37	31	47	39

6-6.1 Octal-to-Decimal Conversion

Octal numbers are converted to their decimal equivalent by multiplying the weight of each position by the digit in that position and adding the products. This conversion technique is illustrated in the following examples.

EXAMPLE 6-22 ━━━━━━━━━━━━━━━━━━━━━━━━━━━━━━━━━━━

Convert the following octal numbers to their decimal equivalent.

(a) 35_8 (b) 100_8 (c) 0.24_8

SOLUTION

(a) $35_8 = 3 \times 8^1 + 5 \times 8^0$
$= 3 \times 8 + 5 \times 1$
$= 29_{10}$

(b) $100_8 = 1 \times 8^2 + 0 \times 8^1 + 0 \times 8^0$
$= 1 \times 64 + 0 + 0$
$= 64_{10}$

(c) $0.24_8 = 2 \times 8^{-1} + 4 \times 8^{-2}$
$= 2 \times (\frac{1}{8})^1 + 4 \times (\frac{1}{8})^2$
$= \frac{2}{8} + \frac{4}{64}$
$= 0.3125_{10}$ ❏

6-6.2 Decimal-to-Octal Conversion

To convert decimal numbers to their octal equivalent, the following procedures are employed:

• Whole-number conversion: repeated division-by-8.
• Fractional-number conversion: repeated multiplication-by-8.

Repeated Division-by-8 Method.   The repeated division-by-8 method of converting decimal numbers to octal applies only to whole numbers, where the procedure is illustrated in the following example.

EXAMPLE 6-23 ━━━━━━━━━━━━━━━━━━━━━━━━━━━━━━━━━━━━━

Convert the following decimal numbers to their octal equivalent.

(a) 245   (b) 175

SOLUTION

(a) 8 | 245   5
(b) 8 | 30   6
(c) 8 | 3   3

The remainders make up the octal number where the first remainder is nearest the octal point; therefore, $245_{10} = 365_8$.

(b) 8 | 175   7
    8 | 21   5
    8 | 2   2
       0

Thus $175_{10} = 257_8$.   ❑

Repeated Multiplication-by-8 Method.   To convert decimal fractions to their octal equivalent requires repeated multiplication by 8, as shown in the following example.

EXAMPLE 6-24 ━━━━━━━━━━━━━━━━━━━━━━━━━━━━━━━━━━━━━

Convert the decimal fraction 0.432 to octal.

SOLUTION

$$0.432 \times 8 = 3.456 \qquad\qquad 3$$

$$0.456 \times 8 = 3.648 \qquad\qquad 3$$

$$0.648 \times 8 = 5.184 \qquad\qquad 5$$

$$0.184 \times 8 = 1.472 \qquad\qquad 1$$

The first carry is nearest the octal point; therefore,

$$0.432_{10} = 0.3351_8$$

The conversion to octal is not precise, since there is a remainder. If greater accuracy is required, we simply continue multiplying by 8 to obtain more octal digits.   ❑

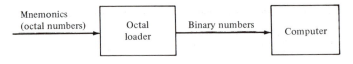

FIGURE 6-3 Binary-to-octal conversion with an octal loader

6-6.3 Octal-to-Binary Conversion

The primary reason for our interest in octal numbers lies in its use in entering and outputting computer data and because of the ease of octal-to-binary conversion. Computers recognize only binary information and may be programmed using only 1's and 0's. This approach, which is called *machine language programming*, makes very efficient use of memory space, but it is very tedious and prone to error. The next higher level of programming, called *assembler language programming*, uses groups of alphabetic characters called *mnemonics*. These mnemonics are applied to an octal loader, which views them as an octal number. The purpose of the octal loader is to convert the mnemonics from an octal number to a binary pattern, as shown in Fig. 6-3.

Converting numbers from octal to binary can be done essentially by inspection. Since there are eight symbols used for counting in the octal system and eight combinations of three binary digits that correspond to these single-digit octal numbers, we can assign a binary three-digit combination to each single-digit octal number, as shown in Table 6-3.

To convert from octal to binary, simply replace each octal digit with the corresponding three-digit binary number, as illustrated in the following example.

EXAMPLE 6-25 ━━━━━━━━━━━━━━━━━━━━━━━━━━━━

Convert the following octal numbers to their binary equivalent.

(a) 247_8 (b) 501_8

TABLE 6-3. / Octal and Binary Number Correspondence

Octal	Binary
0	000
1	001
2	010
3	011
4	100
5	101
6	110
7	111

SOLUTION

	2	4	7	octal
(a)	010	100	111	binary

Thus $247_8 = 010100111_2$.

	5	0	1	octal
(b)	101	000	001	binary

Thus $501_8 = 101000001_2$. ❏

6-6.4 Binary-to-Octal Conversion

Binary-to-octal conversion also makes use of the correspondence between single-digit octal numbers and three-digit binary numbers, shown in Table 6-2. As has been shown, the conversion process is quite straightforward. However, one precaution should be exercised when converting from binary to octal. Grouping the binary digits must begin with the digit nearest the binary point.

EXAMPLE 6-26 ━━━

Convert the binary number 11010101.01101 to its octal equivalent.

SOLUTION

11	010	101.	011	01
3	2	5 .	3	2

Since the outermost groups contain only two digits, we can readily see that a totally different octal number would be obtained if the grouping was done improperly. ❏

6-7
OCTAL ARITHMETIC

Arithmetic operations are performed in the same way in the octal system as in the more familiar decimal system, as can be seen in the following examples of addition and subtraction.

EXAMPLE 6-27 ━━━

Add the octal numbers 512 and 467.

SOLUTION

$$512_8$$
$$+ 467_8$$
$$\overline{1201_8}$$

❏

How we obtain this somewhat strange-looking result can be seen by considering the octal numbers 0 through 12:

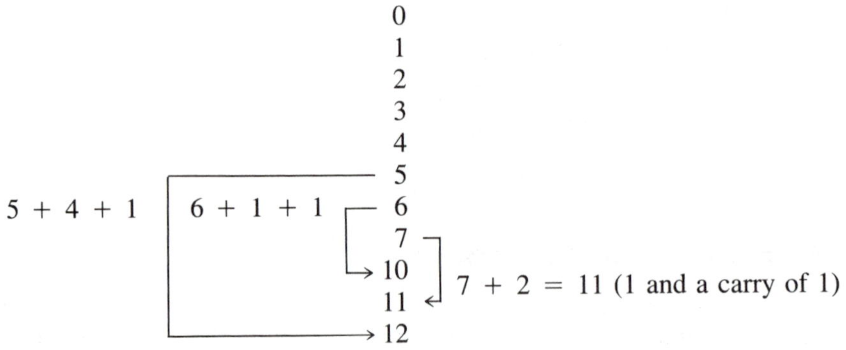

In the units column, 7 plus 2 equals 11. A 1 is carried into the next column; therefore, we have 6 plus 1 plus 1, which equals 10. Again, a 1 is carried into the higher-order column, which gives us a sum of 5 plus 4 plus 1, which equals 12.

EXAMPLE 6-28 ━━━━━━━━━━━━━━━━━━━━

Subtract octal number 467 from octal number 512.

SOLUTION

$$\begin{array}{r} 512_8 \\ -\ 467_8 \\ \hline 23_8 \end{array}$$ ❏

6-8
HEXADECIMAL NUMBER SYSTEM

The hexadecimal number system is a base, or radix, 16 system; thus it uses 16 different symbols in forming numbers. Any 16 symbols could be used, but by convention the decimal digits 0 through 9 and the alphabetic characters A through

TABLE 6-4. / Hexadecimal Conversion Table

Decimal	Binary	Hexadecimal	Decimal	Binary	Hexadecimal
0	0000	0	8	1000	8
1	0001	1	9	1001	9
2	0010	2	10	1010	A
3	0011	3	11	1011	B
4	0100	4	12	1100	C
5	0101	5	13	1101	D
6	0110	6	14	1110	E
7	0111	7	15	1111	F

F are used. The hexadecimal system is widely used in computer systems, since many computers process data in groups that are multiples of four, thus permitting 4-bit binary groups to be represented by a single hexadecimal digit. Table 6-4 shows the single-digit hexadecimal numbers and the corresponding decimal and binary numbers. To distinguish between hexadecimal numbers and decimal or octal numbers, it will sometimes be necessary to subscript the hexadecimal numbers with their base of 16.

Hexadecimal numbers greater in value than F are formed by carrying a 1 to the next higher-order column and combining it with each of the other symbols, as in the decimal system. This gives us a weighted system where the weight of the different positions for the hexadecimal system is the base raised to the appropriate power:

Weight $\qquad 16^3 \quad 16^2 \quad 16^1 \quad 16^0 . \quad 16^{-1} \quad 16^{-2}$

Position notation
(decimal value)

| four thousand ninety sixes | two-fifty sixes | sixteens | units | one-sixteenths | one-two fifty sixes |

6-8.1 Hexadecimal-to-Binary Conversion

Converting a hexadecimal number to an equivalent binary system involves the straightforward process of replacing each hexadecimal symbol with the corresponding four-digit binary group taken from Table 6-4. The procedure is illustrated in the following examples.

EXAMPLE 6-29 ━━━

Convert the following hexadecimal numbers to their binary equivalent.

(a) 7 A 2   (b) 3 D 4. F

SOLUTION

```
        7       A       2       hexadecimal
(a)   0111    1010    0010      binary
```

Thus $7A2_{16} = 011110100010_2$.

```
        3       D       4  .  F
(b)   0011    1101    0100.1111
```

Thus $3D4.F_{16} = 001111010100.1111_2$.

The 0's in the two most significant positions of the binary number are not required and normally would not be included. ❏

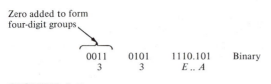

Zero added to form
four-digit groups

0011	0101	1110.101	Binary
3	3	$E .. A$	

FIGURE 6-4

6-8.2   Binary-to-Hexadecimal Conversion

Binary-to-hexadecimal conversion is equally straightforward and is also accomplished directly by using Table 6-4, as shown in the following examples.

EXAMPLE 6-30 ━━━━━━━━━━━━━━━━━━━━━━━━━━━━━━

Convert the following binary numbers to hexadecimal.

(a) 101011010010   (b) 10111010110   (c) 1101011110.101

SOLUTION

	1010	1101	0010	binary
(a)	A	D	2	hexadecimal

Thus $101011010010_2 = AD2_{16}$.

(b) Care must be taken to group the binary digits starting at the binary point.

	101	1101	0110	binary
	5	D	6	hexadecimal

Thus $10111010110_2 = 5D6_{16}$.

(c) Again, care must be taken to group the binary digits starting at the binary point for both the whole numbers and the fractional portion. If there is any confusion regarding grouping, add 0's as required to form four-digit binary groups, as shown in Fig. 6-4.

Thus $1101011110.101_2 = 35E.A_{16}$.   ❏

6-8.3   Hexadecimal-to-Decimal Conversion

Hexadecimal numbers are converted to their decimal equivalent by multiplying the weight of each position by the decimal equivalent of the digit in that position and adding the products. This conversion process is illustrated in the following examples.

EXAMPLE 6-31 ━━━━━━━━━━━━━━━━━━━━━━━━━━━━━━

Convert the following hexadecimal numbers to their decimal equivalent.

(a) 121_{16} (b) $A1C_{16}$

SOLUTION

(a) $121_{16} = 1 \times 16^2 + 2 \times 16^1 + 1 \times 16^0$
$= 1 \times 256 + 2 \times 16 + 1 \times 1$
$= 256 + 32 + 1$
$= 289_{10}$

(b) First, we must express the weight of each symbol as a decimal value, which can be obtained directly from Table 6-4.

$$A = 10_{10}$$

$$1 = 1_{10}$$

$$C = 12_{10}$$

Next, multiply the decimal equivalent of each hexadecimal digit by the weight of the position it occupies.

$$A1C_{16} = 10 \times 16^2 + 1 \times 16^1 + 12 \times 16^0$$

$$= 10 \times 256 + 1 \times 16 + 12 \times 1$$

$$= 2560 + 16 + 12$$

$$= 2588_{10} \qquad \square$$

Hexadecimal numbers can also be converted to their decimal equivalent by converting them to binary and then converting the binary number to decimal. The following example will show this method of conversion.

EXAMPLE 6-32 ━━

Convert the hexadecimal number A3B to its decimal equivalent by first converting it to binary, then converting the binary number to decimal.

SOLUTION

A	3	B	hexadecimal
1010	0011	1011	binary

$$101000111011 = 1 \times 2^{11} + 1 \times 2^9 + 1 \times 2^5 + 1 \times 2^4 + 1 \times 2^3 + 1 \times 2^1 + 1 \times 2^0$$

$$= 2048 + 512 + 32 + 16 + 8 + 2 + 1$$

$$101000111011_2 = 2619_{10} \qquad \square$$

6-8.4 Decimal-to-Hexadecimal Conversion

Decimal numbers are converted to their hexadecimal equivalent using the following procedures:

• Whole-number conversion: repeated division-by-16.
• Fractional-number conversion: repeated multiplication-by-16.

Repeated Division-by-16 Method. To convert a decimal whole number to its hexadecimal equivalent, we repeatedly divide the decimal number by the base to which conversion is being made which, in our case is 16. The following example illustrates the procedure.

EXAMPLE 6-33

Convert the following decimal numbers to their hexadecimal equivalent.

(a) 650   (b) 2588

SOLUTION

Remainder

(a) 16 | 650 $10_{10} = A_{16}$
      16 | 40 $8_{10} = 8_{16}$
           2         $2_{10} = 2_{16}$

Thus $650_{10} = 28A_{16}$.

Remainder

(b) 16 | 2588 $12_{10} = C_{16}$
      16 | 161 $1_{10} = 1_{16}$
      16 | 10 $10_{10} = A_{16}$

Thus $2588_{10} = A1C_{16}$.

$A1C_{16}$ was converted to decimal in Example 6-31; therefore, this example serves to demonstrate the conversion method as well as to check the results of the earlier example. ❏

Repeated Multiplication-by-16 Method. Decimal fractions can be converted to an equivalent hexadecimal fraction by repeated multiplication by 16, as shown in the following example.

EXAMPLE 6-34

Convert the decimal fraction 0.642 to its hexadecimal equivalent.

SOLUTION

$0.642 \times 16 = 10.272$ $10_{10} = A_{16}$

$0.272 \times 16 = 4.352$ $4_{10} = 4_{16}$

$0.353 \times 16 = 5.632$ $5_{10} = 5_{16}$

$0.632 \times 16 = 10.112$ $10_{10} = A_{16}$

Thus $0.642_{10} = 0.A45A_{16}$. ❏

6-9
HEXADECIMAL ARITHMETIC

Addition and subtraction of hexadecimal numbers are performed in essentially the same way as in the more familiar decimal system. Because of our familiarity with the decimal system, hexadecimal arithmetic will be easier if we mentally convert hexadecimal symbols to their decimal equivalent before adding. If the sum of the decimal equivalent values in any column is less than 15_{10}, place the corresponding hexadecimal digit in the column. If the sum of the decimal equivalent values in a column exceed 15_{10}, subtract 6_{10} from the sum, place the hexadecimal equivalent for the difference in the column, and carry a 1 to the next higher-order column.

EXAMPLE 6-35 ━━━━━━━━━━━━━━━━━━━━━━━━━━━━━━━━━━━━━

Add the following hexadecimal numbers.

(a) $25_{16} + 39_{16}$　　(b) $78_{16} + C2_{16}$　　(c) $BD_{16} + AE_{16}$

SOLUTION

(a) Hexadecimal　　　　Decimal Equivalent
$$\begin{array}{ll} 25_{16} & 2_{10} \quad 5_{10} \\ \underline{39_{16}} & \underline{3_{10} \quad 9_{10}} \\ 5E_{16} & 5_{10} \quad (14_{10} = E_{16}) \end{array}$$

(b) Hexadecimal　　　　Decimal Equivalent
$$\begin{array}{ll} 78_{16} & 7_{10} \quad 8_{10} \\ \underline{C2_{16}} & \underline{12_{10} \quad 2_{10}} \\ 13A_{16} & 19_{10} \quad (10_{10} = A_{16}) \\ & \underline{-16_{10}} \\ & 3_{10} = 3_{16} \text{ with carry of } 1 \end{array}$$

(c) Hexadecimal　　　　Decimal Equivalent
$$\begin{array}{ll} B\ D_{16} & 11_{10} \quad 13_{10} \\ \underline{A\ E_{16}} & \underline{10_{10} \quad 14_{10}} \\ 16\ B_{16} & 21_{10} \quad 27_{10} \\ & \underline{1_{10} - 16_{10}} \\ & 22_{10} \quad 11_{10} = B_{16} \text{ with carry of } 1 \\ & \underline{-16_{10}} \\ & 6_{10} = 6_{16} \text{ with carry of } 1 \end{array}$$　❑

　　Hexadecimal subtraction can be performed by applying ordinary rules for subtraction or by the process of subtraction by complementary addition. The following examples illustrate both techniques.

EXAMPLE 6-36 ━━━━━━━━━━━━━━━━━━━━━━━━━━━━━━━━━━━━━

Subtract the following pairs of hexadecimal numbers.

(a) $A5_{16} - 23_{16}$　　(b) $FF_{16} - AA_{16}$　　(c) $235_{16} - 55_{16}$

SOLUTION

(a) $A5_{16}$ (b) FF_{16} (c) 235_{16}
        -23_{16} $-AA_{16}$ -55_{16}
        $\overline{82_{16}}$ $\overline{55_{16}}$ $\overline{1E0_{16}}$

When it is necessary to borrow, as in (c), remember that the weight of the position—16 to the proper power—is borrowed from the next higher-order column.   ❑

To perform hexadecimal subtraction by 2's-complement addition, we convert the number to be subtracted to binary, obtain its 2's complement, convert this number to hexadecimal, and add, as demonstrated in the following example.

EXAMPLE 6-37 ━━━━━━━━━━━━━━━━━━━━━━━━━━━━

Perform the following hexadecimal subtraction by 2's-complement addition: $B7_{16} - 84_{16}$.

SOLUTION

The first step is to convert 84_{16} to its binary equivalent:

$$84_{16} = 10000100_2$$

The 1's complement of 10000100 is 01111011 and its 2's complement is 01111100. The hexadecimal equivalent of 01111100, which is $7C_{16}$, in added to $B7_{16}$ following the rules for hexadecimal addition to obtain the following results:

$$\begin{array}{r} B7_{16} \\ +7C_{16} \\ \hline 33_{16} \end{array}$$

As always in complementary addition, the carry is dropped.   ❑

6-10 SUMMARY

The base, or radix, of a number system tells us how many different symbols are used to form numbers in that system. The familiar decimal system has a base of 10 and thus uses 10 symbols. The binary system, which is of primary interest when studying digital circuitry, has a base of 2 and therefore uses 2 symbols, 0 and 1, to form numbers. Octal and hexadecimal numbers, which have base 8 and base 16, respectively, are used extensively in digital systems to process binary bits in groups of three or four, respectively.

Since digital systems make extensive use of binary, octal, and hexadecimal numbers, it is important for human operators who are familiar with decimal numbers to be proficient in each of the number systems as well as in making conversions between each of the systems.

PROBLEMS

1. Convert the following binary numbers to their decimal equivalent.
   (a) 101   (b) 110   (c) 100   (d) 111   (e) 1001

2. Convert the following binary numbers to their decimal equivalent.
   (a) 1010   (b) 1110   (c) 10101   (d) 10110   (e) 11011

3. Convert the following binary numbers to their decimal equivalent.
   (a) 10100.11   (b) 110101.01   (c) 1110011.011
   (d) 111101.111   (e) 1100001.0001

4. Convert the following decimal numbers to their binary equivalent by using the sum-of-weights method.
   (a) 18   (b) 39   (c) 57   (d) 72   (e) 96

5. Convert the following decimal numbers to their binary equivalent by using the sum-of-weights method.
   (a) 246   (b) 123   (c) 479   (d) 168   (e) 326

6. Convert the following decimal numbers to their binary equivalent by repeated division by 2.
   (a) 15   (b) 27   (c) 61   (d) 84   (e) 97

7. Convert the following decimal numbers to their binary equivalent by repeated division by 2.
   (a) 137   (b) 254   (c) 381   (d) 295   (e) 432

8. Add the following binary numbers.
   (a) 10 + 11   (b) 10 + 10   (c) 11 + 11
   (d) 110 + 10   (e) 101 + 100   (f) 111 + 101

9. Add the following binary numbers.

```
(a)     101     (b)     1011    (c)     101011   (d)     1110101
        110           + 1101          + 110101         + 1011111
      + 111
```

10. Add the following binary numbers.

```
(a)     1011    (b)     1010110  (c)     1010110  (d)     10110110
        1101            1110101          1111011          11010101
      + 1011          + 1001010        + 1011111        + 11010110
```

11. Subtract the following binary numbers.
   (a) 11 − 10   (b) 101 − 11   (c) 111 − 101
   (d) 110 − 11   (e) 111 − 10

12. Subtract the following binary numbers.
   (a) 101011 − 100101   (b) 10101 − 100111
   (c) 100101 − 100011   (d) 10100101 − 10011111
   (e) 11100001 − 10011110

13. Determine the 1's complement of the following binary numbers.

(a) 1010   (b) 1101   (c) 10110   (d) 11011   (e) 101101
(f) 100101

14. Subtract the following binary numbers by the method of 1's-complement addition.

(a) $1011 - 101$ (b) $1101 - 1001$
(c) $11010 - 10101$ (d) $110101 - 101101$

15. Determine the 2's complement of the following binary numbers.

(a) 1101   (b) 1001   (c) 10110
(d) 11010   (e) 11001   (f) 101101

16. Subtract the following binary numbers by the method of 2's-complement addition.

(a) $1001 - 110$ (b) $1110 - 1001$
(c) $11011 - 10010$ (d) $110101 - 100111$

17. Multiply the following binary numbers.

(a) 11×11 (b) 101×10 (c) 110×101
(d) 1010×101 (e) 1011×1101 (f) 11110×10101

18. Divide the following binary numbers.

(a) $110 \div 10$ (b) $1001 \div 11$ (c) $1010 \div 101$
(d) $100011 \div 111$ (e) $110110 \div 1001$

19. Convert the following octal numbers to their decimal equivalent.

(a) 14_8 (b) 36_8 (c) 47_8
(d) 75_8 (e) 236_8 (f) 1432_8

20. Convert the following octal numbers to their decimal equivalent.

(a) 0.43_8 (b) 0.671_8 (c) 0.254_8

21. Convert the following decimal numbers to their octal equivalent.

(a) 29   (b) 68   (c) 125
(d) 243.67   (e) 419.35   (f) 634.58

22. Convert the following octal numbers to their binary equivalent.

(a) 13_8 (b) 27_8 (c) 65_8
(d) 124.375_8 (e) 217.436_8 (f) 777.77_8

23. Convert the following binary numbers to their octal equivalent.

(a) 10110010   (b) 10101101   (c) 1110101
(d) 110101101   (e) 10111.101   (f) 111010.001

24. Add the following octal numbers.

(a) $16_8 + 23_8$ (b) $14_8 + 15_8$ (c) $17_8 + 31_8$
(d) $12_8 + 35_8$ (e) $20_8 + 10_8$ (f) $25_8 + 35_8$

25. Convert the following hexadecimal numbers to their binary equivalent.

(a) 27_{16} (b) $A21_{16}$ (c) $5CD_{16}$
(d) $4F2.C_{16}$ (e) $6B3.E$ (f) $FEF.B2$

26. Convert the following binary numbers to their hexadecimal equivalent.

(a) 111001010010   (b) 101000010011
(c) 101100101.110   (d) 1110101101.001

27. Convert the following hexadecimal numbers to their decimal equivalent.

(a) 39_{16} (b) $B4_{16}$ (c) $6CEA_{16}$
(d) $EA76_{16}$ (e) $F941$ (f) $EA87$

28. Convert the following decimal numbers to their hexadecimal equivalent.

(a) 429   (b) 1758   (c) 2143

29. Add the following hexadecimal numbers.

(a) $28_{16} + 37_{16}$ (b) $84_{16} + 7A_{16}$
(c) $1B5_{16} + 72E_{16}$ (d) $AE6_{16} + 41D_{16}$

30. Subtract the following hexadecimal numbers by the method of 2's-complement addition.

(a) $743_{16} - 521_{16}$ (b) $C7E_{16} - 462_{16}$
(c) $9EF_{16} - 6B5_{16}$ (d) $B8D_{16} - A2C_{16}$

31. Add the following numbers and express the result as a decimal number.

(a) $63_8 + 29_{16}$ (b) $42_8 + 10101101_2$
(c) $FF_{16} + 11010110_2$

REFERENCES

Floyd, Thomas L., *Digital Fundamentals*. Columbus, Ohio: Charles E. Merrill Publishing Company, 1982.

Levine, Morris E., *Digital Theory and Practice Using Integrated Circuits*. Englewood Cliffs, N.J.: Prentice-Hall, Inc., 1978.

Malvino, Albert P., and Donald P. Leach, *Digital Principles and Applications*, 3rd ed. New York: McGraw-Hill Book Company, 1981.

Porat, Dan, and Arpad Barna, *Introduction to Digital Techniques*. New York: John Wiley & Sons, Inc., 1979.

7

Arithmetic Circuits

INSTRUCTIONAL OBJECTIVES

In this chapter we discuss basic arithmetic circuits that are used to add and subtract binary numbers. After completing the chapter, you should be able to:

1. Trace logic signals through an arithmetic circuit to determine the outputs for a given set of inputs.
2. Determine the Boolean expression for the output of a given arithmetic circuit.
3. Define or describe the following terms:
   (a) Modulo-2 addition.
   (b) Half-adder.
   (c) Full adder.
   (d) Half-subtracter.
   (e) Full subtracter.
   (f) End-around carry.
   (g) Look-ahead carry.
   (h) True/complement circuit.
4. Identify an arithmetic circuit from its truth table.
5. Draw the truth table for the arithmetic circuits discussed.
6. Determine the number of half-adders and full adders required to implement a parallel adder to add n bits.
7. Describe the operation of each arithmetic circuit discussed.
8. Design basic adder or subtracter circuits.

SELF-EVALUATION QUESTIONS

The following questions deal with the material presented in this chapter. Read the questions prior to studying the chapter and, as you read through the material, watch for answers to the questions. After completing the chapter, return to this section and evaluate your comprehension of the material by answering questions again.

1. Why is the Exclusive-OR gate so named?
2. What is the difference between a half-adder and a full adder?
3. How many half-adders and full adders are needed to construct a parallel binary adder that can add two 5-bit binary numbers?
4. What is the advantage of subtraction by complementary addition over ordinary binary subtraction?
5. What is the purpose of the look-ahead carry?
6. How many input terminals and output terminals does a full adder have?

7. What is the purpose of the MODE input on the true/complement circuit?
8. What is the symbol for modulo-2 addition?
9. What circuit can be connected to a parallel adder to permit it to function as an adder/subtracter circuit?
10. What is the Boolean equation at the output of an Exclusive-NOR gate?

7-1
INTRODUCTION

In this chapter we discuss digital circuits that electronically perform the arithmetic operations that we studied in Chapter 6. The arithmetic circuits discussed are fabricated from the basic logic gates introduced in Chapter 2, so even though this chapter presents several new concepts, it will serve to pull together the concepts that have been presented to this point.

As we study arithmetic circuits, we should not lose sight of the fact *all* digital circuits respond to logic HIGH and LOW signals. Therefore, when performing arithmetic operations, binary 1's and 0's are viewed as *logic levels* by the logic gates. The gates are interconnected in such a way that the logic levels generate an output that agrees with the arithmetic operations performed with paper and pencil in Chapter 6.

7-2
THE EXCLUSIVE-OR GATE

The Exclusive-OR gate is a very interesting logic device that is used extensively in arithmetic circuits. Strictly speaking, the logic circuit that performs the Exclusive-OR function is not a logic gate but a combination of the basic logic gates discussed in Chapter 2. On the other hand, the circuit has two input terminals and one output terminal, which fits the definition of a logic gate; consequently, it is categorized as a logic gate. The logic diagram for the Exclusive-OR gate is shown in Fig. 7-1. One can easily verify the Boolean algebra expression at the output terminal of the gate by tracing the path of the input signals. Table 7-1 is the truth table for the Exclusive-OR gate. We can conclude from the truth table that the circuit performs the logic proposition of *either–or*, but not both; hence the name Exclusive-OR.

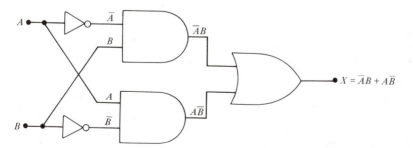

FIGURE 7-1 Exclusive-OR circuit

TABLE 7-1. / Exclusive-OR-Gate Truth Table

A	B	X
0	0	0
0	1	1
1	0	1
1	1	0

Since the logic circuit shown in Fig. 7-1 is categorized as a logic gate, it is assigned a unique Exclusive-OR gate symbol, which is shown in Fig. 7-2. Another reason the Exclusive-OR gate is given a logic gate symbol is due to the fact that because of its wide use, it is manufactured as an integrated circuit, as shown in Fig. 7-3. When working with the Exclusive-OR IC, users see only input and output terminals; thus they view the circuit as a logic gate. In conjunction with the Exclusive-OR logic symbol, the mathematical symbol $\oplus$, implying exclusion, is used to indicate that the output expression describes the Exclusive-OR function.

The response of the Exclusive-OR gate to input logic signals is sometimes called *modulo-2 addition*. This type of addition is performed in the same manner as

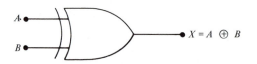
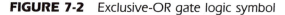

$X = A \oplus B$

FIGURE 7-2 Exclusive-OR gate logic symbol

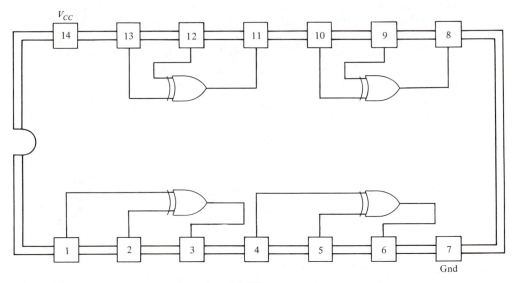

FIGURE 7-3 7486 Quad Exclusive-OR IC

ordinary binary addition except that all carries are discarded, as can be seen by
the following rules for modulo-2 addition:

$$0 + 0 = 0$$
$$0 + 1 = 1$$
$$1 + 0 = 1$$
$$1 + 1 = 0 \qquad \text{(discard the carry)}$$

By observing the rules for modulo-2 addition we can see that the output of the
Exclusive-OR gate is HIGH only when the inputs are different, that is, one HIGH
and one LOW.

There is one other significant difference between the OR gate studied in Chapter
2, sometimes called an Inclusive-OR gate, and the Exclusive-OR gate. Inclusive-
OR gates may have any number of input terminals, whereas Exclusive-OR gates
must have exactly two.

EXAMPLE 7-1 ━━━

Determine the signal level on the IC pins to which the output of each Exclusive-
OR gate in Fig. 7-4 is connected if the input levels are as shown.

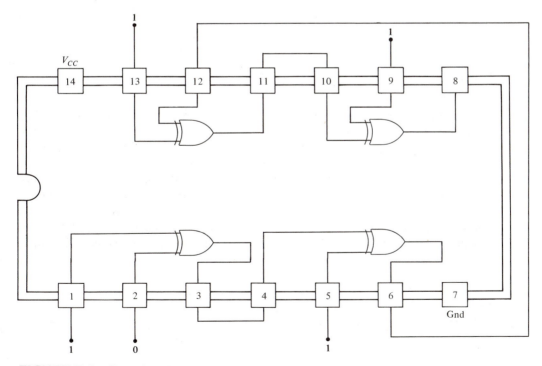

FIGURE 7-4 Exclusive-OR gate circuit

SOLUTION

Starting at pins 1 and 2, we have

$$\text{logic level}_{\text{pin } 3} = 1 \oplus 0 = 1$$

$$\text{logic level}_{\text{pin } 6} = 1 \oplus 1 = 0$$

$$\text{logic level}_{\text{pin } 11} = 1 \oplus 0 = 1$$

$$\text{logic level}_{\text{pin } 8} = 1 \oplus 1 = 0 \qquad \square$$

7-3
EXCLUSIVE-NOR GATE

The Exclusive-NOR gate is essentially an Exclusive-OR gate followed by an inverter, as shown in Fig. 7-5. This circuit implies that if we construct a circuit, using individual logic gates, to function as an Exclusive-NOR gate, it will contain all the logic gates and NOT functions required to construct the Exclusive-OR gate plus an additional NOT circuit. However, the output equation of Fig. 7-5 can be rewritten using De Morgan's theorems as follows:

$$X = \overline{A\bar{B} + \bar{A}B}$$

$$= \overline{\overline{A\bar{B}} \cdot \overline{\bar{A}B}}$$

$$= \overline{A\bar{B}} \cdot \overline{\bar{A}B}$$

$$= \overline{\bar{A} + B} \cdot \overline{A + \bar{B}}$$

$$= (\bar{A} + B)(A + \bar{B})$$

$$= AB + \bar{A}\bar{B}$$

The logic circuit required to implement this Boolean expression is shown in Fig. 7-6. As can be seen, this circuit actually requires the same number of logic gates and inverters as the Exclusive-OR gate circuit of Fig. 7-1. Therefore, we eliminated one inverter by using De Morgan's theorems.

As with the Exclusive-OR gate, the Exclusive-NOR gate is assigned a unique symbol, which is shown in Fig. 7-7. The truth table for the gate, which can be developed from the output equation, is shown in Table 7-2. By observing the truth table, we can see that the output of Exclusive-NOR gates is HIGH whenever both inputs are at the same logic level. For this reason they are sometimes called "coincidence gates."

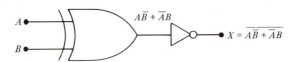

FIGURE 7-5 Exclusive-NOR gate

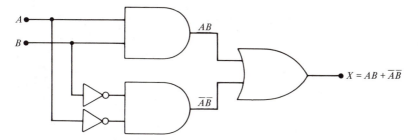

FIGURE 7-6   Exclusive-NOR circuit

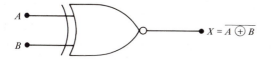

FIGURE 7-7   Exclusive-NOR-gate logic symbol

TABLE 7-2. / Exclusive-NOR-Gate Truth Table

A	B	X
0	0	1
0	1	0
1	0	0
1	1	1

As was the case with Exclusive-OR gates, Exclusive-NOR gates always have two input terminals and are manufactured in IC packages as shown in Fig. 7-8. A more popular IC is the Exclusive-OR/NOR package shown in Fig. 7-9. The operation is easier understood by redrawing the gates in a more standard logic diagram form, as shown in Fig. 7-10. Table 7-3 shows the truth table for the circuit. Terminals A and B together with intermediate point Z function as an Exclusive-OR gate, as we would expect. If input C is LOW, the relationship between terminals A, B, and X is that of an Exclusive-OR gate. If input C is

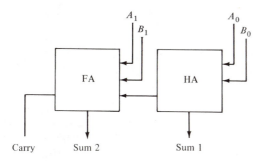

FIGURE 7-8   74LS266 quad exclusive-NOR IC

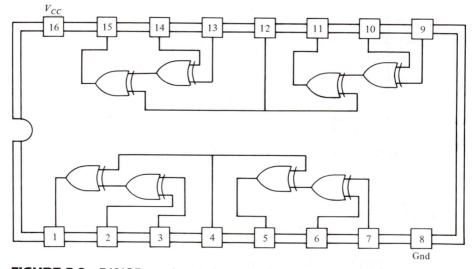

FIGURE 7-9   74S135 quad exclusive-OR/NOR IC

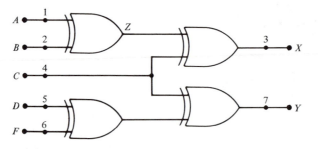

FIGURE 7-10   Exclusive-OR/NOR circuit

TABLE 7-3. / Exclusive-OR/NOR-Gate Truth Table

A	B	Z	C	X
0	0	0	0	0
0	1	1	0	1
1	0	1	0	1
1	1	0	0	0
0	0	0	1	1
0	1	1	1	0
1	0	1	1	0
1	1	0	1	1

HIGH, the relationship between terminals A, B, and X is that of an Exclusive-NOR gate. Therefore, we can see that the logic level of terminal C determines whether the circuit functions as an Exclusive-OR gate or as an Exclusive-NOR gate.

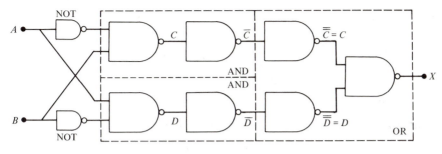

FIGURE 7-11   Figure 7-1 drawn with NAND gates

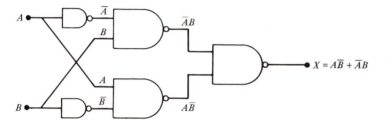

FIGURE 7-12   Exclusive-OR gate using only NAND gates

EXAMPLE 7-2 ▬▬▬▬▬▬▬▬▬▬▬▬▬▬▬▬▬▬▬▬▬▬▬▬▬▬▬

Show how an Exclusive-OR gate can be constructed using NAND gates only.

SOLUTION

We will begin by replacing each logic gate and inverter in Fig. 7-1 with its equivalent constructed with NAND gates as shown in Fig. 7-11. Since the logic signals at point C and $\overline{\overline{C}}$ are the same as are those at D and $\overline{\overline{D}}$, four NAND gates can be omitted. Therefore, our required circuit is as shown in Fig. 7-12. The output equation is put in the desired form by use of De Morgan's law as follows:

$$X = \overline{\overline{A\overline{B}} \cdot \overline{\overline{A}B}}$$

$$= \overline{\overline{A\overline{B}}} + \overline{\overline{\overline{A}B}} = A\overline{B} + \overline{A}B \qquad \square$$

7-4
ADDER CIRCUITS

One of the primary applications of Exclusive-OR gates is in adder circuits, including half-adders, full adders, and parallel binary full adders.

7-4.1   Half-Adders

The most basic arithmetic circuit is the half-adder, which is shown in Fig. 7-13. As can be seen, it can be constructed with an Exclusive-OR gate and an AND

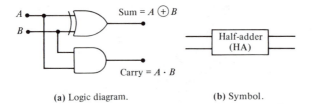

(a) Logic diagram.   (b) Symbol.

FIGURE 7-13   Half-adder: (a) logic diagram; (b) symbol

**TABLE 7-4. / Half-Adder
Truth Table**

A	B	Carry	Sum
0	0	0	0
0	1	0	1
1	0	0	1
1	1	1	0

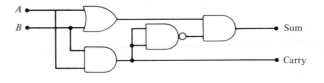

(a)

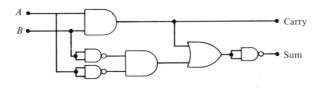

(b)

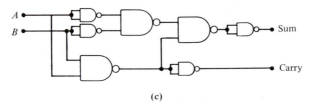

(c)

FIGURE 7-14   Half-adder circuits

gate. This basic circuit is capable of adding two binary digits, where the output of the Exclusive-OR gate corresponds to the *sum,* or the least significant bit, and the output of the AND gate corresponds to the *carry,* or the most significant bit.

Since half-adders have two input terminals, there are four combinations of

1's and 0's which they must be able to add. These combinations and the resulting output signals are as follows:

1. When $A = 0$ and $B = 0$:

$$\text{sum} = A \oplus B = 0 \oplus 0 = 0$$

$$\text{carry} = A \cdot B = 0 \cdot 0 = 0$$

2. When $A = 1$ and $B = 0$:

$$\text{sum} = 1 \oplus 0 = 1$$

$$\text{carry} = 1 \cdot 0 = 0$$

3. When $A = 0$ and $B = 1$:

$$\text{sum} = 0 \oplus 1 = 1$$

$$\text{carry} = 0 \cdot 1 = 0$$

4. When $A = 1$ and $B = 1$:

$$\text{sum} = 1 \oplus 1 = 0$$

$$\text{carry} = 1 \cdot 1 = 1$$

These conditions agree with the results obtained with binary arithmetic, as shown below.

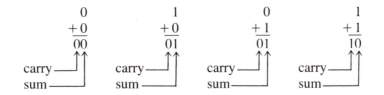

Table 7-4 is the truth table for the half-adder.

There are many other ways besides that shown in Fig. 7-13 to interconnect logic gates so that they function as half-adders. Three other half-adder circuits are shown in Fig. 7-14. Half-adders are certainly very basic arithmetic circuits whose individual capabilities are quickly exceeded in any kind of practical computational system. However, they are a first step toward circuits that are capable of more difficult arithmetic operations.

7-4.2 Full Adders

When adding two multiple-bit binary numbers, a carry is very often generated, which prohibits the use of half-adders. For example, when we add the numbers,

```
        1   ←carry from LSB column
      1 0 1
      1 1 1
    ─────────
    1 1 0 0
```

there is a carry from the least significant bit column, which necessitates adding 3 bits in the next-higher-order column.

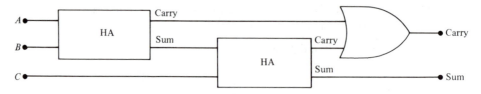

FIGURE 7-15 Full adder from half-adders

To add 3 bits electronically requires a logic circuit called a *full adder*. We can construct a full adder by using two half-adders and an OR gate connected as shown in Fig. 7-15. Since each half-adder consists of an Exclusive-OR gate and an AND gate, we can easily draw the logic diagram for the full adder. This, as well as the symbol for the full adder, are shown in Fig. 7-16. Tracing the input signals through the logic circuit permits us to obtain a Boolean expression for the sum and carry terminals. The Boolean expression at the sum terminal is developed as follows:

$$
\begin{aligned}
\text{sum} &= A\bar{B} + \bar{A}B \oplus C \\
&= (A\bar{B} + \bar{A}B)\bar{C} + \overline{(A\bar{B} + \bar{A}B)}C \\
&= A\bar{B}\bar{C} + \bar{A}B\bar{C} + (\overline{A\bar{B}} \cdot \overline{\bar{A}B})C \\
&= A\bar{B}\bar{C} + \bar{A}B\bar{C} + \left[(\bar{A} + \bar{\bar{B}})(\bar{\bar{A}} + \bar{B})\right]C \\
&= A\bar{B}\bar{C} + \bar{A}B\bar{C} + \left[(\bar{A} + B)(A + \bar{B})\right]C \\
&= A\bar{B}\bar{C} + \bar{A}B\bar{C} + ABC + \bar{A}\bar{B}C
\end{aligned}
$$

By using the Boolean expressions for the sum and carry terminals, the full-adder truth table can be developed as shown in Table 7-5.

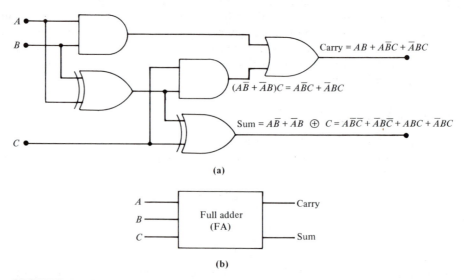

(a)

(b)

FIGURE 7-16 Full adder: (a) logic diagram; (b) symbol

TABLE 7-5. / Full-Adder Truth Table

A	B	C	Carry	Sum
0	0	0	0	0
0	0	1	0	1
0	1	0	0	1
0	1	1	1	0
1	0	0	0	1
1	0	1	1	0
1	1	0	1	0
1	1	1	1	1

EXAMPLE 7-3 ▬▬▬▬▬▬▬▬▬▬▬▬▬▬▬▬▬▬▬▬▬▬▬▬▬▬▬

Show that the correct sum and carry values will be seen at the output terminals of the full adder shown in Fig. 7-15 when the following combinations of binary bits are applied to the input terminals.

(a) $1 + 0 + 0$ (b) $1 + 1 + 0$ (c) $1 + 1 + 1$

SOLUTION

By adding the binary digits, we can readily see what binary numbers should be observed at the output terminals

$$
\begin{array}{lll}
\text{(a)} \quad
\begin{array}{r}
1 \\
+\,0 \\
+\,0 \\
\hline
01
\end{array}
&
\text{(b)} \quad
\begin{array}{r}
1 \\
+\,1 \\
+\,0 \\
\hline
10
\end{array}
&
\text{(c)} \quad
\begin{array}{r}
1 \\
+\,1 \\
+\,1 \\
\hline
11
\end{array}
\end{array}
$$

Figure 7-15 is redrawn in Fig. 7-17 with the desired combinations of binary inputs

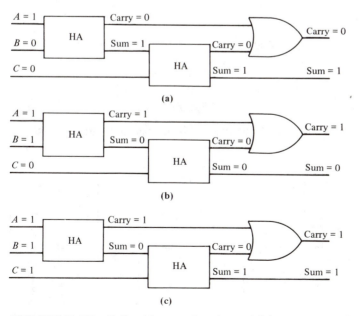

FIGURE 7-17 *Full adders performing addition*

and the resulting binary values at intermediate points and at the sum and carry terminals.   ❏

7-4.3 Parallel Binary Adders

The full-adder circuit discussed in the preceding section can add only three single binary digits; therefore, additional circuitry is required to add two multiple-digit binary numbers. To perform such addition, we can connect half-adders and full adders as shown in Fig. 7-18. The circuit shown, which is called a *parallel binary adder*, is capable of adding two 4-bit binary numbers. In the figure, the numbers $A_3 A_2 A_1 A_0$ and $B_3 B_2 B_1 B_0$ are being added. The sum is

$$\begin{array}{r} A_3\ A_2\ A_1\ A_0 \\ +\,B_3\ B_2\ B_1\ B_0 \\ \hline S_4\ S_3\ S_2\ S_1\ S_0 \end{array}$$

Since there are only 2 bits to be added in the least significant bit column, only a half-adder is required. However, in any other column there may be a carry from the preceding column; therefore, we must use a full adder in each column except the first. To see how the parallel adder works, consider the following example.

EXAMPLE 7-4 ━━━

Determine the binary value at the sum terminals when the following binary numbers are applied to the input terminals of the parallel adder shown in Fig. 7-19: $1101 + 1011$.

SOLUTION

We can see by using ordinary binary arithmetic that the sum is

$$\begin{array}{r} 1101 \\ +\,1011 \\ \hline 11000 \end{array}$$

which agrees with the results obtained with our parallel adder.   ❏

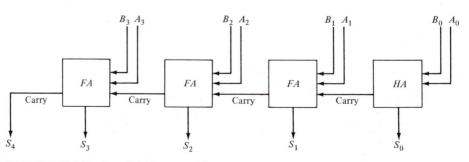

FIGURE 7-18 Parallel binary adder

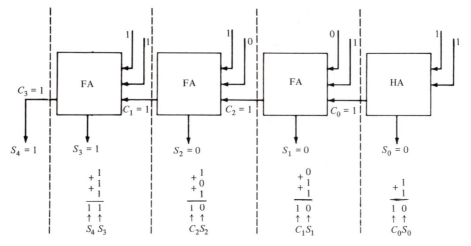

FIGURE 7-19 Addition with parallel binary adder

The parallel adder circuit shown in Fig. 7-19 can add two binary numbers whose maximum size is four digits. To add larger numbers, more full adders can be connected to the end of the circuit opposite the half-adder. An additional full adder is required for each additional digit.

We can add more than two multiple-digit binary numbers by adding two and then adding the sum of these to the third number. For example, consider the following three binary numbers.

$$
\text{add three numbers}\;\left\{\;
\begin{array}{l}
\quad 110 \\
+\,101 \\
\underline{+\,011} \\
\quad 1110
\end{array}
\qquad
\begin{array}{l}
\left.\begin{array}{l}
\quad 110 \\
\underline{+\,101}
\end{array}\right\}\;\text{add two numbers} \\
\quad 1011 \\
\left.\begin{array}{l}
+\;\underline{\;011} \\
\;\;1110
\end{array}\right\}\;\text{add third number}
\end{array}
\right.
$$

The circuit for adding three numbers is shown in Fig. 7-20.

If we wish to add more than three multiple-digit numbers, we must add another parallel adder for each additional number. Although this is a valid technique for

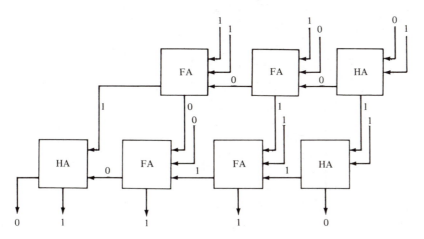

FIGURE 7-20 Adding three numbers with parallel binary adders

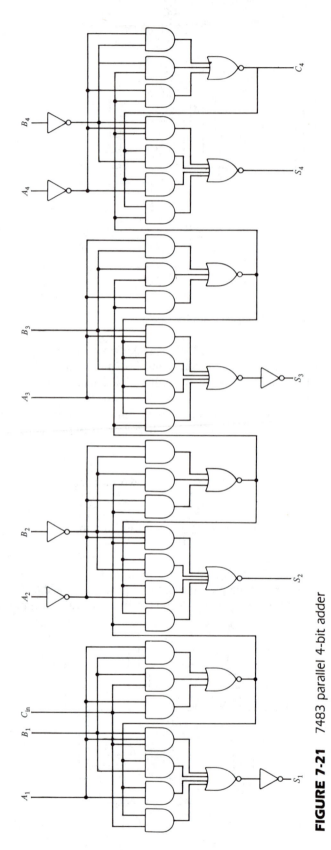

FIGURE 7-21 7483 parallel 4-bit adder

adding binary numbers, modern calculators and computers use parallel adders in conjunction with circuits called registers and accumulators to add columns of numbers. This method of adding is discussed in Chapter 13.

7-4.4 Integrated-Circuit Adders

In most practical computational circuits, multiple-digit numbers are added; therefore, most IC adders are designed to add two 4-bit numbers. However, 1-bit as well as 2-bit adders are also available in the TTL 7400 series.

The 7483 is an MSI parallel adder that is capable of adding two 4-bit binary numbers. The circuit, which is shown in Fig. 7-21, is designed for medium- to high-speed operation and incorporates a *look-ahead* carry. Look-ahead carry means that rather than propagate the carry-out through each adder, the outputs of the gates that OR A_n and B_n and that AND A_n and B_n are applied to the logic gates associated with the carry-out. If these logic levels are such that a carry-out would be generated after all the binary numbers propagate through the complete parallel adder circuit, the carry-out is generated with very little propagation delay. Typical propagation delay for the 7483 4-bit parallel adder with look-ahead carry is about 12 ns. This is about the same as the propagation delay experienced by the signals at the sum terminals; therefore, the carry-out signal appears at approximately the same time as, or even sooner than, the digits at the sum terminals.

7-5
SUBTRACTER CIRCUITS

Binary subtraction is the process of finding the difference between two binary numbers. The types of logic circuits that perform binary subtraction are very similar to the adder circuits already discussed. In fact, by using complementary addition to perform subtraction, the same circuits can be used for both addition and subtraction.

7-5.1 Half-Subtracters

Half-subtracters are the most basic logic circuits for performing binary subtraction. As with half-adders, only two binary bits are involved. The subtraction process generates a difference term D and a borrow term B_o according to the following arithmetic processes:

$$\text{minuend } (A) \searrow \quad \swarrow \text{subtrahend } (B)$$
$$0 - 0 = 0 \text{ difference, } \quad \text{borrow} = 0$$
$$1 - 0 = 1 \text{ difference, } \quad \text{borrow} = 0$$
$$0 - 1 = 1 \text{ difference, } \quad \text{borrow} = 1$$
$$1 - 1 = 0 \text{ difference, } \quad \text{borrow} = 0$$

Inspection of the difference column reveals that the difference always equals the Exclusive-OR of the inputs A and B. Therefore, we can write a Boolean equation for the difference output as

$$D = A \oplus B$$

TABLE 7-6. / Half-Subtracter Truth Table

A	B	Difference	Borrow
0	0	0	0
0	1	1	1
1	0	1	0
1	1	0	0

Further observation of the arithmetic processes reveals that a borrow is generated for only one set of conditions, when $A = 0$ and $B = 1$. Therefore, we can write the Boolean equation for the borrow output as

$$B_o = \overline{A} \cdot B$$

Using these Boolean expressions, we can develop the truth table for the half-subtracter (Table 7-6). The logic circuit that is required to implement the difference and borrow expressions for the half-subtracter, as well as the half-subtracter symbol, is shown in Fig. 7-22. The following example illustrates subtraction as performed by the half-subtracter.

EXAMPLE 7-5 ━━━━━━━━━━━━━━━━━━━━━━━━━━━━━━━━━━━━━━

Show that the correct difference and borrow values will be seen at the output terminals of the half-subtracter shown in Fig. 7-22 when the following combinations of binary bits are applied to the input terminals.

(a) $1 - 0$ (b) $0 - 1$

SOLUTION

Figure 7-22 is redrawn in Fig. 7-23 with the desired combinations of inputs and the resulting binary values at the difference and borrow terminals.   ❏

The binary digit at the difference terminal is the solution to the subtraction problem. As you will see in the next section, the borrow bit becomes important when performing subtraction of multiple-digit numbers with parallel full subtracters.

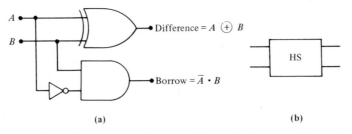

(a)   (b)

FIGURE 7-22 Half-subtracter: (a) logic diagram; (b) symbol

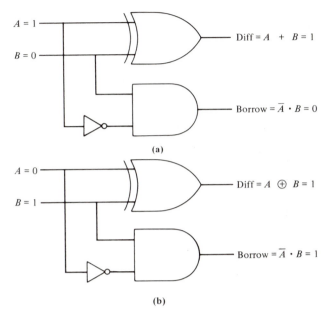

(a)

(b)

FIGURE 7-23 Subtraction with half-subtracters

7-5.2 Full Subtracters

As in adder circuits, a *full subtracter* is required for all bits, other than the least significant bit, when doing subtraction of multiple-digit numbers. Each subtracter state, including the half-subtracter for the least significant bit, generates a *borrow-out* signal if it must borrow from a more significant state. Each stage except the half-subtracter for the least significant bit may receive a *borrow-in* signal from the preceding stage. Therefore, full subtracters must have three input terminals, which are A, B, and borrow-in (B_i), and two output terminals, which are a difference output (D) and a borrow-out (B_o).

We can construct a full subtracter by using two half-subtracters and an OR gate as shown in Fig. 7-24. Since each half-subtracter is made up of an Exclusive-OR gate, an AND gate, and an inverter, as shown in Fig. 7-22, we can draw the logic diagram for the full subtracter. This, as well as the symbol for the full subtracter, is shown in Fig. 7-25. By tracing the input signals through the logic circuit, we

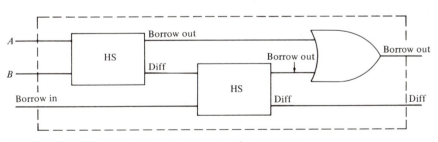

FIGURE 7-24 Full subtracter from half-subtracters

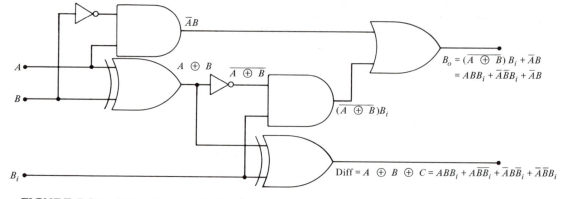

FIGURE 7-25 Full subtracter logic diagram

can obtain a Boolean expression for the difference and borrow-out terminals. The final Boolean expression at the difference output is developed as follows:

$$D = A \oplus B \oplus B_i$$

$$= A\overline{B} + \overline{A}B \oplus B_i$$

$$= (A\overline{B} + \overline{A}B)\overline{B}_i + (\overline{A\overline{B} + \overline{A}B})B_i$$

$$= A\overline{B}\overline{B}_i + \overline{A}B\overline{B}_i + [(\overline{A} + B)(A + \overline{B})]B_i$$

$$= A\overline{B}\overline{B}_i + \overline{A}B\overline{B}_i + ABB_i + \overline{A}\overline{B}B_i$$

The final expression for the borrow-out terminal is developed as follows:

$$B_o = \overline{(A \oplus B)}_i + \overline{A}B$$

$$= (\overline{A\overline{B} + \overline{A}B})B_i + \overline{A}B$$

$$= [(\overline{A} + B)(A + \overline{B})]B_i + \overline{A}B$$

$$= ABB_i + \overline{A}\overline{B}B_i\overline{A}B$$

Using the Boolean expressions for the difference and borrow-out terminals, we can develop the truth table for the full subtracter (Table 7-7).

TABLE 7-7. / Full-Subtracter Truth Table

A	B	B_i	Difference	Borrow-Out
0	0	0	0	0
0	0	1	1	1
0	1	0	1	1
0	1	1	0	1
1	0	0	1	0
1	0	1	0	0
1	1	0	0	0
1	1	1	1	1

EXAMPLE 7-6 ━━━━━━━━━━━━━━━━━

Draw a block diagram to show what is required to perform the following subtraction problem.

$$A_1 A_0$$
$$- \underline{B_1 B_0}$$

SOLUTION

A half-subtracter is required for the least significant bits, whereas a full subtracter is required for the most significant bits. The blocks are interconnected as shown in Fig. 7-26. ❏

EXAMPLE 7-7 ━━━━━━━━━━━━━━━━━

Show that the correct results are obtained when the following subtraction problem is performed using the circuit shown in Fig. 7-27. ❏

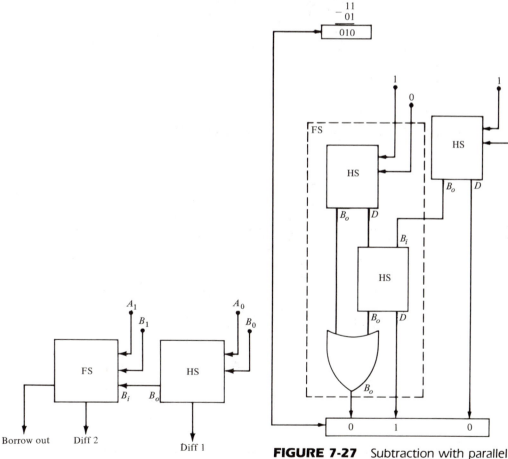

FIGURE 7-26 Parallel full subtracter

FIGURE 7-27 Subtraction with parallel full subtracter

7-6
SUBTRACTION BY 1'S-COMPLEMENT ADDITION

Most modern computational systems perform subtraction by 1's-or 2's-complement addition, since less total circuitry is required. Once the complement of the subtrahend is obtained, the problem becomes an addition problem; therefore, the same circuitry as that used for addition is used for subtraction. The 1's complement of the subtrahend is easily obtained by the use of inverters, as can be seen in Fig. 7-28. However, recall that when doing subtraction by 1's-complement addition, an end-around carry is involved. To accomplish this with arithmetic circuits requires another parallel adder, connected as shown in Fig. 7-29.

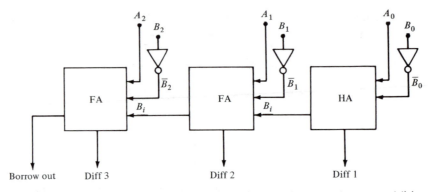

FIGURE 7-28 Circuit for doing subtraction by 1's-complement addition

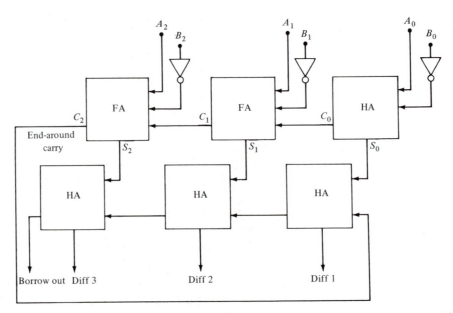

FIGURE 7-29 Complete circuit for doing subtraction by 1's-complement addition

EXAMPLE 7-8 ━━━━━━━━━━━━━━━━━━━━━━━━━━━━━━━

Show that the correct results are obtained when the following subtraction problem is performed as a 1's-complement addition problem using the circuit shown in Fig. 7-29.

$$\begin{array}{r} 111 \\ -\ \underline{101} \end{array}$$

SOLUTION

For purposes of comparison, the solution to the problem is

Binary Subtraction	1's-Complement Addition
$\begin{array}{r} 111 \\ -\ \underline{101} \\ 010 \end{array}$	$\begin{array}{r} 111 \\ +\ \underline{010} \\ 1001 \\ +\ \underline{\ 11} \\ 010 \end{array}$

Applying the binary numbers to the adder circuit shown in Fig. 7-30 yields the same results.

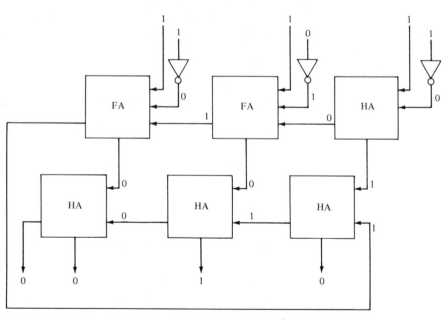

FIGURE 7-30   Subtraction by 1's-complement addition

7-7
SUBTRACTION BY 2'S-COMPLEMENT ADDITION

The 2's-complement system offers several attractive characteristics with regard to performing arithmetic operations related to signed numbers and handling the signs that will be discussed in Chapter 13. Another important characteristic is the ease

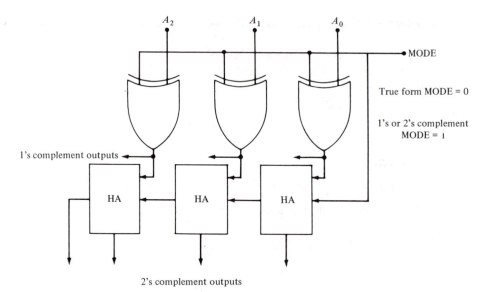

FIGURE 7-31   True/complement circuit

with which systems that perform both addition and subtraction can be constructed. Such systems incorporate the circuit shown in Fig. 7-31, which is called a *true/complement circuit*, which, incidentally, illustrates another important application of Exclusive-OR gates. To illustrate how the true/complement circuit works, consider the following example.

EXAMPLE 7-9 ━━━━━━━━━━━━━━━━━━━━━━━━━━━━━━━━━━━

Use the true/complement circuit to obtain the 2's-complement of 111 as shown in Fig. 7-32.

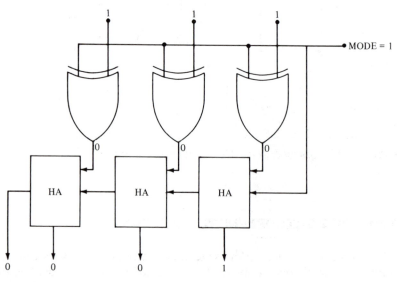

FIGURE 7-32   Obtaining 2's complement of 111

SOLUTION

To obtain the 2's complement, the MODE input must be a logic 1. If we connect the outputs of the true/complement circuit to one set of inputs of a parallel binary adder, as shown in Fig. 7-33, we have an adder/subtracter circuit. Actually, the half-adders in Fig. 7-33 are no longer needed to obtain the 2's complement of the number $B_3B_2B_1B_0$ in the subtract mode. If we connect the circuit as shown in Fig. 7-34, we obtain the 2's complement by adding the carry-in, which equals 1 in the subtract mode, to the 1's complement of $B_3B_2B_1B_0$. The circuit provides

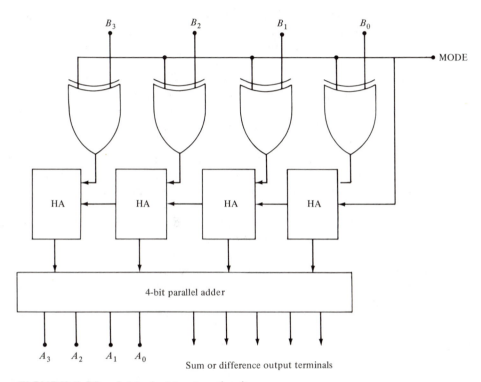

FIGURE 7-33 Adder/subtracter circuit

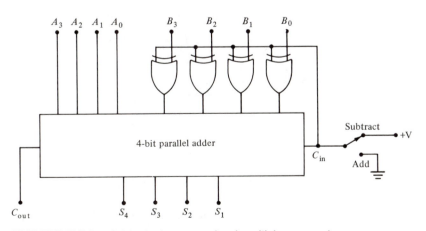

FIGURE 7-34 Adder/subtracter circuit utilizing carry-in

an output equal to $(A_3A_2A_1A_0) \pm (B_3B_2B_1B_0)$. For the moment, the most significant bit (the carry-out) of the result will be discarded if the MODE input is logic 1, which indicates subtraction, but is part of the result in an addition problem (when MODE = 0). This bit will be dealt with in Chapter 13 when we deal with the treatment of signed numbers in arithmetic circuits.   ❏

7-8
SUMMARY

Logic gates respond to HIGH and LOW logic levels; however, by equating binary 1 to logic HIGH and binary 0 to logic LOW the gates are, in effect, responding to binary data. By properly interconnecting these logic gates, they function as arithmetic circuits capable of providing the sum or difference of binary data.

Exclusive-OR gates, which are used extensively in adder and subtracter circuits, generate the sum or difference output by performing modulo-2 addition. This kind of addition is the same as ordinary binary addition except that we discard any carries. The symbol for modulo-2 addition is $\oplus$.

The primary adder circuits of interest are half-adders, full adders, and parallel binary adders. Half-adders have two input terminals and two output terminals and can add two binary bits. Full adders, which may consist of two half-adders and an OR gate, have three input terminals and two output terminals and can add three binary bits. By connecting half-adders and full adders in parallel, we can make a parallel binary adder, which is capable of adding numbers with many digits.

Subtraction can be performed by using special subtracter circuits or by using adder circuits with complemented numbers. Most modern computational systems perform subtraction by complementary addition and, by doing so, use the circuits that are used for addition. A wide range of arithmetic circuits are available in integrated-circuit form.

PROBLEMS

1. Show that each circuit in Fig. 7-14 is a half-adder by writing the output equations and developing the truth table for each.

2. Show how a half-adder can be fabricated using only NOR gates.

3. Draw the output waveforms for the following circuit.

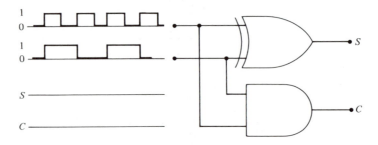

4. Draw the logic diagram of a full-adder using only NAND gates.

5. Draw the logic diagram of a half-subtracter using only NAND gates.

6. Determine what kind of circuit (adder, subtracter, etc.) is shown below by writing the Boolean expressions for the output terminals and developing the truth table.

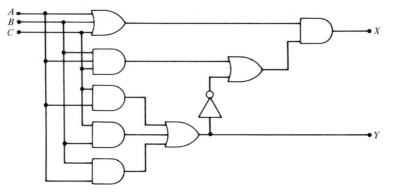

7. Determine what kind of circuit (adder, subtracter, etc.) is shown below by writing the Boolean expressions for the output terminals and developing the truth table.

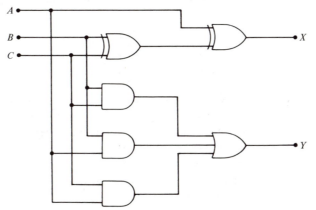

8. Use the 1's-complement adder of Fig. 7-29 to perform the subtraction $110 - 011$.

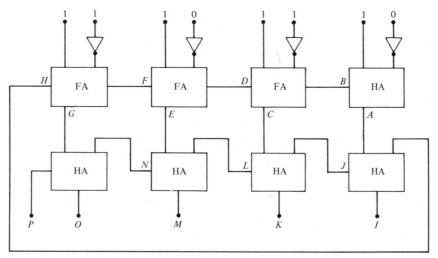

9. With the binary numbers applied to the input terminals of the arithmetic circuit as shown below, indicate the binary number that exists at points *A* through *F*.

10. Draw the complete logic diagram for the following parallel adder block diagram.

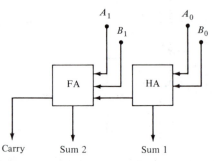

11. Design a full adder using AND, OR, and Exclusive-OR gates by synthesizing the sum and carry equations shown below. Show any Boolean algebra performed, the truth table, minterms for the sum and carry, and the logic diagram.

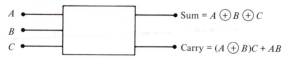

12. The typical propagation delay for the look-ahead carry for a parallel binary adder is about 12 ns. Draw the logic diagram for the parallel adder shown below and determine the propagation delay between terminal B_0 and the output carry terminal assuming that A_0 plus B_0 generate a carry and using the following propagation delay times: AND gates, 12 ns; OR gates, 10 ns; inverters, 14 ns.

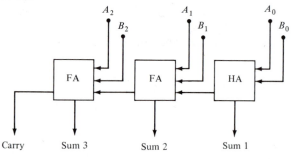

REFERENCES

Deem, William, Kenneth Muchow, and Anthony Zeppa, *Digital Computer Circuits and Concepts.* Reston, Va.: Reston Publishing Co., Inc., 1980.

Greenfield, Joseph D., *Practical Digital Design Using ICs,* 2nd ed. New York: John Wiley & Sons, Inc., 1983.

Malvino, Albert P., and Donald P. Leach, *Digital Principles and Applications,* 3rd ed. New York: McGraw-Hill Book Company, 1981.

Nashelsky, Louis, *Introduction to Digital Technology,* 3rd ed. New York: John Wiley & Sons, Inc., 1983.

Multivibrators

INSTRUCTIONAL OBJECTIVES

In this chapter you are introduced to multivibrators and some of their applications in digital electronics. After completing the chapter, you should be able to:

1. List the different types of multivibrators discussed.
2. Describe the difference between combinational and sequential logic circuits.
3. Describe the difference between latches and flip-flops.
4. Show how flip-flops are constructed with NAND gates or NOR gates.
5. Define or describe the following terms:
    (a) Timing diagram.
    (b) Synchronous operation.
    (c) Asynchronous operation.
    (d) Single-rail device.
    (e) Double-rail device.
    (f) Race condition.
    (g) Setup time.
    (h) Hold time.
    (i) Hysteresis.
6. Describe the purpose of the direct SET and CLEAR inputs on flip-flops.
7. Describe the difference between edge triggering and level triggering.
8. Describe the major advantage of *JK* flip-flops over *RS* flip-flops.
9. State what type of multivibrator a Schmitt trigger is classified as and list two other names by which it is known.
10. Compute the frequency and duty cycle of an astable multivibrator.
11. Compute the duration of the output pulse of a monostable multivibrator.
12. Draw the output waveforms when a set of input waveforms are given for any of the bistable multivibrators discussed.
13. Describe the major advantage of master-slave flip-flops.
14. List the two modes of operation of 555 timers.

SELF-EVALUATION QUESTIONS

The following questions deal with the material presented in this chapter. Read the questions prior to studying the chapter and, as you read through the material, watch for answers to the questions. After completing the chapter, return to this section and evaluate your comprehension of the material by answering the questions again.

1. How many stable states does each of the following multivibrators have?
    (a) Astable.

201

        (b) Bistable.

        (c) Monostable.

2. Define a sequential logic circuit.

3. What does the term *synchronous operation* mean?

4. What is the major advantage of *JK* flip-flops over *RS* flip-flops?

5. What conditions create a ''race problem'' with a flip-flop?

6. Describe the difference between a latch and a flip-flop.

7. By what other name is a Schmitt trigger known?

8. Which kind of flip-flop provides single-rail data?

9. What are the two modes of operation of 555 timers?

10. What is setup time?

8-1
INTRODUCTION

Multivibrators are regenerative circuits with two active devices. By using positive feedback, one of the devices is made to conduct while the other is cut off. There are three basic types of multivibrators in use: *bistable, monostable,* and *astable*. Although each type has important applications in digital circuits and is discussed in this chapter, major emphasis will be on bistable multivibrators.

The development of integrated circuits has tended to reduce the importance of a detailed study of the interval circuitry of all logic devices, including multivibrators. Therefore, in this chapter we focus on the operation, characteristics, and applications of multivibrators.

8-2
BISTABLE MULTIVIBRATORS

The logic circuits that have been discussed to this point are categorized as combinational logic circuits, since there was no memory associated with the circuits. Multivibrators fall in the category of *sequential* logic circuits, since their output depends on both the present state of their output as well as on their input signals. Sequential logic circuits have memory and therefore complement combinational logic circuits in most logic systems. The most widely used *memory* element is the bistable multivibrator.

Bistable multivibrators are usually called *flip-flops*. Such circuits have two output states, logic 1 and logic 0. Both states are said to be *stable states*, which means that whichever state the output is in, it will remain in that state until it is caused to change to the other state by an input signal called a *trigger*. The general symbol for a flip-flop is shown in Fig. 8-1. The inputs are for establishing initial conditions (Q HIGH, $\overline{Q}$ LOW, etc.) and/or a pulse train input from a clock. The outputs, which are generally designated as Q and $\overline{Q}$, are the complements of each other. When Q is HIGH, $\overline{Q}$ is LOW, or vice versa. It is very important to recognize that when an input signal causes a flip-flop output to be in a certain state, the output will remain in that state even after the input signal is removed. This is a characteristic of a *memory* device. Several types of bistable multivibrators are discussed in the following paragraphs.

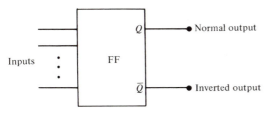

FIGURE 8-1   *General flip-flop symbol*

8-3
RS FLIP-FLOPS

RS flip-flops are the most basic bistable multivibrator. Circuits that function as *RS* flip-flops can be fabricated from discrete components or with logic gates: NAND or NOR gates. Since the vast majority of the flip-flops in use today are fabricated as ICs, the form of the actual circuit is of little concern. We shall build our discussion of the operation and characteristics of flip-flops around circuits constructed with logic gates.

8-4
RS LATCHES

Latches are a type of bistable multivibrator that are normally placed in a separate category from flip-flops. The logic for an *RS* latch is identical to the logic circuit for an *RS* flip-flop, assuming that the same kind of gates are used. Both are regenerative circuits with two stable states. The main difference between latches and flip-flops lies in the method used to cause each to change its output state.

Latches are *level*-controlled circuits, whereas flip-flops respond to a *transition* of a triggering input. As such, latches may be thought of as bistable multivibrators that are made to change states by manually operating a mechanical switch, whereas flip-flops are made to change states by a train of pulses called *clock pulses*.

Our primary concern in this chapter is with regard to flip-flops; therefore, no major distinction will be made between flip-flops and latches in subsequent paragraphs.

8-5
NOR-GATE *RS* FLIP-FLOPS

Figure 8-2 shows an *RS* flip-flop constructed with NOR gates. Note that the output of NOR gate 1 serves as one of the inputs to gate 2, and vice versa. This kind of connection is referred to as *feedback*. The circuit operates as a flip-flop because of nonsymmetry between the two NOR gates. If *every* component in both gates were *identical*, both gates would attempt to turn ON at the same instant when dc voltage is applied. Since the gates are not likely to be identical, one will turn ON an instant before the other. The fact that one gate is ON prohibits the other from turning ON. To analyze the operation of a flip-flop, one must assume that one

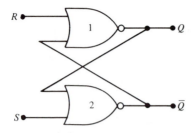

FIGURE 8-2 NOR-gate *RS* flip-flop

gate (or other active device) is ON and the other gate OFF. It makes no difference which is assumed to be ON, as each will be ON, or OFF, half the time.

If we assume that the output of gate 2 in Fig. 8-2 is LOW, the output of gate 1 must be HIGH. If Q is HIGH, the upper input of gate 2 is HIGH. Also, if $\overline{Q}$ is LOW, the lower input of gate 1 is LOW. If these signal levels exist, Q can remain HIGH only if R equals 0. If $\overline{Q}$ is LOW and the upper input of gate 2 is HIGH, it makes no difference whether the signal at S equals 1 or 0, $\overline{Q}$ will remain LOW. On the other hand, if the signal at R is changed to a logic 1, output Q will be forced LOW while $\overline{Q}$ must go HIGH. When $\overline{Q}$ is HIGH, the lower input of gate 1 is HIGH; therefore, it makes no difference whether the signal at R equals 1 or 0, Q will remain LOW.

We can use the foregoing relationships and the redrawn logic diagram for the *RS* flip-flop shown in Fig. 8-3 to develop the truth table for an *RS* flip-flop. We will work toward the truth table for the complete circuit by considering first the truth tables for the individual NOR gates 1 and 2, which are shown in Table 8-1. These truth tables are drawn as though the gates are independent of each other, which is obviously not the case. The truth tables show Q and $\overline{Q}$ as inputs; however, their primary role is as output signals, where they are dependent on the inputs at R and S. Therefore, our primary concerns are the R and S inputs. From Table 8-1a we can see that any time R equals 1, Q equals 0, but whenever R equals 0, Q may be either 0 or 1. From Table 8-1b we can see that any time S equals 1, $\overline{Q}$ equals 0, but whenever S equals 0, $\overline{Q}$ may equal either 0 or 1. Furthermore, the third row in both tables is not possible, since Q and $\overline{Q}$ cannot be 0 at the same time. Therefore, our primary interest lies with the first and fourth rows of the tables.

From the fourth row of Table 8-1a we see that when R equals 1, Q equals 0; therefore, $\overline{Q}$ must equal 1. From the first row of Table 8-1b, when Q equals 1, S equals 0. Also, from the fourth row of Table 8-1b, when S equals 1, $\overline{Q}$ equals 0 and Q equals 1. Finally, from the first row of Table 8-1a, Q equals 1 when R

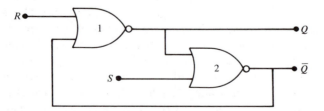

FIGURE 8-3 Redrawn *RS* flip-flop

TABLE 8-1. / Truth Tables for NOR Gates of Fig. 8-3

R	$\overline{Q}$	Q
0	0	1
0	1	0
1	0	0
1	1	0

(a)

S	Q	$\overline{Q}$
0	0	1
0	1	0
1	0	0
1	1	0

(b)

equals 0. Combining these conditions, we can develop Table 8-2, which is the truth table for the *RS* flip-flop fabricated with NOR gates. The *R* and *S* inputs are identified as such by virtue of their effect on the output at *Q*. When the RESET (*R*) input equals 1, it resets *Q* to 0. When the SET (*S*) input equals 1, it sets *Q* to 1. The operation of the NOR-gate *RS* flip-flop is summarized in the following statements.

1. Applying a 1 to the SET input sets *Q* HIGH. If *Q* is already HIGH, no change occurs.
2. Applying a 0 to the SET input causes no change regardless of the level of *Q*.
3. Applying a 1 to the RESET input resets *Q* LOW. If *Q* is already LOW, no change occurs.
4. Applying a 0 to the RESET input causes no change regardless of the level of *Q*.
5. Applying a 1 to both inputs simultaneously is not allowed.

To clarify the operation of sequential logic circuits, it is often beneficial to construct *timing diagrams* that show various circuit voltage levels as a function of time. The timing diagram for the NOR-gate *RS* flip-flop is shown in Fig. 8-4. As can be seen, whenever a pulse is applied to the SET input, *Q* is set HIGH unless it is already HIGH, and whenever a pulse is applied to the RESET input, *Q* is reset LOW unless it is already LOW. Any time *Q* goes HIGH, $\overline{Q}$ goes LOW, and vice versa.

TABLE 8-2. / NOR-Gate *RS* Flip-Flop Truth Table

R	S	Q	$\overline{Q}$	Command
1	0	0	1	Reset
0	1	1	0	Set
0	0	No change		Remember
1	1	Not allowed		Indeterminate

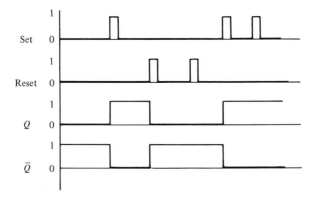

FIGURE 8-4   Timing diagram for NOR-gate *RS*
flip-flop

EXAMPLE 8-1 ━━━

An *RS* flip-flop can be constructed from the 7402 quad two-input NOR gate. Show
the wiring diagram if pin 2 on the IC is SET and pin 6 is RESET. Draw the timing
diagram if the SET and RESET inputs go HIGH for 0.1 ms at the following times.

$$\text{SET:} \quad 1 \text{ ms, } 4 \text{ ms, } 5.5 \text{ ms, } 7 \text{ ms}$$
$$\text{RESET: } 2 \text{ ms, } 3 \text{ ms, } \quad 5 \text{ ms, } 8 \text{ ms}$$

Q is initially LOW and $\overline{Q}$ is HIGH.

SOLUTION

The wiring and timing diagrams are shown in Fig. 8-5.   ❏

8-6
NAND-GATE *RS* FLIP-FLOPS

The discussion thus far has dealt with *RS* flip-flops fabricated with NOR gates.
Such circuits can also be built with NAND gates, as shown in Fig. 8-6. As before,
if we consider each NAND gate individually and combine their truth tables, we
obtain the truth table for the NAND-gate *RS* flip-flop. The individual truth tables
are shown in Table 8-3. We can see from Table 8-3a that any time S equals 0,
Q equals 9; however, when S equals 1, Q may be either 1 or 0. From Table 8-
3b we can see that whenever R equals 0, $\overline{Q}$ equals 1, but when $R = 1$, $\overline{Q}$ may
be either 1 or 0. This means we set Q HIGH or reset Q LOW with a logic 0 on
the SET or RESET inputs, respectively. The complete truth table for the NAND-
gate *RS* flip-flop is shown in Table 8-4.

The operation of the NAND-gate *RS* flip-flop is summarized in the following
statements.

1. Applying a 0 to the SET input sets Q HIGH. If Q is already HIGH, no
   change occurs.
2. Applying a 1 to the SET input causes no change regardless of the level
   of Q.

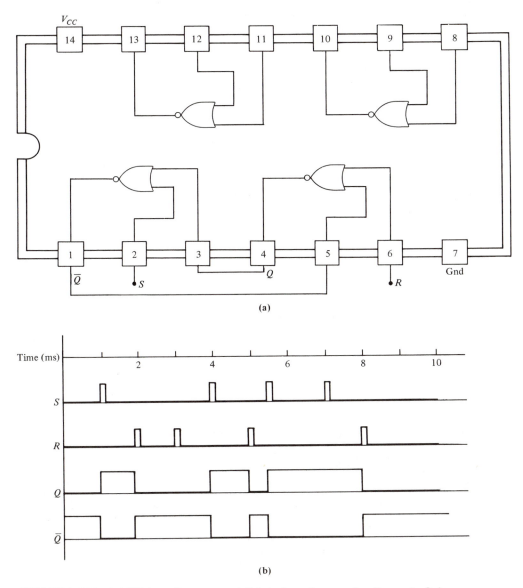

FIGURE 8-5 (a) Wiring diagram and (b) timing diagram for Example 8-1

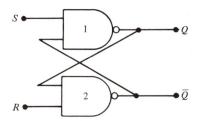

FIGURE 8-6 *RS* flip-flop fabricated with NAND gates

TABLE 8-3. / Truth Tables for NAND Gates of Fig. 8-6

S	$\overline{Q}$	Q
0	0	1
0	1	1
1	0	1
1	1	0

(a)

R	Q	$\overline{Q}$
0	0	1
0	1	1
1	0	1
1	1	0

(b)

TABLE 8-4. / NAND-Gate *RS* Flip-Flop Truth Table

R	S	Q	$\overline{Q}$	Command
0	1	0	1	Reset
1	0	1	0	Set
1	1	No change		Remember
0	0	Not allowed		Indeterminate

3. Applying a 0 to the RESET input rests Q LOW. If Q is already LOW, no change occurs.
4. Applying a 1 to the RESET input causes no change regardless of the level of Q.
5. Applying a 0 to both inputs simultaneously is not allowed.

The timing diagram for the NAND-gate *RS* flip-flop, which is shown in Fig. 8-7, will help clarify the operation of the circuit. Note that when the SET input goes to logic 0, the output Q is set HIGH unless it is already HIGH, and when the RESET input goes to logic 0, the output is reset LOW unless it is already LOW. Whenever Q goes HIGH, $\overline{Q}$ goes LOW, and vice versa.

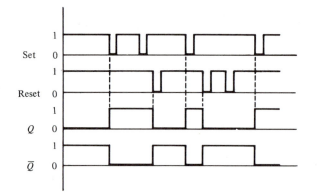

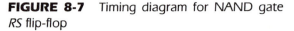

FIGURE 8-7   Timing diagram for NAND gate *RS* flip-flop

8-7
RS FLIP-FLOP SYMBOLS

As we have seen, *RS* flip-flops may be fabricated with either NAND or NOR gates; however, they will be set or reset by opposite logic levels. To distinguish between NAND- and NOR-gate *RS* flip-flops, the general flip-flop symbol of Fig. 8-1 is modified slightly, as shown in Fig. 8-8. The symbol and partial truth table in Fig. 8-8a is for an *RS* flip-flop fabricated with NOR gates, and Fig. 8-8b shows the symbol and partial truth table for the NAND-gate flip-flop. The circles at the inputs can be associated with the NAND-gate symbol shown in Fig. 8-6.

Example 8-2 ━━━━━━━━━━━━━━━━━━━━━━━━━━━━━━━━━

The output of each of the *RS* flip-flops shown in Fig. 8-9 is initially in the LOW state ($Q = 0$). Draw the timing diagram for each flip-flop.

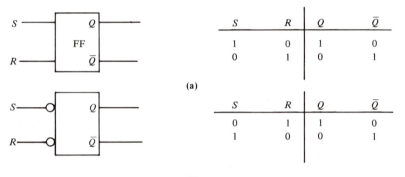

S	R	Q	$\bar{Q}$
1	0	1	0
0	1	0	1

(a)

S	R	Q	$\bar{Q}$
0	1	1	0
1	0	0	1

(b)

FIGURE 8-8 Symbol for *RS* flip-flop: (a) assembled with NOR gates; (b) assembled with NAND gates

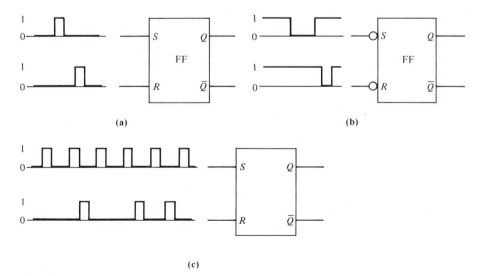

FIGURE 8-9 Illustrations for Example 8-2

SOLUTION

The timing diagram for each flip-flop is shown in Fig. 8-10.

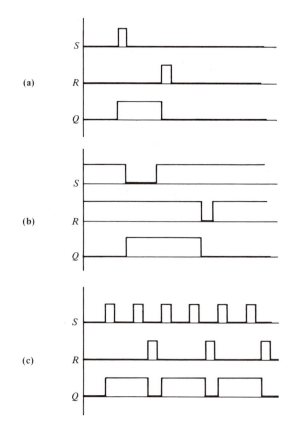

FIGURE 8-10   Timing diagrams for Example 8-2   ❏

8-8
CLOCKED CIRCUITS

Most digital systems operate in a *synchronous* mode, which means that the operation of the system is synchronized by a *master clock signal*. The master clock generates periodic pulses that are transmitted throughout the system. Such systems are referred to as *clocked systems* or *triggered systems*.

The clock usually generates a square-wave pulse train; however, the actual trigger pulse seen by many circuits within the system is a very short duration pulse, usually on the order of a few nanoseconds. Trigger pulses can be obtained by connecting the pulse train from the master clock to a pulse-narrowing circuit, as shown in Fig. 8-11. A negative-going pulse can be obtained by replacing the AND gate with a NAND gate or by using another inverter after the AND gate.

Flip-flops are designed to change states due to a pulse transition from 0 to 1 or from 1 to 0, but not both. The points of transition are shown in Fig. 8-12. The transition from 0-to-1 is called the *leading edge* or *positive-going edge* of the

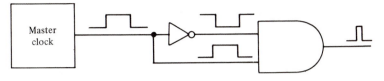

FIGURE 8-11 Pulse-narrowing circuit

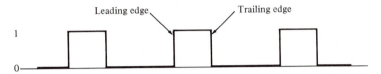

FIGURE 8-12 Pulse transition points

pulse, and the 1-to-0 transition is called the *trailing edge* or *negative-going edge* of the pulse.

8-9
CLOCKED *RS* FLIP-FLOPS

When using flip-flops, it is often desirable to establish the logic levels of R and S without actually setting or resetting the flip-flop the moment these levels are established. Most digital systems operate in this manner by using clocked, or triggered, flip-flops. Such flip-flops have a third input called the clock, or trigger, input, which, as its name implies, triggers the flip-flop. This causes the output to establish itself at the logic level dictated by the preestablished logic levels of R and S. However, the change occurs on the clock pulse rather than when the levels of R and S are established.

Flip-flops may be designed so that they trigger on either the leading or trailing edge of the clock pulse. Figure 8-13 shows the logic symbol and logic diagram for an *RS* flip-flop that changes states on the leading (positive-going) edge of the clock pulse. The NOR gates in Fig. 8-13b form a basic, unclocked *RS* flip-flop such as that discussed in the preceding section. This part of the circuit is generally called a *latch*, since it responds to what basically amounts to a dc level. The incorporation of the AND gates provides us with a means of triggered operation. A timing diagram for a leading-edge triggered *RS* flip-flop is shown in Fig. 8-14.

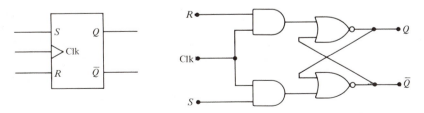

FIGURE 8-13 Leading-edge-triggered clocked *RS* flip-flop: (a) symbol; (b) logic circuit

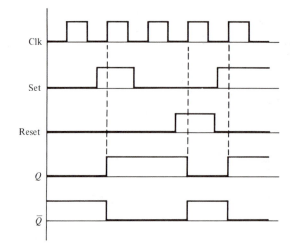

FIGURE 8-14 Timing diagram for leading-edge-triggered *RS* flip-flop

By observing the timing diagram, we can analyze the operation of a clocked *RS* flip-flop as follows:

1. Initially, both data inputs (inputs *R* and *S*) are LOW and the flip-flop is reset (*Q* is LOW).

2. Since *R* and *S* are both LOW at the leading edge of the first clock pulse, the flip-flop is not triggered; therefore, *Q* remains LOW.

3. Prior to the second clock pulse, the *S* input is set HIGH; therefore, the flip-flop is triggered on the leading edge of the second clock pulse. The output at *Q* is now HIGH. This is equivalent to storing a binary 1 in the flip-flop.

4. At the leading edge of the third clock pulse, both *R* and *S* are LOW; therefore, the flip-flop "remembers" (no change in output occurs).

5. Prior to the fourth clock pulse, the *R* input is set HIGH; therefore, *Q* is reset LOW at the leading edge of the clock pulse. This is equivalent to storing a binary 0 in the flip-flop.

6. Prior to the fifth clock pulse, the *S* input is set HIGH and the *R* input is reset LOW; therefore, the flip-flop changes states, which sets *Q* HIGH.

7. No change in the output occurs on the sixth clock pulse, since *R* and *S* are the same as at the beginning of the preceding clock pulse.

The effect of the clock pulse is to transfer the data at the *R* and *S* inputs into the flip-flop for temporary storage by the data storage block consisting of the two NOR gates. The *R* and *S* input levels should be set prior to the leading edge of the clock pulse and should not change states during a clock pulse. The possibility of this occurring can be significantly reduced by using a pulse-narrowing circuit such as the one shown in Fig. 8-11.

The change of state from a LOW output to a HIGH state, or vice versa, by a flip-flop is often referred to as a *toggle*. Flip-flops may be designed to toggle on the trailing edge of the clock pulse as well as on the leading edge. The logic

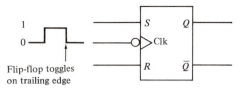
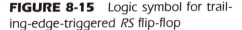

1

0

Flip-flop toggles
on trailing edge

FIGURE 8-15 Logic symbol for trail-ing-edge-triggered *RS* flip-flop

symbol for a trailing-edge triggered *RS* flip-flop is shown in Fig. 8-15. The tri-angular symbol on the CLOCK input is called a *dynamic input indicator* and is used to indicate that the flip-flop is *transition triggered,* or *edge triggered,* rather than *level triggered.* The circle at the CLOCK input denotes the use of NAND gates in the data storage block that toggle to a HIGH state on a 1-to-0 transition.

The CLOCK input shown in Fig. 8-15 is often designated as the TRIGGER input. When this terminology is used, the flip-flop is referred to as an *RST* (reset–set–trigger) flip-flop. In either case, the function of the clock pulse is to "clock," "gate," "strobe," or "trigger" the flip-flop. Clocked *RS* flip-flops have two data input lines, *R* and *S*, which accept what is often called *double-rail* data.

8-10
MASTER-SLAVE *RS* FLIP-FLOPS

In many applications, the clocked *RS* flip-flop has a rather serious limitation in that if the input data change at the instant the flip-flop toggles, incorrect data may be stored by the flip-flop. This problem is referred to as a *race* condition and may be encountered wherever flip-flops are used.

The race problem can be solved by using *master-slave flip-flops,* which provide isolation between the input and output by effectively disconnecting the input sig-nals at the instant the clock pulse arrives. This type of the flip-flop is made up of two sections, called a *master* section and a *slave* section, as shown in Fig. 8-16. Both the master and the slave sections are leading-edge-triggered, clocked *RS* flip-flops. As can be seen, the slave section is controlled by outputs of the master section rather than by external inputs. The master section will be set to the state determined by the *R* and *S* inputs on the leading edge of the clock pulse. The outputs of the master section now serve as inputs to the slave section; therefore, the state of the master section will be transferred to the slave section on the trailing edge of the clock pulse. This is because the clock pulse is inverted; therefore, the

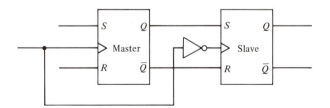

FIGURE 8-16 *RS* master-slave flip-flop

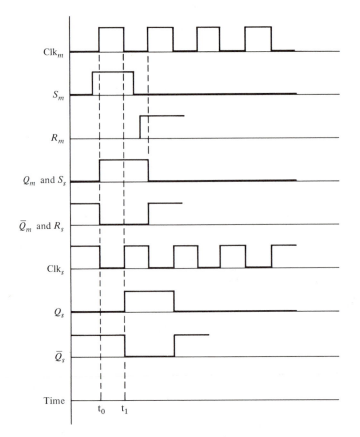

FIGURE 8-17 Timing diagram for master-slave *RS* flip-flop

1-to-0 transition at the clock pulse's trailing edge becomes a 0-to-1 transition at the slave CLOCK input.

The operation of the master-slave (M/S) flip-flop shown in Fig. 8-16 is best explained by considering separately the events that occur at the leading edge and at the trailing edge of the clock pulse. These events can be seen on the timing diagram for the circuit shown in Fig. 8-17.

LEADING-EDGE EVENTS

1. At t_0, the leading edge of the clock pulse triggers the master. The slave is disabled, since its CLOCK input sees a pulse transition to 0.

2. The output of the master (Q_m) goes *HIGH* while $\overline{Q}_m$ goes LOW. These outputs provide set and reset inputs to the slave (S_s and R_s).

TRAILING-EDGE EVENTS

1. At t_1, the negative transition of the clock pulse is inverted. This is seen as a leading-edge clock pulse by the slave; therefore, its output (Q_s) goes HIGH.

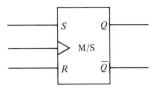

FIGURE 8-18 Logic symbol for master-slave *RS* flip-flop

2. The logic level of *S* at the leading edge of the clock pulse is now stored by the slave.

The timing associated with the operation of an *RS* master-slave flip-flop, whose logic symbol is shown in Fig. 8-18, is actually somewhat more complicated than described above in that, to avoid any race problems, the logic gates are designed so that the master is disabled before the slave is enabled.

8-11
SYNCHRONOUS AND ASYNCHRONOUS INPUTS

The *R* and *S* inputs associated with *RS* flip-flops are called *synchronous inputs* because transfer of data from these inputs to the output of the flip-flop occurs on the triggering edge of the clock pulse; that is, data transfer is synchronized with the clock.

Many integrated circuits also have *asynchronous inputs*. These inputs, which are normally called *preset* and *clear,* affect the state of the flip-flop independent of the clock. These inputs, which are shown in Fig. 8-19, are connected directly to the slave flip-flop and therefore have priority over all other inputs. The flip-flop shown is preset or cleared by connecting a positive voltage, corresponding to a logic 1, to the PRESET or CLEAR input. For example, if the CLEAR input is set HIGH, the output is cleared (set LOW) regardless of the *S, R,* or CLOCK inputs. When digital systems are turned on, flip-flop outputs may be either HIGH or LOW. These inputs are typically used to clear all flip-flops at turn-on or to

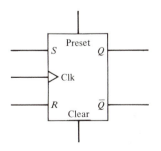

FIGURE 8-19 *RS* flip-flop with preset and clear inputs

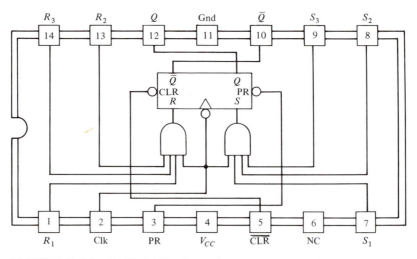

FIGURE 8-20   7471 *RS* flip-flop with preset and clear inputs

preset the state of the flip-flop. Attempting to preset and clear a flip-flop at the same time is not logical and therefore is not allowed.

A commercially produced *RS* master-slave flip-flop with preset and clear inputs is shown in Fig 8-20. This particular device has AND gates on the *R* and *S* inputs and active-LOW CLEAR and PRESET inputs. Therefore, HIGH input data on the *S* input set *Q* HIGH, while a LOW applied to the CLEAR or PRESET input either clears or presets the output.

8-12
D FLIP-FLOPS

RS flip-flops have associated with them two undesirable combinations of input conditions: if *R* and *S* are both LOW or both HIGH at the same time. These combinations cause a ''no change'' or a ''not allowed'' condition to exist. Additionally, since *RS* flip-flops are double-rail devices (two data inputs), two signals are necessary to drive the device, which is often undesirable.

D flip-flops provide a very attractive solution to the problems associated with *RS* flip-flops. The possibility of identical inputs is eliminated by using a single line that is connected directly to the *S* input. Connection to the *R* input is through an inverter as shown in Fig. 8-21a. Since the *D*, or *data*, flip-flop has only one data line, we have *single-rail* data.

D flip-flops are edge-triggered devices, as can be seen from the logic symbol, whose output assumes the state of the *D* input at the proper edge of the clock

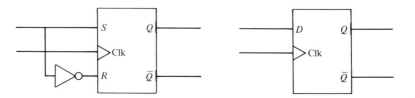

FIGURE 8-21   *D* flip-flop: (a) fabricated from *RS* flip-flop; (b) logic symbol

pulse. Figure 8-22 shows the timing diagram for a leading-edge triggered D flip-flop. Since the output waveform at Q is identical to the waveform on the D input, except delayed by some Δt, D flip-flops are sometimes referred to as *delay flip-flops*. Since there is only one input, the truth table for the D flip-flop is quite simple. It is shown in Table 8-5.

D flip-flops are one of the two primary categories of flip-flops produced by TTL manufacturers, the other being JK flip-flops. One of the most widely used D flip-flops in the TTL series is the 7474. This is a dual-D, positive-edge-triggered flip-flop with preset and clear inputs. The pin layout and truth table are shown in Fig. 8-23.

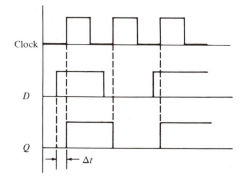

FIGURE 8-22 Timing diagram for a leading-edge-triggered D flip-flop

TABLE 8-5. / D Flip-Flop Truth Table

D	Q	Comments
0	0	Q follows D
1	1	On clock pulse

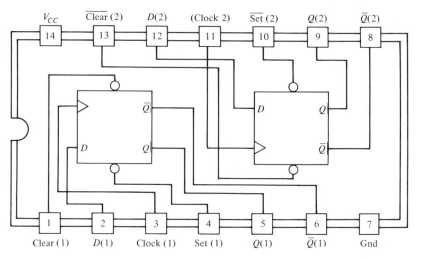

FIGURE 8-23 7474 dual-D flip-flop with leading-edge triggering

8-13
T Flip-Flops

A flip-flop that is very useful in applications where pulses are counted is the T, or *toggle, flip-flop*. Figure 8-24 shows a symbol that is commonly used for T flip-flops. As can be seen, the device has a single input, labeled "clock" or "toggle." Actually, T flip-flops are generally not available as such, but rather are obtained by modifying some other type of flip-flop, often the RS flip-flop. The modifications involve cross-coupled feedback, shown in Fig. 8-25. If the flip-flop is in the reset state, $\overline{Q}$ is HIGH; therefore, S is HIGH. When a clock pulse arrives, the output is set HIGH; therefore, R is HIGH. The next clock pulse will reset the output LOW. By operating in this manner, the output changes states, or toggles, on every clock pulse, which provides the timing diagram shown in Fig. 8-26, for trailing-edge-triggered devices. Observing the waveforms, we can readily see that there are half as many positive-going pulses at the output as there are clock pulses, or, stated differently, the output frequency is exactly one-half the input frequency. This inherent divide-by-2 capability makes T flip-flops very useful in applications such as frequency dividers and electronic counters.

8-14
JK FLIP-FLOPS

JK flip-flops are the most useful and versatile flip-flops available. They can function as clocked *RS* flip-flops or as T flip-flops as well as performing other more

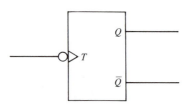

FIGURE 8-24 T flip-flop symbol

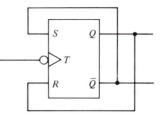

FIGURE 8-25 Modified *RS* flip-flop that functions as T flip-flop

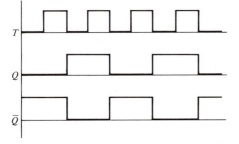

FIGURE 8-26 Timing diagram for T flip-flop

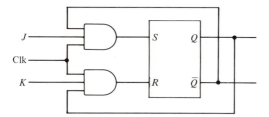

FIGURE 8-27   Basic *JK* flip-flop

specialized functions. Beyond this, their most important characteristic is that they have no invalid combinations of inputs, in that the disallowed combinations associated with *RS* flip-flops is a toggle condition with *JK* flip-flops.

One way to fabricate a *JK* flip-flop is to add a pair of three-input AND gates at the inputs of an *RS* flip-flop, as shown in Fig. 8-27. The *J* and *K* inputs are called *control inputs* because they control the state of the flip-flop in the same way that the *R* and *S* inputs control an *RS* flip-flop. *JK* flip-flops operate as follows.

1. When $J = 1$ and $K = 1$, the *JK* flip-flop behaves the same as a *T* flip-flop. The output changes states on the leading edge of every clock pulse.
2. When $J = 1$ and $K = 0$, the flip-flop will set *Q* HIGH on the first clock pulse if it is not already HIGH. Subsequent pulses will have no affect.
3. When $J = 0$ and $K = 1$, the flip-flop will reset *Q* LOW on the first clock pulse if it is not already LOW. Subsequent pulses will have no affect.
4. When $J = 0$ and $K = 0$, all clock pulses are blocked by the AND gates; therefore, no change occurs in the output state.

These operations are shown in tabular form in Table 8-6, which is the truth table for the *JK* flip-flop.

TABLE 8-6. / *JK* Flip-Flop Truth Table

J	K	Q
0	0	No change
0	1	0
1	0	1
1	1	Toggles

8-15
JK MASTER-SLAVE FLIP-FLOPS

Each of the different types of flip-flops discussed to this point have applications; however, the workhorse of flip-flops is the *JK* master-slave flip-flop. This very versatile device can be used in virtually every application where the flip-flops already discussed are used without the race problems or indeterminate states associated with other flip-flops.

A functional block diagram for the *JK* master-slave flip-flop is shown in Fig. 8-28. As can be seen, the circuit consists of two flip-flops in cascade, with their

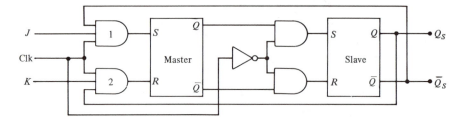

FIGURE 8-28   *JK* master-slave flip-flop

clock inputs connected in a complementary manner. The operation of the circuit
is summarized as follows.

1. Let $J = 0$ and $K = 0$.
   The output of both three-input AND gates will be LOW, which will block
   all clock pulses to the master flip-flop; therefore, no changes occur in the
   output state of the circuit.
2. Let $J = 1$ and $K = 0$.
   a. Assume that $Q_s = 1$ and $Q_s = 0$. When the clock input goes HIGH, the
      inputs to AND gates 1 and 2 in Fig. 8-28 are

   $$\text{Gate 1:} \quad J = 1, \text{Clk} = 1, \overline{Q}_s = 0$$

   $$\text{Gate 2:} \quad K = 0, \text{Clk} = 1, Q_s = 1$$

   Neither AND gate has a HIGH on all its inputs. This condition blocks
   all clock pulses; therefore, no changes occur in the output state of the
   circuit.
   b. Assume that $Q_s = 0$ and $\overline{Q}_s = 1$. When the clock input goes HIGH, the
      inputs to the AND gates are

   $$\text{Gate 1:} \quad J = 1, \text{Clk} = 1, \overline{Q}_s = 1$$

   $$\text{Gate 2:} \quad K = 0, \text{Clk} = 1, Q_s = 0$$

   Since all inputs to AND gate 1 are HIGH, the Q output of the master
   flip-flop will be set HIGH at the leading edge of the clock pulse, and the
   output of the slave, Q_s, will be set HIGH at the trailing edge of the clock
   pulse. Subsequent clock pulses cause no change.
3. Let $J = 0$ and $K = 1$.
   a. Assume that $Q_s = 0$ and $\overline{Q}_s = 1$. When the clock input goes HIGH, the
      inputs to AND gates 1 and 2 are

   $$\text{Gate 1:} \quad J = 0, \text{Clk} = 1, \overline{Q}_s = 1$$

   $$\text{Gate 2:} \quad K = 1, \text{Clk} = 1, Q_s = 0$$

   Since neither AND gate has a HIGH on all its inputs, all clock pulses are
   blocked; therefore, no changes in the output state of the circuit occur.
   b. Assume that $Q_s = 1$ and $\overline{Q}_s = 0$. When the clock input goes HIGH, the
      inputs to AND gates 1 and 2 are

   $$\text{Gate 1:} \quad J = 0, \text{Clk} = 1, \overline{Q}_s = 0$$

   $$\text{Gate 2:} \quad K = 1, \text{Clk} = 1, Q_s = 1$$

Since all inputs to AND gate 2 are HIGH, the Q output of the master flip-flop will be reset LOW at the leading edge of the clock pulse and the output of the slave, Q_s, will be set LOW at the trailing edge of the clock pulse. Subsequent clock pulses cause no change.

4. Let $J = 1$ and $K = 1$.
   a. Assume that $Q_s = 1$ and $\overline{Q}_s = 0$. When the clock input goes HIGH, the inputs to AND gates 1 and 2 are

$$\text{Gate 1:} \quad J = 1,\ \text{Clk} = 1,\ \overline{Q}_s = 0$$

$$\text{Gate 2:} \quad K = 1,\ \text{Clk} = 1,\ Q_s = 1$$

Since all inputs to AND gate 2 are HIGH, the Q output of the master flip-flop will be reset LOW at the leading edge of the clock pulse, and the output of the slave, Q_s, will be reset LOW at the trailing edge of the clock pulse.

   b. Assume that $Q_s = 0$ and $\overline{Q}_s = 1$. When the clock input goes HIGH, the inputs to AND gates 1 and 2 are

$$\text{Gate 1:} \quad J = 1,\ \text{Clk} = 1,\ \overline{Q}_s = 1$$

$$\text{Gate 2:} \quad K = 1,\ \text{Clk} = 1,\ Q_s = 0$$

Since all inputs to AND gate 1 are HIGH, the Q output of the master flip-flop will be set HIGH at the leading edge of the clock pulse and the output of the slave, Q_s, will be set HIGH at the trailing edge of the clock pulse. Each subsequent clock pulse will cause the output of the circuit to change states, or toggle.

The operation described above is shown graphically in the timing diagram of Fig. 8-29. It should be pointed out that while the foregoing description of the operation of master-slave flip-flops and the following waveforms show the changing states at the leading or trailing edges of the clock pulse, they are *level-triggered* devices. They change states when the pulse transition is at a certain level but are not triggered by the transition.

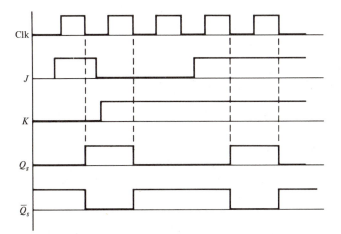

FIGURE 8-29   Timing diagram for *JK* master-slave flip-flop

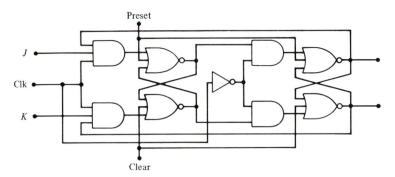

FIGURE 8-30 *JK* master-slave flip-flop with preset and clear inputs

Most *JK* master-slave flip-flops that are commercially available incorporate one additional feature, the asynchronous *preset* and *clear* inputs. These inputs, which are shown in Fig. 8-30, permit flip-flops to be set to a particular state at turn-on or to establish a desired condition.

8-16
INTEGRATED-CIRCUIT *JK* MASTER-SLAVE FLIP-FLOPS

JK master-slave flip-flops are available in most of the popular logic families with the widest selection in TTL. Some of the most common TTL *JK* master-slave flip-flops are the 7473, 7476, and 74107.

The 7473 is a dual flip-flop with individual *J*, *K*, clock, and clear inputs; however, it has no preset input. The maximum toggle frequency is 20 MHz and the maximum supply current per chip is 10 mA. The package is a 14-pin DIP with a nonstandard pin-out.

The 7476 is also a dual flip-flop; however, the package is a 16- rather than a 14-pin DIP. With 16 pins, the flip-flop has individual *J*, *K*, clock, preset, and clear inputs. The maximum toggle frequency is 20 MHz with a maximum supply current of 10 mA.

The 74107 has the same inputs as the 7473 but provides a standard pin-out configuration, as shown in Fig. 8-31.

8-17
LEVEL TRIGGERING VERSUS EDGE TRIGGERING

There are two primary methods of triggering flip-flops, level triggering and edge triggering. Level triggering is suitable for synchronous operation; however, when incoming data are not synchronized with the clock, edge-triggered devices should be used. For example, a pushbutton mounted on the front panel for some type of digital instrument may be depressed at any time by an instrument user. It would be unreasonable to expect this action to be synchronized with the clock. Master-slave devices place this undesirable restriction on us and would therefore be unsuitable for this type of application. Edge-triggered devices overcome this restriction, since triggering occurs only in response to pulse transition. Steady state conditions are not responded to by edge-triggered devices.

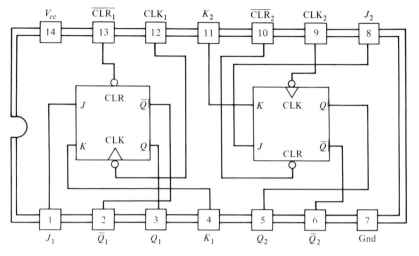

FIGURE 8-31   74107 dual *JK* flip-flop with clear input

8-18
SWITCHING CHARACTERISTICS

As we have seen, flip-flops are basically regenerative circuits that can be constructed with logic gates. Since there is *propagation delay* associated with logic gates, there is also propagation delay associated with flip-flops. The amount of delay in a flip-flop is the sum of the delays associated with the affected logic gates and will vary depending on the particular type of flip-flop.

There are several categories of propagation delay that are encountered in the operation of flip-flops:

1. The *turn-off delay* (t_{PLH}) measured from the *clock pulse* triggering edge to the LOW-to-HIGH transition of the output.
2. The *turn-on delay* (t_{PHL}) measured from the *clock pulse* triggering edge to the HIGH-to-LOW transition of the output.
3. The *turn-off delay* (t_{PLH}) measured from the *preset* input to the LOW-to-HIGH transition of the output.
4. The *turn-on-delay* (t_{PHL}) measured from the *preset* input to the HIGH-to-LOW transition of the output.
5. The *turn-off delay* (t_{PLH}) measured from the *clear* input to the LOW-to-HIGH transition of the output.
6. The *turn-on-delay* (t_{PHL}) measured from the *clear* input to the HIGH-to-LOW transition of the output.

The concept of propagation delay is illustrated in Fig. 8-32, which shows the turn-off delay (t_{PLH}) measured from the triggering edge of the clock pulse to the LOW-to-HIGH transition of the output for a *T* flip-flop.

Two other important flip-flop characteristics are the setup and hold times. The *setup time*, t_S, is the amount of time that the voltage level at the control inputs must remain constant prior to the triggering edge of the clock pulse. The *hold time*, t_H, is the amount of time that the voltage level at the control inputs must remain constant after the triggering edge of clock pulse. For master-slave flip-

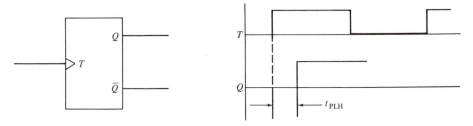

FIGURE 8-32 Propagation delay through a *T* flip-flop

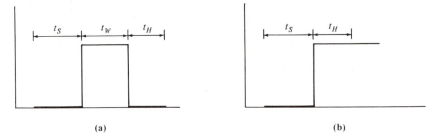

(a)   (b)

FIGURE 8-33 Setup and hold times: (a) master-slave; (b) edge triggered

flops, the control voltages must also remain constant during the clock pulse; therefore, it is usually desirable to use short duration, or low duty cycle, clock pulses when using master-slave flip-flops. The idea of setup and hold times for both master-slave and edge-triggered flip-flops is illustrated in Fig. 8-33. For master-slave flip-flops, the setup time is approximately equal to the pulse duration; however, the hold time is very short as is the hold time for the edge-triggered flip-flops.

8-19
ELECTRIC CHARACTERISTICS

Some of the major electrical parameters associated with the operation of the flip-flops are listed in Table 8-7.

8-20
THE SCHMITT TRIGGER

Schmitt triggers are a somewhat specialized application of bistable multivibrators. Basically, the circuit has two stable states, as do all bistable multivibrators, and is a *level-sensitive* circuit. It switches output states at two distinct voltage levels called the *lower trigger level* (LTL) and the *upper trigger level* (UTL).

Schmitt triggers are often categorized by circuit function as *voltage comparators* or *voltage-level detectors*, since they compare two voltages or detect when a voltage crosses a specified level. They find many applications in digital systems; however, virtually every application involves level detection or signal conditioning. Schmitt triggers provide an excellent mechanism for pulse restoration of

TABLE 8-7. / Typical Electrical Characteristics for Flip-Flops

Symbol	Parameter		Test Conditions	Limits			Units
				Mini-mum	Typical	Maxi-mum	
V_{CC}	Supply voltage			4.75	5.0	5.25	V
V_{IH}	HIGH-level input voltage			2.0			V
V_{IL}	LOW-level input voltage					0.8	V
V_{OH}	HIGH-level output voltage		V_{CC} = min, V_{IH} = min, V_{IL} = max	2.4	3.4		V
V_{OL}	LOW-level output voltage		V_{CC} = min, V_{IH} = min, V_{IL} = max		0.2	0.4	V
I_{IH}	HIGH-level input current	R, S, J, K, or D	V_{CC} = max, V_{in} = 2.4 V			40	μA
		Clear or Preset				80	μA
		Clock				80	μA
		All inputs	V_{CC} = max, V_{in} = 5.5 V			1	mA
I_{IL}	LOW-level input current	R, S, J, K, or D	V_{CC} = max, V_{in} = 0.4 V			−1.6	mA
		Clear or Preset				−3.2	mA
		Clock				−3.2	mA
I_{OH}	HIGH-level output current					−400	μA
I_{OL}	LOW-level output current					16	mA
I_{OS}	Short-circuit output current		V_{CC} = max	−18		−57	mA
I_{CC}	Total supply current		V_{CC} = max		10	40	mA

deteriorated square waves or for obtaining a very good quality square-wave output from some other type of waveform at the input. For example, suppose that a square wave has noise or ringing associated with it or it has a long rise time. If we apply such signals to a Schmitt trigger, as shown in Fig. 8-34, the output will be a clean square pulse with fast rise time.

As can be seen, the circuit changes states when the leading edge of the input waveform equals the voltage level at the upper trigger point. It changes states again at the lower trigger point, thus generating a pulse whose time duration equals that from the UTL to the LTL, which is identified as t on both the input and output waveforms.

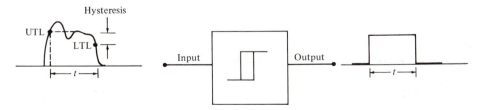

FIGURE 8-34 Schmitt trigger

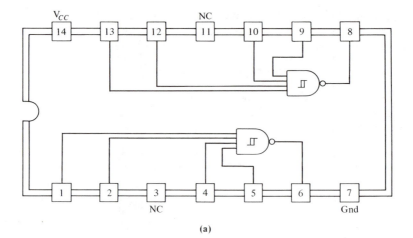

(a)

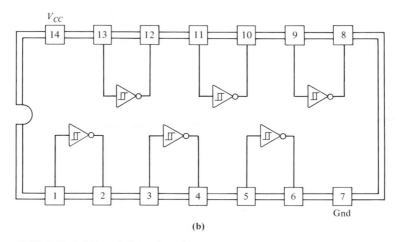

(b)

FIGURE 8-35   *IC Schmitt triggers*

The voltage difference between the upper and lower trigger levels in Fig. 8-34 is identified as hysteresis, where

$$\text{hysteresis} = \text{UTL} - \text{LTL}$$

Hysteresis can be reduced to near 0 V; however, some hysteresis is desirable because it helps ensure very fast switching action.

There are a number of IC circuits available that provide Schmitt trigger operation. Two of the more widely used are the 7413 dual four-input NAND Schmitt triggers and the 7414 hex Schmitt triggers shown in Fig. 8-35. Several CMOS Schmitt trigger ICs are also readily available.

8-21
MONOSTABLE MULTIVIBRATORS

Monostable multivibrators are regenerative circuits that have one stable state and one momentary, or quasi-stable, state. The devices, so named because of their one stable state, are often referred to as ''one-shots'' or ''single-shots.''

Monostable multivibrators remain in their stable state until triggered. When triggered into the quasi-stable state, the output Q in Fig. 8-36 is set HIGH and remains HIGH for a precise period of time. This period of time is proportional to the RC time constant associated with the resistor and capacitor in Fig. 8-36. These components are usually wired external to the multivibrator package to facilitate changing them to adjust pulse width. Monostable multivibrators are almost always edge-triggered devices, which makes the width of the input pulse unimportant to the operation of the circuit, as shown in Fig. 8-37.

There are two basic types of monostable multivibrator circuits that operate in the same way except when certain types of pulse train inputs are applied. The most common type is a *nonretriggerable* device, which, once triggered, ignores any other trigger pulses until the output returns to its stable state. On the other

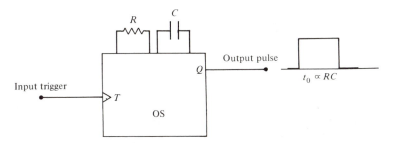

FIGURE 8-36   Monostable multivibrator

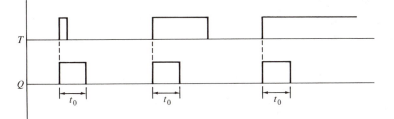

FIGURE 8-37   Input and output pulses of a monostable multivibrator

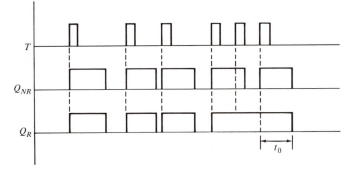

FIGURE 8-38   Response of retriggerable and nonretriggerable monostable multivibrators to identical inputs

hand, *retriggerable* devices reinitiate the timing sequence if another trigger pulse is applied before the end of an existing output pulse. When such a device is retriggered, the time to the end of the output pulse equals t_0 and is measured with respect to the time of retriggering. Figure 8-38 illustrates how retriggerable and nonretriggerable monostable multivibrators respond to identical input conditions.

There are several IC monostable multivibrator packages commercially available. One of the most widely used in the TTL series is the 74121. The circuit diagram for this nonretriggerable device is shown in Fig. 8-39. As can be seen, the timing resistance consists of an internal fixed resistance and an external resistance that is variable. The duration of the output pulse *for this device* is given as

$$t_0 = 0.7RC$$

where R is the total *external* resistance and C is an externally mounted variable capacitor.

The device is triggered to produce an output pulse by applying one of the following combinations of inputs:

1. When A_1 or A_2 or both are LOW and B goes HIGH, an output pulse is obtained.
2. When B is HIGH and A_i or A_2 goes LOW while the other remains HIGH, an output pulse is obtained.
3. When B is HIGH and A_1 and A_2 go LOW at the same time, an output pulse is obtained.

These conditions are summarized in Table 8-8, which is the truth table for the 74121.

If the inputs are continuously HIGH or LOW, as shown in the first four lines of the truth table, no output pulse is produced. The truth table entries marked

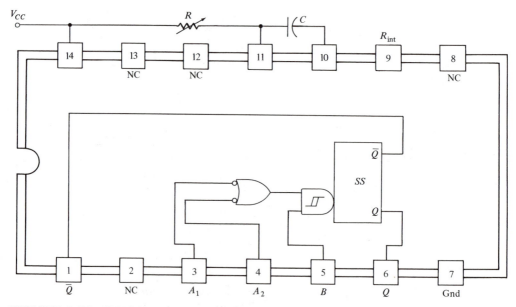

FIGURE 8-39   74121 monostable multivibrator

TABLE 8-8. / Truth Table for the 74121 Monostable Multivibrator

A_1	A_2	B	Q	$\overline{Q}$
0	$\times$	1	0	1
$\times$	0	1	0	1
$\times$	$\times$	0	0	1
1	1	$\times$	0	1
1	$\downarrow$	1	⊓	⊔
$\downarrow$	1	1	⊓	⊔
$\downarrow$	$\downarrow$	1	⊓	⊔
0	$\times$	$\uparrow$	⊓	⊔
$\times$	0	$\uparrow$	⊓	⊔

with an $\times$ are called *don't-care* conditions or are referred to as "irrelevant" since, with the other inputs as shown, setting the don't-care inputs to either 1 or 0 will not cause the output to change. An output pulse is obtained by transition of the inputs, as indicated by the arrows in the table.

EXAMPLE 8-3 ━━━━━━━━━━━━━━━━━━━━━━━━━━━━━━━━

If an external resistance of 28 kΩ is used with the 74121, what value of capacitance is required to produce an output pulse of 4-s duration?

SOLUTION

The output pulse is expressed as

$$t_0 = 0.7RC$$

Solving for C yields

$$C = \frac{t_0}{0.7R}$$

$$= \frac{4}{(0.7)(28 \times 10^3)} = 204 \ \mu F$$

❏

8-22
ASTABLE MULTIVIBRATORS

Astable multivibrators are regenerative circuits that have two output states but are stable in neither. Such circuits operate in one state momentarily, then switch to the other state, where they operate for a period of time before switching back to the first state. This continuous switching between the two states, as a predictable rate, produces a rectangular output pulse with fast rise times. Since no input signal is required to cause such circuits to change states, astable multivibrators are often called *free-running multivibrators*.

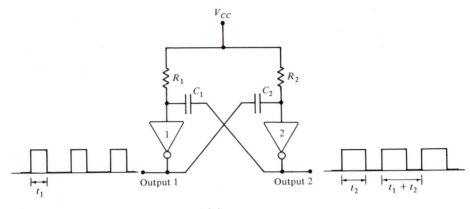

FIGURE 8-40   Basic astable multivibrator

A basic astable multivibrator that uses two inverters is shown in Fig. 8-40. The frequency of operation of the astable circuit is given as

$$f = \frac{1}{t_1 + t_2}$$

$$= \frac{1}{0.7R_1C_1 + 0.7R_2C_2}$$

$$= \frac{1.4}{R_1C_1 + R_2C_2}$$

If the resistors and capacitors are of equal value, the output frequency is

$$f = \frac{1}{2T} = \frac{1}{2(0.7RC)} = \frac{1}{1.4RC}$$

$$= \frac{0.7}{RC}$$

Astable multivibrators are useful in digital systems as sources of clock pulses for timing purposes.

EXAMPLE 8-4 ━━

A synchronous digital system incorporates several edge-triggered bistable and monostable multivibrators that are to be triggered every 250 ns by a pulse from an astable multivibrator. If 5-kΩ resistors are used in the astable circuit, what value must the capacitors have if they are equal in value?

SOLUTION

The pulse train from the astable multivibrator must have a period of 250 ns, as shown in Fig. 8-41. The frequency is computed as

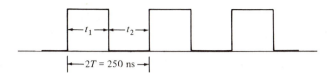

FIGURE 8-41   Illustration for Example 8-4

$$f = \frac{1}{2T} = \frac{1}{250 \text{ ns}} = 4 \text{ MHz}$$

and the required capacitance is therefore

$$C = \frac{0.7}{Rf}$$

$$= \frac{0.7}{(5 \times 10^3)(4 \times 10^6)} = 35 \text{ pF}$$ ❏

8-23
THE 555 TIMER

A linear IC that has found many applications in digital systems is the 555 timer. This very useful and versatile device was first introduced by Signetics Corporation; however, it is presently being manufactured by virtually every IC manufacturer. A block diagram of the 555 timer is shown in Fig. 8-42. The pin assignments are for a standard eight-pin package.

The 555 timer has two basic modes of operation—astable and monostable—with the particular mode of operation determined by external wiring. When wired for the astable mode, the device is a TTL-compatible pulse generator that can serve as a clock for timing purposes in digital systems. When wired for the

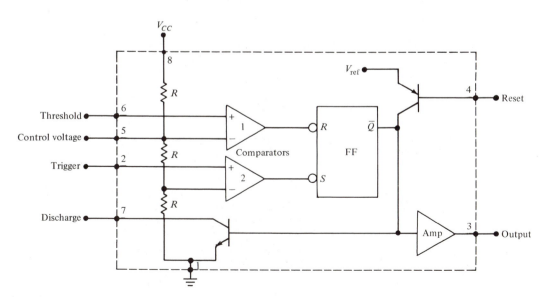

FIGURE 8-42   555 timer

monostable mode, a pulse at the trigger input turns the timer on for a period of time set by an externally connected resistor and capacitor, thus producing a TTL-compatible output pulse.

The external connections that cause the 555 timer to function as an astable multivibrator are shown in Fig. 8-43. We can see that for astable operation, the trigger and threshold inputs are connected together. The capacitor C charges toward $\frac{2}{3}V_{CC}$ through R_A and R_B and discharges through R_B; therefore, the frequency and duty cycle of the output waveform are established in the selection of these resistors. The waveforms at terminal 6 and at the output terminal are shown in Fig. 8-44. The time constant for charging the capacitor with the output in the HIGH state is

$$t_H = 0.7(R_A + R_B)C$$

The time constant for discharging the capacitor with the output in the LOW state is

$$t_L = 0.7R_BC$$

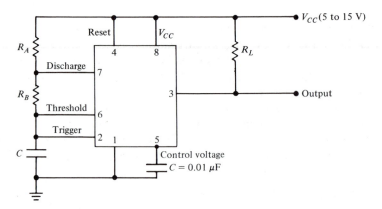

FIGURE 8-43   555 timer wired to function as an astable multivibrator

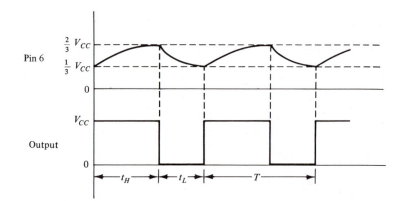

FIGURE 8-44   555 timer waveforms for astable operation

The period for the output waveform is the sum of the HIGH and LOW times expressed as

$$T = t_H + t_L = 0.7(R_A + 2R_B)C$$

Therefore, the output frequency is

$$f = \frac{1}{T} = \frac{1.4}{(R_A + 2R_B)C}$$

The duty cycle is the percentage of the period during which the output is HIGH and is given as

$$\text{duty cycle} = \frac{t_H}{T} = \frac{t_H}{t_H + t_L} = \frac{R_A + R_B}{R_A + 2R_B}$$

The 555 timer functions as a nonretriggerable monostable multivibrator when connected as shown in Fig. 8-45. Unlike the astable circuit, a trigger pulse is

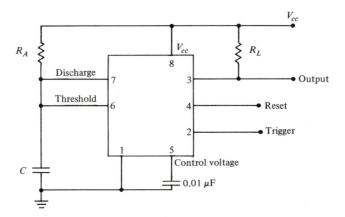

FIGURE 8-45 555 timer connected to function as a one-shot

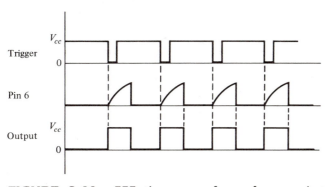

FIGURE 8-46 555 timer waveforms for one-shot operation

required to initiate circuit operation to produce an output pulse. The waveforms associated with the 555 timer when connected to function as a one-shot are shown in Fig. 8-46. Once the circuit is triggered by the negative-going trigger pulse, the timing interval determined by R_A and C and expressed as

$$t_P = 1.1R_A C$$

will complete even if the other trigger pulses occur.

8-24
APPLICATIONS

Applications of multivibrators in electronic circuits are interesting and varied, with their scope limited only by one's imagination. The following paragraphs are examples of but a few of the numerous uses for these devices.

8-24.1   Switch Debouncers

Toggle and pushbutton switches are frequently used in digital circuits to permit users to control some aspect of the circuit's operation. These mechanical devices almost always produce contact bounce when their contacts are closed or opened. When viewed in real time, the waveform out of the switch is an irregular analog waveform. However, when this signal is applied to a logic gate whose response is much faster than the rate of contact bounce, the logic gate responds as though the analog waveform generated by contact bounce were a pulse train of logic 1's and 0's as shown in Fig. 8-47. *RS* latches are widely used in switch contact debouncer applications. The circuit can be fabricated as shown in Fig. 8-48.

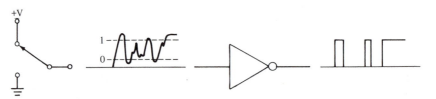

FIGURE 8-47   Incorrect logic signal due to contact bounce

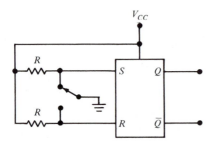

FIGURE 8-48   *RS* latch used in contact debounce circuit

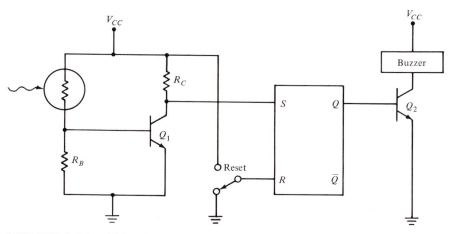

FIGURE 8-49 *RS* latch used in alarm system

8-24.2 Alarm System

The ability to store momentary information makes *RS* flip-flops useful alarm systems where the signal that activates the alarm may be of short duration: for example, a burglar passing through a laser beam. Figure 8-49 shows a simple alarm which, once activated, would continue to sound the alarm until the flip-flop is reset. The *RS* flip-flop is initially reset ($Q = 0$), so the alarm is deactivated. When light is directed onto the photocell, its resistance is low; therefore, transistor Q_1 is biased ON, which applies a LOW to the SET input. The output Q is LOW and thus transistor Q_2 is OFF; therefore, the alarm is deactivated, as was previously stated. If the light beam is interrupted, the resistance increases abruptly to a high value which biases Q_1 off. Therefore, the voltage at the SET input goes high, which sets the output at Q HIGH. This biases Q_2 ON, which sounds the buzzer.

8-24.3 Pressure Monitoring System

The cabin pressure in an airplane or a space vehicle such as the Space Shuttle is constantly monitored. The logic diagram shown in Figure 8-50 shows how a flip-

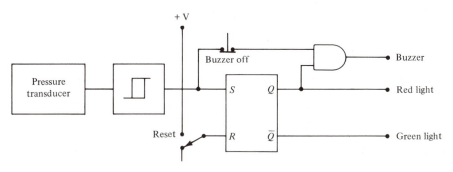

FIGURE 8-50 Schmitt trigger and *RS* latch used in pressure-monitoring system

flop might be used in this application. If we assume that the transducer output voltage is 0 V when the pressure is safe and increases as pressure decreases, we can design the circuit so that the analog output of the pressure transducer is less than the trigger point of the Schmitt trigger when the pressure is safe. Then the alarm will sound when the Schmitt trigger changes states due to decreasing pressure. The buzzer can be turned off manually, but the red light will stay on until the flip-flop can be reset as a result of an increase in pressure. The green light is on when the pressure is satisfactory.

8-25
SUMMARY

Multivibrators are regenerative circuits that utilize two active devices, of which one is conducting while the other is cut off. Multivibrators are classified as either bistable, monostable, or astable circuits according to the number of stable operating states.

The most widely used type of multivibrator in digital circuitry is the bistable multivibrator, or flip-flop, which serves as a basic memory element. There are several types of flip-flops, such as *RS, T, D,* and *JK* devices, that enjoy wide industrial and commercial use in a broad range of applications, but regardless of the application, flip-flops serve as memory elements. Flip-flops that are designed to change states only on a clock pulse are referred to as clocked flip-flops and are said to operate in a synchronous mode. The advantage of synchronous operation is that the operation of an entire digital system can be coordinated by a single clock.

Timing diagrams are widely used to provide a graphical representation of pulse waveforms with respect to time or with respect to another waveform. Propagation delay, which is a characteristic of all flip-flops, should be accounted for in timing diagrams if it is likely to affect the operation of the circuit; otherwise, it should not be shown.

Schmitt triggers, which are classified as bistable multivibrators, are widely used in digital systems for pulse restoration, level detection, comparison, and many other applications.

Astable multivibrators have two output states, both of which are unstable, or quasi-stable, states. This type of circuit requires no external triggering signal to cause it to change states and is therefore referred to as a free-running multivibrator. These devices are widely used in digital systems for timing and synchronizing purposes.

Monostable multivibrators also have two output states, one stable and one unstable, or momentary, state. The device is placed in its momentary state by an external trigger signal and remains in this state for a period of time determined by an externally connected *RC* network. The time in the momentary state is usually adjustable from a few nanoseconds to several minutes for nonretriggerable devices. Retriggerable one-shots can be made to operate in their unstable state for an indefinite period of time by periodic retriggering.

PROBLEMS

1. If the following waveforms are applied to the *RS* flip-flop, draw the output waveforms seen at Q and $\overline{Q}$. Assume that the flip-flop is initially reset.

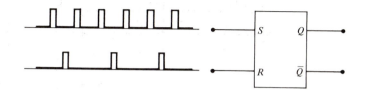

2. If the following waveforms are applied to the clocked *RS* flip-flop, draw the waveform seen at Q. Assume that the flip-flop is initially reset.

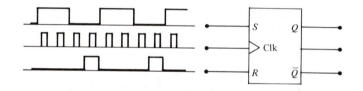

3. If the following waveforms are applied to the clocked *RS* flip-flop, draw the waveforms seen at Q and $\overline{Q}$. Assume that the flip-flop is initially set.

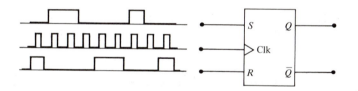

4. Draw the output waveform at X and Y for the following circuit. Assume that both flip-flops are initially reset.

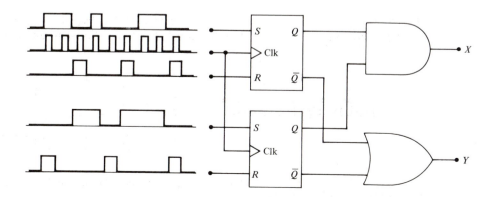

5. Draw the output waveform at Q for both edge-triggered RS flip-flops shown below. Assume that both flip-flops are initially reset.

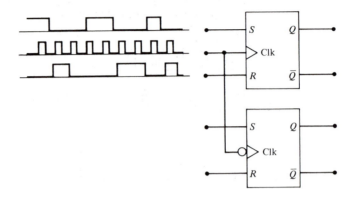

6. Draw the output waveform at Q if the following waveforms are applied to the D flip-flop. Assume that the flip-flop is initially reset.

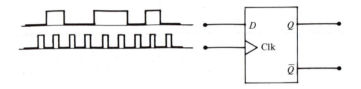

7. The following waveforms are applied to the leading-edge triggered JK flip-flop. Draw the resulting waveform at Q if the flip-flop is initially reset.

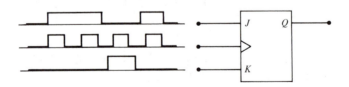

8. Draw the waveform that will be seen at $\overline{Q}$ of T flip-flop 2 when the pulse train shown below is applied to flip-flop 1 if both flip-flops are initially reset.

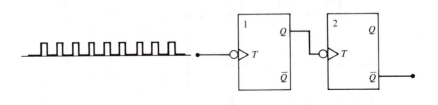

9. Draw the waveform that will be seen at $\overline{Q}$ if the following set of waveforms are applied to the JK flip-flop.

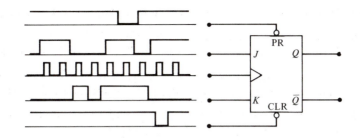

10. A leading-edge-triggered D flip-flop has the following specifications on its data sheet:

$$t_S = 20 \text{ ns}$$

$$t_H = 5 \text{ ns}$$

$$t_{PLH} = 30 \text{ ns}$$

$$t_{PHL} = 15 \text{ ns}$$

(a) How far ahead of the triggering edge of the clock pulse must data be applied to ensure correct storage?

(b) How long after the triggering edge of the clock pulse must the data remain stable to ensure correct storage?

(c) How much time elapses after the triggering edge of the clock pulse before the output changes from LOW to HIGH?

(d) How much time elapses after the triggering edge of the clock pulse before the output changes from HIGH to LOW?

11. A pulse train with a frequency of 400 kHz and a 20 percent duty cycle is applied to a leading-edge-triggered D flip-flop as shown. If t_1 and t_2 are equal, how many nanoseconds is the output delayed with respect to the input data?

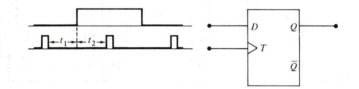

12. If a 1-MHz square-wave pulse train is applied to the clock input terminal of a JK master-slave flip-flop, what is the approximate setup time?

13. If the UTL for the Schmitt trigger shown below is 9.0 V and the LTL is 6.0 V, draw the output waveform that will be observed due to the input waveform shown.

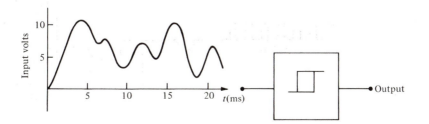

14. How much hysteresis does the Schmitt trigger in Problem 13 have?

15. Determine the pulse width of the output of a 74121 monostable multivibrator if the external resistor is 7.5 kΩ and the external capacitance is 1200 pF.

16. An output pulse of 10 s is to be generated by a 74121 one-shot. What is the value of the capacitor that is required if the external resistance is 3.9 kΩ?

17. Draw the output waveforms that should be observed at X and Y in the following circuit. Each one-shot is a 74121.

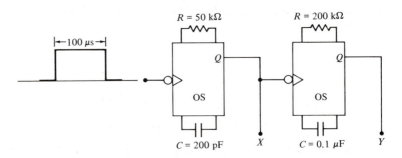

18. Draw the waveforms that will be observed at points A through F in the following circuit.

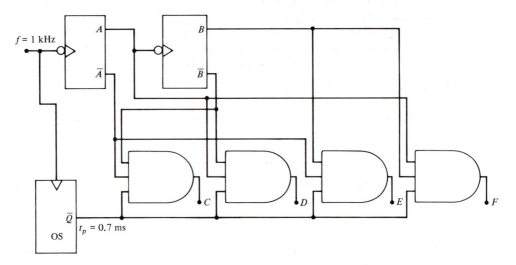

19. Draw the output waveform at $\overline{Q}$ for the following circuit where the one-shot is a 74121.

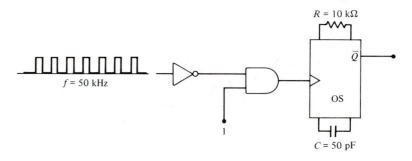

20. Design an alarm system that will activate a buzzer for 5 s if either of the front doors on your car are opened. The buzzer should stay activated even if the door is closed within 5 s.

21. A 555 timer is to be wired to operate as an astable multivibrator with a frequency of 200 kHz. If R_A equals 1.2 kΩ and C equals 0.01 μF, what value must R_B have?

22. A 555 timer is to be wired to operate as an astable multivibrator with a frequency of 10 kHz and a duty cycle of 25 percent. If a 0.002-μF capacitor is used, determine the value of R_A and R_B.

REFERENCES

The Logic Databook. Santa Clara, Calif.: National Semiconductor Corp., 1982.

Mims, Forrest M., *Engineer's Notebook, A Handbook of Integrated Circuit Applications*. Forth Worth, Tex.: Radio Shack, 1980.

Strangio, C. E., *Digital Electronics, Fundamental Concepts and Applications*. Englewood Cliffs, N.J.: Prentice-Hall, Inc., 1980.

The TTL Data Book. Dallas, Tex.: Texas Instruments, Inc., 1976.

The TTL Logic Data Manual. Sunnyvale, Calif.: Signetics Corp., 1982.

Williams, Gerald E., *Digital Technology*. Chicago: Science Research Associates, Inc., 1982.

Counters

9

INSTRUCTIONAL OBJECTIVES

In this chapter you are introduced to counters and a few of their many applications in digital systems. After completing the chapter, you should be able to:

1. Determine the modulus of a counter.
2. Draw the timing diagram for a given counter.
3. Determine the number of flip-flops required to scale the frequency by a desired scale factor.
4. Show how feedback is used to obtain a mod-N counter.
5. Show how lower-modulus counters are used to fabricate higher-modulus counters.
6. Describe how a down counter can be constructed.
7. Compute the maximum frequency of an asynchronous counter.
8. Design a counter without cascading lower-modulus counters.
9. Decode a counter.
10. Compute the pulses stored by the counter during a given number of gate pulses.
11. Compute the input voltage of a digital voltmeter that uses a VCO when given a voltage–frequency relationship.
12. Define or describe the following terms:
    (a) Asynchronous counter.
    (b) Modulus.
    (c) Synchronous.
    (d) State diagram.
    (e) Decoding gates.
    (f) Gating pulse.
    (g) Scalar.

SELF-EVALUATION QUESTIONS

The following questions deal with the material presented in this chapter. Read the questions prior to studying the chapter and, as you read through the material, watch for answers to the questions. After completing the chapter, return to this section and evaluate your comprehension of the material by answering the questions again.

1. How do synchronous and asynchronous counters differ?
2. What is the major shortcoming of asynchronous counters?
3. How is the modulus of a counter changed?
4. How many count states does a binary ripple counter made up of n flip-flops in cascade have?
5. What is the purpose of gating a counter?
6. How is a counter decoded, and what is the purpose of decoding?
7. What modification is required to make a ''down'' counter from an ''up'' counter?
8. What is the best way to construct a higher-modulus counter?
9. What is the purpose of a scalar circuit?
10. What three moduli counters can be readily obtained with a 7490 IC chip?

9-1
INTRODUCTION

The flip-flops discussed in Chapter 8 are the heart of the counters to be discussed in this chapter. Counting circuits, which are a vital part of most digital systems, are among the most useful and versatile type of digital circuits. They find application in pulse counting; for frequency division; for measuring time, period, and frequency; in digital multimeters and other test instruments; as well as for many other industrial and commercial uses.

As their name implies, counters are used to register count. The binary ripple counter is the simplest and most straightforward counter and counts in straight binary. We begin with a discussion of how flip-flops are used to construct binary ripple counters. Feedback techniques, which are used to obtain counters of various moduli, are discussed next. From these counters of various module, counters having any number of desired count states can be implemented. Characteristics and applications related to the various counters are also discussed.

9-2
BINARY RIPPLE COUNTERS

Electronic counters are implemented by interconnecting flip-flops in such a way that they, in effect, count trigger pulses. Binary ripple counters are constructed by connecting flip-flops in cascade as shown in Fig. 9-1. Any type of flip-flop connected in a toggle mode can be used; however, JK flip-flops are by far the most widely used type in counter applications and will therefore be used in all discussions in this chapter.

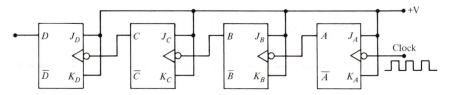

FIGURE 9-1 *Four-stage binary ripple counter*

We begin our discussion of binary ripple counters by considering the operation of the four-stage counter of Fig. 9-1, and start by making the assumption that each flip-flop is initially reset. When the first clock pulse is applied, flip-flop A is triggered on the trailing edge of the pulse. This sets the flip-flop so that its output is HIGH; therefore, it has, in effect, "counted" the first clock pulse. The other three flip-flops were not affected by the first clock pulse, and are therefore still at logic 0. We can express the outputs conditions as

$$DCBA = 0001$$

The second clock pulse causes a flip-flop A to toggle, which resets its output LOW. This negative transition toggles flip-flop B, which sets its output HIGH. This positive transition at the output of flip-flop B has no effect on flip-flop C; therefore, the output states of the flip-flops are

$$DCBA = 0010$$

The third clock pulse toggles flip-flop A; therefore, its output is again set HIGH. No other flip-flop toggles; therefore, the output states are

$$DCBA = 0011$$

The fourth clock pulse causes flip-flop A to toggle again; therefore, its output is reset LOW. This negative transition toggles flip-flop B, which also resets its output LOW. The negative transition at the output of flip-flop B causes flip-flop C to toggle, which sets its output HIGH and gives us the output states

$$DCBA = 0100$$

No doubt you have observed that the output states provide us with a binary number that corresponds to the number of clock pulses counted. If we continue to observe the output states, we shall find that the binary numbers produced by the toggling action continues to increase as a straight binary count until all flip-flops are in their HIGH state. With four flip-flops, this corresponds to binary 15 (1111_2). The next clock pulse will reset each flip-flop; therefore, we can see that the number of pulses which can be counted with n flip-flops, without resetting all the flip-flops, is

$$N = 2^n - 1 \tag{9-1}$$

where N is the number of pulses that can be counted and n is the number of flip-flops. Each of the flip-flops will reset on the next clock pulse, that is, when

$$N = 2^n$$

which corresponds to the total number of count states for flip-flops in cascade.

EXAMPLE 9-1 ━━━

What is the largest number that can be counted with five flip-flops that are connected as a binary ripple counter?

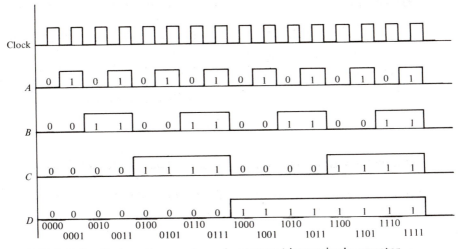

FIGURE 9-2 Timing diagram for a four-state binary ripple counter

SOLUTION

The largest number that can be counted with a five-stage binary ripple counter is

$$N = 2^5 - 1$$
$$= 32 - 1 = 31 \qquad \square$$

The waveforms generated by applying 16 clock pulses to the four-state binary ripple counter of Fig. 9-1 are shown in Fig. 9-2, together with the corresponding count states.

By observing these waveforms, we can see that each flip-flop toggles with its clock input experiences a negative-going transition. Flip-flop A toggles on every clock pulse, and flip-flop B toggles on every pulse generated by flip-flop A. However the frequency at the output of flip-flop A is only one-half of the clock frequency. In fact, we can see that the frequency at the output of each flip-flop is one-half the frequency at its input, so we see that flip-flops act as frequency dividers. Quantitatively, we can say that

$$f_n = \frac{f_{clk}}{2^n} \qquad (9\text{-}2)$$

where

$$f_{clk} = \text{clock frequency}$$
$$f_n = \text{frequency after division by } n \text{ flip-flops}$$
$$n = \text{number of flip-flops}$$

EXAMPLE 9-2

A 2-MHz clock signal is applied to a five-stage binary ripple counter. What is the frequency at the output of the fifth flip-flop?

SOLUTION

The output frequency is computed as

$$f_n = \frac{f_{clk}}{2^n}$$

$$f_5 = \frac{2 \text{ MHz}}{2^5} = \frac{2 \text{ MHz}}{32} = 62.5 \text{ kHz}$$

❏

EXAMPLE 9-3 ━━━━━━━━━━━━━━━━━━━━━━━━━━━━━━━━━━━━

If it is desired to divide an 80-kHz clock signal with flip-flops to obtain a 5-kHz signal, how many flip-flops are required?

SOLUTION

Solving the expression

$$f_n = \frac{f_{clk}}{2^n}$$

for n, we obtain

$$2^n = \frac{f_{clk}}{f_n}$$

$$= \frac{80 \text{ kHz}}{5 \text{ kHz}} = 16$$

$$n(\ln 2) = \ln 16$$

$$n = \frac{\ln 16}{\ln 2} = \frac{2.77}{0.693} = 4$$

❏

9-3
USING FEEDBACK WITH BINARY COUNTERS

When flip-flops are connected in cascade, as shown in Fig. 9-1, so that the output of one triggers the next, the counter is called a ripple counter or *asynchronous counter.* If we cascade n flip-flops, we obtain 2^n discrete combinations of output states. Since this includes the condition when all flip-flop outputs return to 0, the largest number of pulses that can be generated with n flip-flops is one less than 2^n, as has already been stated.

The progression of a binary counter through its 2^n states is referred to as a straight binary count or *natural count,* and the number of natural count states is called the *modulus* of the counter. Since the natural count equals 2^n, the modulus of binary ripple counters will be 2, 4, 8, 16, 32, and so on, depending on the number of flip-flops in cascade. However, it is often desirable to construct counters that have moduli not equal to 2^n. For example, you may need a counter with a modulus of 5 or 10 or 12. Lower-modulus counters can always be obtained by using higher-modulus counters and causing them to skip natural count states. Such counters are said to have a *modified count*. The modified count is always less than 2^n, where n is the number of flip-flops required. For example, a counter with a modulus of 5, generally referred to as a mod-5 counter, requires three flip-flops, since 8 is the lowest natural count greater than the desired modified count of 5.

EXAMPLE 9-4

How many flip-flops are needed to construct each of the following counters?

(a) Mod-3   (b) Mod-7   (c) Mod-12

SOLUTION

(a) The lowest natural count state greater than 3 is 4. If 2^n equals 4, then n equals 2; therefore, two flip-flops are required to construct a mod-3 counter.

(b) The lowest natural count state greater than 7 is 8. If 2^n equals 8, then n equals 3; therefore, three flip-flops are required to construct a mod-7 counter.

(c) The lowest natural count state greater than 12 is 16. If 2^n equals 16, then n equals 4; therefore, four flip-flops are required to construct a mod-12 counter.   ❏

The problem of obtaining a modified count must now be addressed. This is accomplished by using feedback to cause the counter to skip, or omit, some of its natural count states by resetting the flip-flops early. For example, if we use three flip-flops to construct a mod-5 counter, we must use feedback to cause three of the natural count states to be skipped. The count states that are skipped are arbitrary; however, consideration should be given to minimizing the amount of hardware required. With regard to the use of feedback, consider the three-stage counter shown in Fig. 9-3. Since the counter consists of three flip-flops, it has eight natural count stages, which are shown in Table 9-1. Suppose that we wish to implement a mod-7 counter. Since only one count state is to be skipped, it would seem rather natural to allow the counter to progress through its natural count sequence, but reset it one count early. Since the flip-flop outputs are a unique combination of 1's and 0's on the seventh count, as is the case on each count, we can use logic levels to reset each flip-flop. If the flip-flop outputs at A, B, and C are applied to a NAND gate, as shown in Fig. 9-4, the signal at the output of the gate can be used to clear the flip-flops. On count 7, the output of

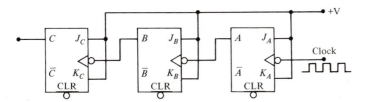

FIGURE 9-3   Three-stage binary ripple counter

**TABLE 9-1. / Truth Table
for a Three-Stage
Binary Counter**

Count	C	B	A
0	0	0	0
1	0	0	1
2	0	1	0
3	0	1	1
4	1	0	0
5	1	0	1
6	1	1	0
7	1	1	1
0	0	0	0

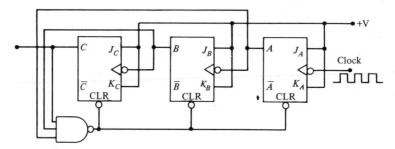

FIGURE 9-4   Binary ripple counter with feedback resulting in
mod-7 counter

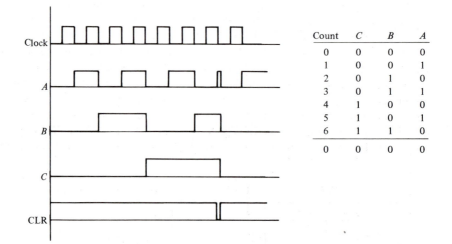

FIGURE 9-5   Timing diagram and truth table for the mod-7 counter
shown in Fig. 9-4

each flip-flop goes HIGH only momentarily; therefore, it essentially skips this state, thus providing us with a mod-7 counter. The truth table and the waveforms for the circuit are shown in Fig. 9-5.

9-4
HIGHER-MODULUS ASYNCHRONOUS COUNTERS

Most practical applications of counters involve higher-modulus counters such as mod-10, mod-12, or even mod-60. Although it is possible to design such counters directly, it is generally more convenient to cascade two or more lower-modulus counters for which there are some fairly standard circuit configurations available. For example, a mod-10 counter can easily be implemented by cascading a mod-2 and a mod-5 counter as shown in Fig. 9-6. The mod-2 counter requires a single flip-flop, while the mod-5 counter requires three flip-flops. Figure 9-7 shows a logic diagram for the mod-10 counter shown in Fig. 9-6. It should be emphasized that this is only one of many ways to implement either the mod-5 or the mod-10 counter. Mention should also be made of the fact that since the JK flip-flops are TTL devices, the J and K inputs will float HIGH when left disconnected; therefore, the flip-flops will function in a toggle mode, which is poor practice. If the J and K inputs are to be HIGH, connect them HIGH. In the toggle mode, flip-flop A toggles on every clock pulse. In fact, the entire counter functions as a binary ripple counter in counting from 0000 through 1001. On the next count state, which is 1010, the flip-flops that make up the mod-5 counter are reset. There is no need

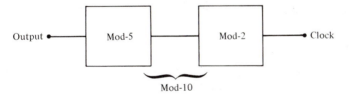

FIGURE 9-6 Mod-10 counter fabricated with lower-modulus counters

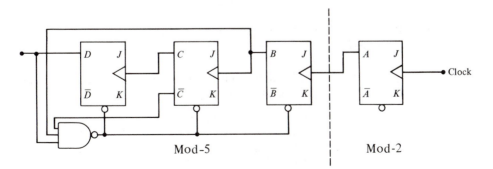

FIGURE 9-7 Logic diagram of the mod-10 counter shown in Fig. 9-6

TABLE 9-2. / Truth Table for the Counter in Fig. 9-7

Count	D	C	B	A
0	0	0	0	0
1	0	0	0	1
2	0	0	1	0
3	0	0	1	1
4	0	1	0	0
5	0	1	0	1
6	0	1	1	0
7	0	1	1	1
8	1	0	0	0
9	1	0	0	1
	1	0	1	0

to reset flip-flop A, since it toggles to 0 anyway, as can be seen in the truth table for the counter, which is shown in Table 9-2.

The mod-10 counter of Fig. 9-7 could just as well be constructed with the mod-5 and mod-2 counters interchanged. Ten unique but different count states would still be obtained.

EXAMPLE 9-5

If mod-2 through mod-7 counters are available, show how the following higher-modulus counters can be constructed using these counters.

(a) Mod-15   (b) Mod-24   (c) Mod-60

SOLUTION

(a) A Mod-15 counter can be constructed by cascading a mod-3 and a mod-5 counter, as shown in Fig. 9-8.

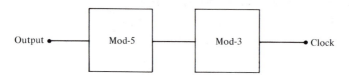

FIGURE 9-8 *Mod-15 counter using mod-5 and mod-3 counters*

(b) A Mod-24 counter can be constructed by cascading a mod-2, a mod-3, and a mod-4 counter, as shown in Fig. 9-9.

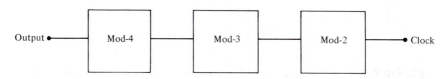

FIGURE 9-9 Mod-24 counter using lower-modulus counters

(c) A Mod-60 counter can be constructed by cascading a mod-2, a mod-5, and a mod-6 counter, as shown in Fig. 9-10. ❏

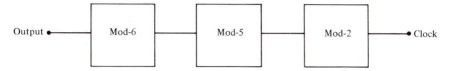

FIGURE 9-10 *Mod-60 counter using lower-modulus counters*

In each case in Example 9-5, the desired number of unique count states will be obtained regardless of the order in which the lower-modulus counters are connected; however, the order determines the unique *set* of count states.

9-5
DOWN COUNTERS

The counters considered to this point have counted upward, that is, from a lower number to a higher number. There are applications where down counters are very useful; for example, the countdown to an event. Down counters are used in microwave ovens and in many sports events to indicate the remaining time, as well as in aerospace applications to indicate the time remaining until launch.

The binary ripple counters already discussed are easily modified to function as down counters. The only change is that the output of each flip-flop is taken at the complemented output, as shown in Fig. 9-11. The counter circuit operates in the same way as the counters already discussed. The only difference is the waveforms being used to provide a display of count. In normal operation, the flip-flops in Fig. 9-11 are reset prior to applying the first clock pulse. If A, B, and C are reset to logic 0, then $\overline{A}$, $\overline{B}$ and $\overline{C}$ are set to logic 1. As clock pulses are applied, the count states shown in Table 9-3a are generated at outputs A, B, and C. Table 9-3b shows the count states that are generated at outputs $\overline{A}$, $\overline{B}$, and $\overline{C}$.

Another way of constructing a down counter is shown in Fig. 9-12. The inverted output of each flip-flop is used to drive the next flip-flop and the pulse count is the signals taken from outputs A, B, and C. The waveforms for the counter are shown in Fig. 9-13.

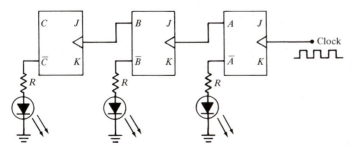

FIGURE 9-11 *Down counter*

TABLE 9-3.

Count	C	B	A		$\overline{C}$	$\overline{B}$	$\overline{A}$	Count
0	0	0	0		1	1	1	7
1	0	0	1		1	1	0	6
2	0	1	0		1	0	1	5
3	0	1	1		1	0	0	4
4	1	0	0		0	1	1	3
5	1	0	1		0	1	0	2
6	1	1	0		0	0	1	1
7	1	1	1		0	0	0	0

(a)   (b)

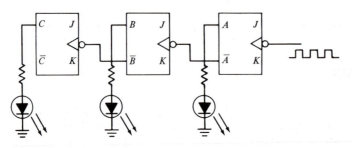

FIGURE 9-12   Down counter using complemented outputs

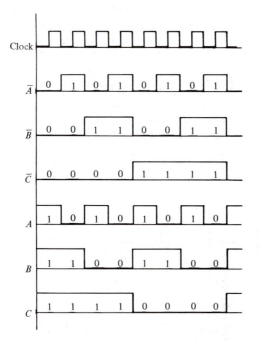

FIGURE 9-13   Timing diagram for down counter of Fig. 9-12

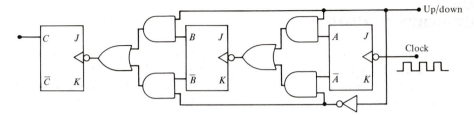

FIGURE 9-14 Up/down counter

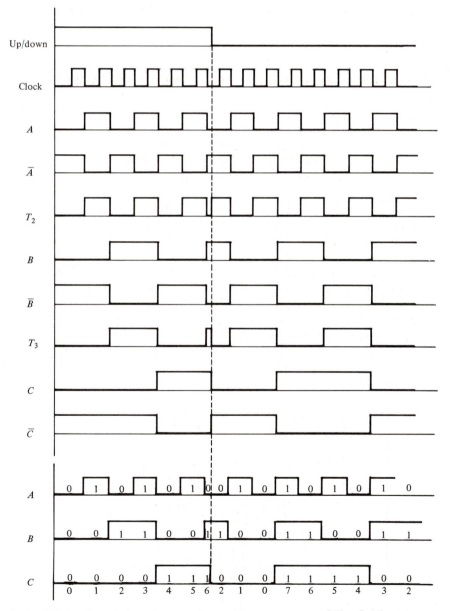

FIGURE 9-15 Timing diagram for up/down counter of Fig. 9-14

9-6
UP/DOWN COUNTERS

There are applications where it is necessary for counters to count either up or down through some number of count states and then reverse direction. For example, on the countdown of a rocket a down count to launch is desirable, whereas after launch an up count of elapsed time from launch is desired.

By combining up and down counter circuits, as shown in Fig. 9-14 on p. 253, we obtain a counter that is capable of counting either up or down. If the up/down line is HIGH, the counter counts in the up direction, whereas if the up/down line is LOW, the count sequence is toward zero. Waveforms for the up/down counter are shown in Fig. 9-15. The waveforms are taken at A, B, and C.

9-7
PROPAGATION DELAY IN ASYNCHRONOUS COUNTERS

Counters of any desired modulus are easy to construct using binary ripple counters with feedback, and by cascading lower-modulus counters; however, such counters have one major disadvantage in many applications—propagation delay. The propagation delay of asynchronous counters is the cumulative delay of each flip-flop and is a major limiting factor with regard to the rate at which counters can be clocked.

If the propagation delay of each of n flip-flops is t_{pd}, then the total propagation delay for the n-stage counter is $n \times t_{pd}$. The cumulative effect of propagation delay is shown in the waveforms of Fig. 9-16. Since the propagation of delay of flip-flops is fixed, we can see that if the period of the clock is less than $n \times t_{pd}$, the counter will malfunction. The waveforms in Fig. 9-17 shows what happens when a pulse train, whose period is 100 ns, is applied to a ripple counter constructed of flip-flops with a propagation delay of 50 ns. Since 3×50 ns exceeds the period of the clock waveform, the counter will not count properly. As can be seen, there is a 150-ns delay after the trailing edge of the fourth clock pulse before the output at C goes HIGH. By this time, the output of flip-flop A has toggled HIGH, giving us a count of 101_2; therefore, the count state of 100_2 was skipped.

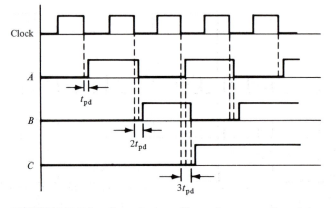

FIGURE 9-16 Cumulative effect of propagation delay

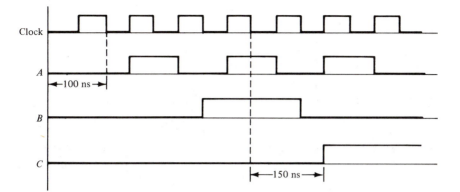

FIGURE 9-17 *Timing diagram showing missed pulse count due to propagation delay*

For proper operation of the counter, the period of the clock should be greater than $n \times t_{pd}$; that is,

$$T_{clk} \geq n \times t_{pd} \qquad (9\text{-}3)$$

Since the clock frequency is the reciprocal of the period, we can express Eq. 9-3 in terms of the clock frequency as

$$f_{clk} = \frac{1}{n \times t_{pd}}$$

We can see that as the number of flip-flops increases, the maximum operating frequency of the counter decreases.

EXAMPLE 9-6

What is the maximum operating frequency of a four-stage binary ripple counter if each flip-flop has a 40-ns propagation delay?

SOLUTION

The maximum operating frequency is

$$f_{clk} = \frac{1}{4 \times 40 \text{ ns}} = 6.25 \text{ MHz}$$

❏

9-8
SYNCHRONOUS BINARY COUNTERS

The propagation delay and the associated problems that limit the usefulness of ripple counters can be overcome by using *synchronous*, or *parallel*, counters. Parallel counters are constructed so that each flip-flop is triggered directly by the

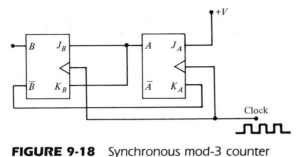 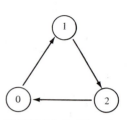

FIGURE 9-18   *Synchronous mod-3 counter*   **FIGURE 9-19**   State diagram for mod-3 counter

clock; thus all flip-flops toggle simultaneously. A two-stage parallel counter is shown in Fig. 9-18. Note that whereas ripple counters progress through the desired number of count states in a natural binary count sequence and are reset by applying the correct combination of flip-flop output signals to the RESET input, synchronous counters use signals applied to the J and K inputs to generate the desired count states. This makes it possible to generate irregular count sequences with synchronous counters.

The counter shown in Fig. 9-18 is a mod-3 counter, which means that it has three unique count states. These are often represented by a drawing called a *state diagram*. The state diagram for the mod-3 counter of Fig. 9-18 is shown in Fig. 9-19. Since the count state does not have to progress as a natural binary count, there are many count sequences that could be used. The actual count states are a function of the signals fed back to the J and K inputs. Since the feedback paths that are needed to generate a desired count sequence are generally not obvious, an example is in order to shed more light on the subject.

EXAMPLE 9-7 ━━━━━━━━━━━━━━━━━━━━━━━━━━━━━━

Design a mod-3 counter with count states 01, 10, and 11.

SOLUTION

The state diagram is shown in Fig. 9-20. The next step is to tabulate the relationship between the J and K signal levels and dynamic output states. We will assume that our flip-flops behave as shown in Table 9-4. The relationships show what levels J and K must be to cause the output to stay either LOW or HIGH or

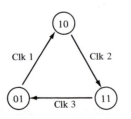

FIGURE 9-20   State diagram for Example 9-7

TABLE 9-4. / Flip-Flop Behavior

Flip-Flop Output	$0 \to 0$	$0 \to 1$	$1 \to 0$	$1 \to 1$
J	0	1	$\times$	$\times$
K	$\times$	$\times$	1	0

to change states from LOW to HIGH or from HIGH to LOW. For example, J must be HIGH and K may be either HIGH or LOW (don't care) when the FF output is to change stages from 0 to 1. ❏

From the relationships above, we can develop a state table (Table 9-5) showing the necessary levels of the J and K inputs to cause the flip-flop outputs to respond as required. The $\times$ entries are don't-care states.

The question that we need to address now is: Where do we get the required J and K inputs when they are needed? The answer comes in part from observation and in part from the use of combinational logic. We can see that output $\overline{B}$ is at logic 1 when K_A needs to be at logic 1; therefore, the necessary signal for K_A can be obtained by connecting $\overline{B}$ to K_A. Since two of the three conditions for J_A and J_B are don't-care states, the only required state is easily obtained by connecting these terminals directly to logic 1, as is in fact done with J_A. For the purpose of illustration, J_B will be connected to either output A or $\overline{B}$. We can see that A is at logic 1 when J_B must be at logic 1; therefore, A can be connected to J_B. The only remaining control terminal is K_B. We can see in Table 9-5 that K_B must be at logic 0 when A is at logic 0 and at logic 1 when A is at logic 1; therefore, we can connect K_B to the output of flip-flop A. Table 9-5 is redrawn in Table 9-6 to include the source of the required inputs for the J and K terminals. The logic diagram for the counter is shown in Fig. 9-21. It should be emphasized that the counter must be preset to one of its three count states because if it is initially preset to count state 00, it will "lock" in this state.

TABLE 9-5. / State Table

A	↺ 1 →	0 →	1 ↻
B	↺ 0 →	1 →	1 ↻
J_A	$\times$	1	$\times$
K_A	1	$\times$	0
J_B	1	$\times$	$\times$
K_B	$\times$	0	1

TABLE 9-6. / State Table and Boolean Expressions for J and K Inputs

A:	1	0	1	J and K Inputs
B:	0	1	1	
J_A	$\times$	1	$\times$	1
K_A	1	$\times$	0	$\overline{B}$
J_B	1	$\times$	$\times$	A
K_B	$\times$	0	1	A

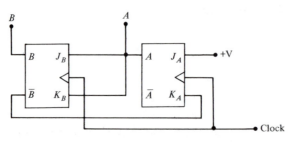

FIGURE 9-21   Mod-3 counter for Example 9-7

When constructing mod-N counters, where N is different from the natural count state, there are always omitted states. When power is first applied to a counter, it may set up one of these omitted states and "lock-up" unless steps are taken to prevent this. This possibility can be precluded by applying signals to the set or reset inputs at the time power is applied to place the counter in one of the count states of its state diagram. Self-correcting, or steering circuits, which steer counters into their normal count-state sequence, will be discussed in Chapter 10.

9-9
DECADE COUNTERS

Decade counters are one of the most important categories of counters because of their wide range of applications. Decade counters are mod-10 counters—that is, they have 10 unique count states—which appeals to human beings, who are used to counting in the decimal number system.

Since a mod-10 counter requires four flip-flops, there will be 6 omitted states, which, as before, may be any 6 of the 16 natural count states. The state diagram for a decade counter is shown in Fig. 9-22.

A decade counter implemented by cascading a mod-2 and a mod-5 counter is shown in Fig. 9-6. As stated earlier, it is generally easier to implement higher-modulus counters by cascading lower-modulus counters. However, this may not always be possible. The following example shows how a decade counter can be implemented without using lower-modulus counters.

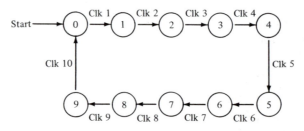

FIGURE 9-22   State diagram for mod-10 (decade) counter

TABLE 9-7. / Flip-Flop Behavior

Flip-Flop Output	0→0	0→1	1→0	1→1
J	0	1	×	×
K	×	×	1	0

EXAMPLE 9-8

Design a decade counter with count states 0000 through 1001 without using lower-modulus counters. Assume that the relationship between the J and K signal levels and the output for the flip-flops to be used are as shown in Table 9-7.

SOLUTION

The state table showing the count states (read horizontally) and the required levels of J and K for each flip-flop is shown in Table 9-8.

TABLE 9-8. / State Table

A: 0→1→0→1→0→1→0→1→0→1
B: 0→0→1→1→0→0→1→1→0→0
C: 0→0→0→0→1→1→1→1→0→0
D: 0→0→0→0→0→0→0→0→1→1

Boolean Terms for J and K Inputs

											Boolean Term
J_A	1	×	1	×	1	×	1	×	1	×	$\overline{A}$
K_A	×	1	×	1	×	1	×	1	×	1	A
J_B	0	1	×	×	0	1	×	×	0	0	$A\overline{D}$
K_B	×	×	0	1	×	×	0	1	×	×	A
J_C	0	0	0	1	×	×	×	×	0	0	AB
K_C	×	×	×	×	0	0	0	1	×	×	AB
J_D	0	0	0	0	0	0	0	1	×	×	ABC
K_D	×	×	×	×	×	×	×	×	0	1	A

❏

The Boolean terms can be obtained by considering all logic 1 entries for the J and K inputs for each flip-flop with the flip-flop outputs used as required to obtain the necessary 1's. A better approach for higher-modulus counters is to use a Karnaugh map, as will be done to complete this example.

A Karnaugh map must be used for each J and K input. Cells corresponding to addresses where there are 1's and 0's required on the J or K inputs should be identified. All other cells can be treated as don't-care conditions and filled with an ×. For example, there are five cells in the Karnaugh map for J_A, which must contain a 1; all other cells can be treated as don't cares. After each map is filled, all 1's must be enclosed, together with ×'s, if desired, for simplification; however, no 0's are to be enclosed. The required Karnaugh maps are shown in Fig. 9-23.

The Boolean terms read from the maps provide us with the necessary signals for the J and K inputs. The complete logic diagram for the counter can now be drawn as shown in Fig. 9-24. It should be readily apparent that the mod-10 counter

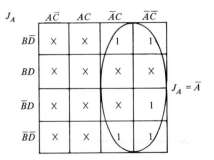

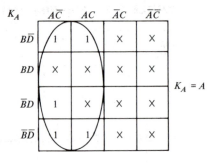

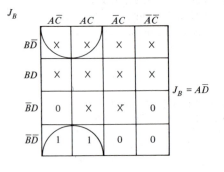

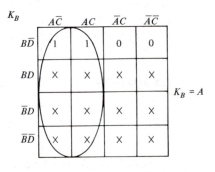

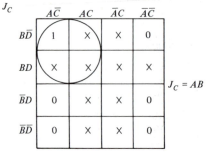

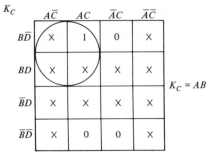

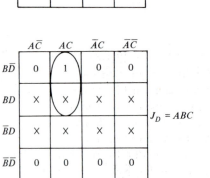

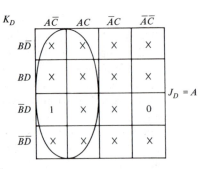

FIGURE 9-23 Karnaugh maps for Example 9-8

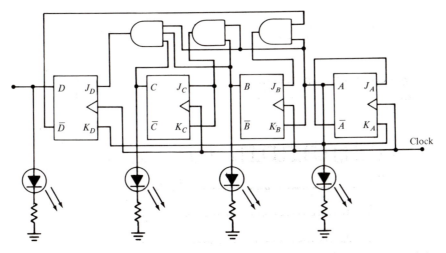

FIGURE 9-24   Decade counter designed without using lower-modulus counters

shown in Fig. 9-7 is a much easier circuit to construct when needing a decade counter. The light-emitting diodes (LEDs) shown connected to the output of the flip-flops in Fig. 9-24 provide a convenient way to indicate count.

9-10
COUNTER DECODING

The decade counter shown in Fig. 9-24 has 10 unique count states, binary 0000 through binary 1001. Each count state is displayed, in binary, by the LEDs connected to the output of each flip-flop. However, this is not a particularly convenient way to count! A more convenient way to indicate count would be to use 10 LEDs, or other indicating device, with only one ON at a time. With 10 LEDs, if only one is ON at a time, the ON LED corresponds to the decimal number of pulses counted.

Applying flip-flop outputs to AND gates to turn ON one of 10 LEDs is referred to as *counter decoding*. The AND gates are called *decoding* gates and their inputs depend on the 10 count states of the decade counter. As an example, suppose that we have a counter whose count states are as shown in Table 9-9. The minterms

TABLE 9-9. / Count States

Count	D	C	B	A	Minterms for Decoding Gates
0	0	0	0	0	$\bar{D}\bar{C}\bar{B}\bar{A}$
1	0	0	0	1	$\bar{D}\bar{C}\bar{B}A$
2	0	0	1	0	$\bar{D}\bar{C}B\bar{A}$
3	0	0	1	1	$\bar{D}\bar{C}BA$
4	0	1	0	0	$\bar{D}C\bar{B}\bar{A}$
5	0	1	0	1	$\bar{D}C\bar{B}A$
6	0	1	1	0	$\bar{D}CB\bar{A}$
7	0	1	1	1	$\bar{D}CBA$
8	1	0	0	0	$D\bar{C}\bar{B}\bar{A}$
9	1	0	0	1	$D\bar{C}\bar{B}A$

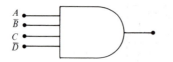

FIGURE 9-25   Decoding gate

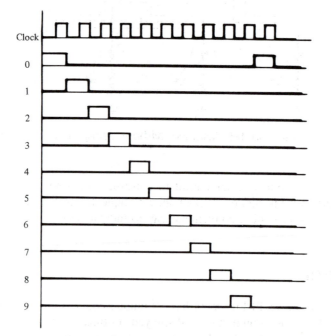

FIGURE 9-26   Output waveforms from decoding gates used with decade counter

show the sources of signals to turn ON a single LED. For example, if outputs A, $\overline{B}$, C, and $\overline{D}$ are applied to a four-input AND gate as shown in Fig. 9-25, its output should be HIGH only on the sixth count state, which is binary 5.

If the flip-flop outputs are connected to the decoding gates in accordance with the minterms of Table 9-9, the waveforms shown in Fig. 9-26 can be observed at the output of the decoding gates. As can be seen, only one decoding gate output is HIGH at any point in time; therefore, only the LED which that gate is driving will be ON.

EXAMPLE 9-9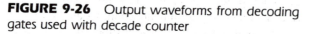

Determine the inputs for the gates that are necessary to decode the counter shown in Fig. 9-27.

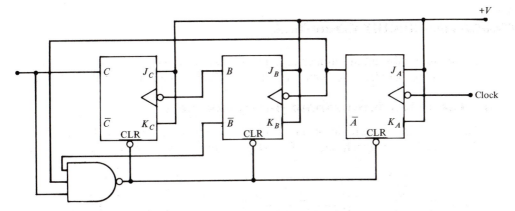

FIGURE 9-27 Counter for Example 9-9

**TABLE 9-10. /
Minterms for
Decoding**

Count	Minterm
000	$\overline{C}\,\overline{B}\,\overline{A}$
001	$\overline{C}\,\overline{B}A$
010	$\overline{C}B\overline{A}$
011	$\overline{C}BA$
100	$C\overline{B}\,\overline{A}$

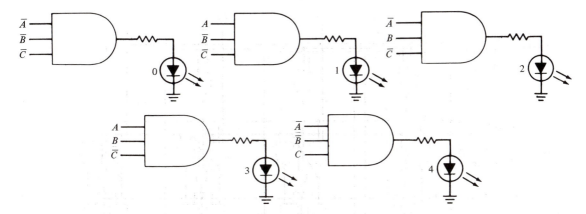

FIGURE 9-28 Decoding gates for Example 9-9

SOLUTION

The counter counts in straight binary until count 101, when it is reset. The counter is therefore a mod-5 counter with count states 000, 001, 010, 011, and 100. The required minterms for decoding are given in Table 9-10.

The decoding gates are shown in Fig. 9-28. ❑

9-11
INTEGRATED-CIRCUIT COUNTERS

There are a substantial number of counters manufactured as IC packages. In the following paragraphs we examine several of the more widely used of these devices.

9-11.1 The 74163 Synchronous Binary Counters

The 74163 is an example of an integrated-circuit binary counter. Pin and logic diagrams for this synchronous, four-stage counter are shown in Fig. 9-29. This

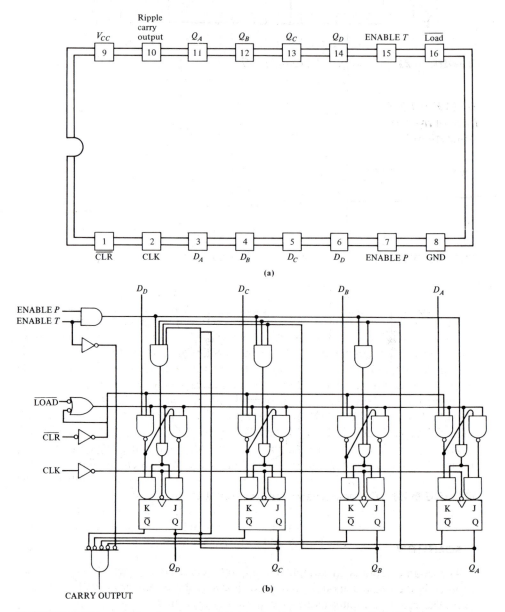

FIGURE 9-29 (a) Pin diagram and (b) logic diagram for the 74163 binary counter

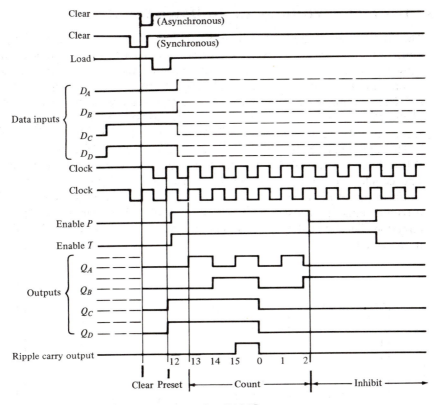

FIGURE 9-30 Timing diagram for 74163 counter

counter has several additional features in addition to the functions generally associated with basic synchronous binary counters. For example, it can be preset to any 4-bit binary number by applying the proper logic levels to the *data inputs*. By applying a LOW to the *load input,* the outputs will agree with the input data after the next clock, which allows the counter sequence to be started with any 4-bit binary number. There are also two *enable inputs,* labeled P and T, which must be HIGH for the counter to cycle through its count states. If either is LOW, the counter is disabled. The counter also has *carry look-ahead* circuitry, which makes it possible to cascade counters without additional gating requirements. The terminals used to accomplish look-ahead carry are the two *enable inputs* and the *carry output.* Both *enable inputs* must be HIGH and *enable T* is applied to the *carry output,* which goes HIGH on the last count state.

A timing diagram for the 74163 counter is shown in Fig. 9-30. The diagram shows the counter being preset to binary 1100 and then progressing through the remaining count states to binary 1111. In analyzing the waveforms, note that a logic 0 on the *clear* input causes all outputs (A, B, C, and D) to go low. Next, a logic 0 on the *load* input *presets* the counter so that input data (D_A, D_B, D_C, and D_D) can be entered into the counter on the next clock pulse. On the next three clock pulses the counter completes its count sequence. It then resets to binary 0000 and continues its normal count sequence.

9-11.2 The 74191 Synchronous Up/Down Counters

The 74191 is an example of an integrated-circuit up/down counter. The pin configuration and logic symbols for this four-stage synchronous binary counter are shown in Fig. 9-31. This counter is fully programmable; that is, the outputs may be preset either LOW or HIGH by placing a LOW on the *load* input and entering the desired data at the data inputs. The output changes independent of the level of the clock input, which allows the counter to be used as a mod-*N* counter by simply modifying the number of count states with the presets.

The direction of counting is controlled by a simple *up/down count control* line. When this line is LOW, the counter counts up, and when HIGH, it counts down. The counters are easily cascaded by connecting the *ripple clock* output to the *enable* input of the next counter if parallel clocking is used, or to the *clock* input if parallel enabling is used. A HIGH on the *enable* line inhibits counting in either direction. The count state that exists at the time the *enable* line goes HIGH remains in the counter.

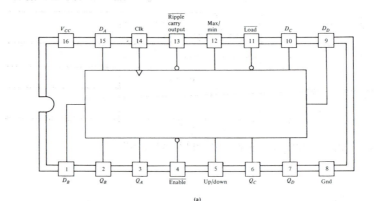

(a)

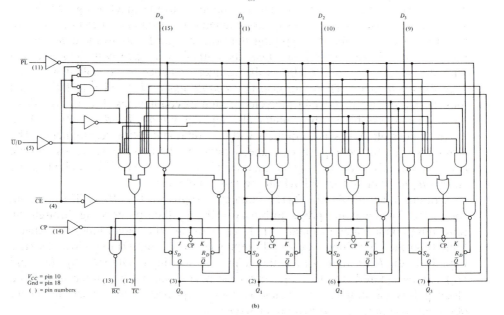

(b)

FIGURE 9-31 (a) Pin configuration and (b) logic diagram for the 74191 up/down counter

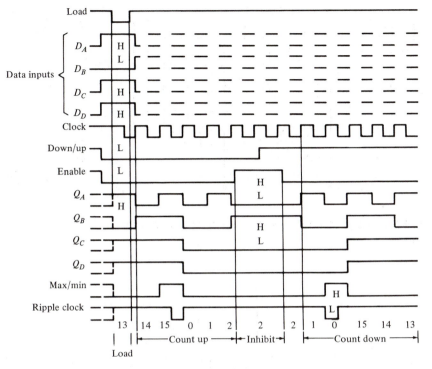

FIGURE 9-32 Timing diagram for the 74191 counter

A timing diagram for the 74191 is shown in Fig. 9-32. The diagram shows the counter being loaded with binary 1101. The counter counts up to binary 1111, resets to 0000, and begins another count sequence. At 0010, the *enable* line is set HIGH, which inhibits counting. Also, the *up/down count* line is set HIGH; therefore, when the *enable* line is reset LOW, the counter will count down.

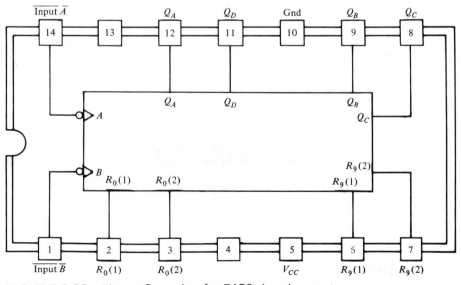

FIGURE 9-33 Pin configuration for 7490 decade counter

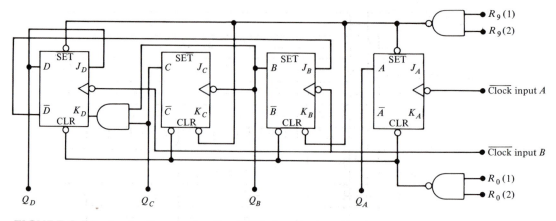

FIGURE 9-34   Logic diagram for the 7490 decade counter

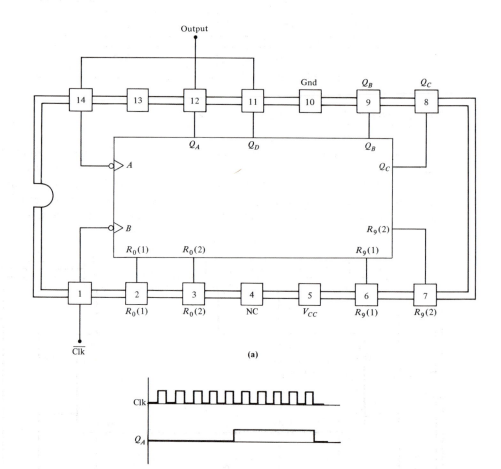

FIGURE 9-35   (a) 7490 connections to operate as a divide-by-10 counter; (b) the corresponding waveforms

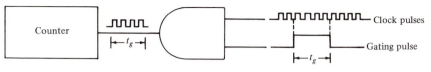

FIGURE 9-36 AND gate used to "gate" a counter

9-11.3 The 7490 Asynchronous Decade Counters

The 7490 is an example of an integrated-circuit decade counter. The pin config-uration for the device is shown in Fig. 9-33 on p. 267. The 7490 is a four-stage ripple-type decade counter consisting of four master-slave JK flip-flops that are interconnected to provide a mod-2 and a mod-5 section. The logic diagram for the counter is shown in Fig. 9-34. With no external connection between Q_A and input B, the mod-2 and mod-5 counters operate independently. When Q_A is con-nected externally to input B, the mod-2 and mod-5 counters are cascaded; thus we have a decade counter. If external connections to the counter are made as shown in Fig. 9-35a, the circuit operates as a divide-by-10 counter; that is, one output pulse is generated for every 10 clock pulses as shown in the timing diagram of Fig. 9-35b.

9-12
GATING COUNTERS

Gating a counter means that we establish the necessary conditions for clock pulses to be applied to the counter for a specific period of time. The conditions that are necessary for the counter to count the clock pulses during this specific time interval can be established by using an AND gate, as shown in Fig. 9-36. If we apply clock pulses to one input of the AND gate and a *gating pulse* to the other AND-gate input, we obtain an output signal only when both inputs are present. Since the gating pulse is essentially a dc level during the time interval t_g, the output is a train of clock pulses. The number of output pulses equals the number of clock pulses that occur during the time the gating pulse is present, t_g. Before and after the gating pulse, the AND gate is disabled; therefore, the counter is disabled as well.

EXAMPLE 9-10 ━━

Four 7490 decade counters are cascaded as shown in Fig. 9-37. If the count states of each 7490 are a straight binary count from 0000 to 1001, determine the reading of the four-stage counter after one gate pulse.

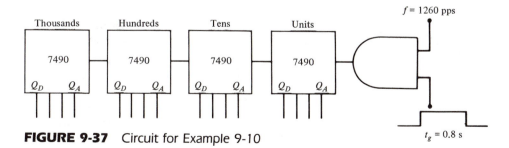

FIGURE 9-37 Circuit for Example 9-10

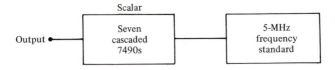

FIGURE 9-38 Cascaded decade counters used as a scalar

SOLUTION

The number of pulses counted by the four-stage counter is given as

$$\text{pulses counted} = \frac{f}{t_g}$$

$$= \frac{1260 \text{ pulses/s}}{0.8 \text{ s}} = 1575 \text{ pulses}$$

The units counter will indicate 5, the tens counter 7, the hundreds counter 5, and the thousands counter 1. Therefore, the counter circuit output will be

$$0\ 0\ 0\ 1 \quad 0\ 1\ 0\ 1 \quad 0\ 1\ 1\ 1 \quad 0\ 1\ 0\ 1 \qquad ❑$$

Since the accuracy of the count depends on the accuracy of the gating pulse, it is usually desirable to derive the gating pulse from a very accurate higher-frequency source and divide down the frequency by using a *scalar* circuit. Since a single flip-flop divides the frequency of its input signal by 2, we can cascade flip-flops to divide, or *scale*, the input frequency as required. This is illustrated in Fig. 9-38. If the 7490s are wired as shown in Fig. 9-35, the accurate 5-MHz signal is successively divided by 10. After being divided by 10 seven times, the output frequency is 0.5 Hz; therefore, its period is 2 s. Since we are interested only in the positive half-cycle, we have an accurate 1-s gating pulse.

9-13
APPLICATIONS

There are many very interesting and practical applications for the counters discussed in this chapter. A few examples are discussed in the following paragraphs. As these are discussed, you will undoubtedly think of other applications.

9-13.1 Time-Interval Measurement

Counters are very useful for measuring the time interval between two events. One application for this type of measurement is to determine the reaction time of a person. This can be done with the circuit shown in Fig. 9-39. At the instant the person doing the testing presses the "start" switch, the LED at the output of the AND gate comes ON, and the counter begins to register count. The instant the person being tested realizes that the LED is ON, he or she should press the "stop" switch, which disables the AND gate and the counter. With a clock frequency of 1 kHz, one count per millisecond is registered by the counter. Therefore, n pulses counted correlates directly as n milliseconds reaction time. This may have appli-

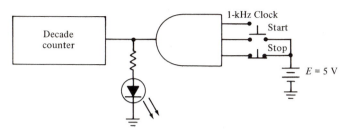

FIGURE 9-39 Counter used for time-interval measurement

cations with regard to evaluating the effects of alcohol, drugs, fatigue, or other parameters that may affect reaction time.

Another application of time-interval measurement is to determine the closing time of relay contacts. When voltage is applied to the coil or a relay, some finite time elapses before the contacts close. This time interval can be measured with the circuit shown in Fig. 9-40. When the switch shown in the drawing is closed, voltage is applied to the relay coils and to the lower input of the AND gate. The counter begins to register count immediately. After some finite time, typically in the millisecond region, the normally closed (NC) relay contacts open, thus disabling the counter. Since the clock frequency is 1 kHz, the counter output is again read directly in milliseconds.

9-13.2 Digital Voltmeters

Another very important application of counters is in digital voltmeters. By modifying the gated counter of Fig. 7-36, we obtain a basic digital voltmeter. The modification involves adding a voltage-controlled oscillator (VCO) to the gated counter circuit as shown in Fig. 9-41. The VCO functions as a voltage-to-frequency converter. If there is a linear relationship between the VCO input voltage and output frequency, the output of the AND gate can be applied to the counter to provide an indication of the voltage being measured.

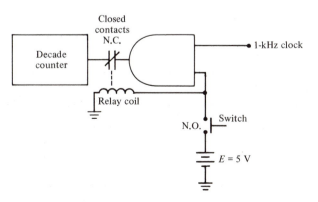

FIGURE 9-40 Counter used to measure relay contact response time

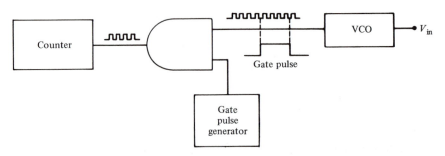

FIGURE 9-41   Counter used in basic digital voltmeter circuit

EXAMPLE 9-11 ▬▬▬▬▬▬▬▬▬▬▬▬▬▬▬▬▬▬▬▬▬▬▬▬▬▬▬▬▬▬▬▬▬▬▬

The relationship between the input voltage V_{in} and the output frequency f for the VCO in Fig. 9-41 is given as

$$V_{in} = \frac{f}{80}$$

If 560 pulses pass through the AND gate during a 0.1-s gate pulse, what is the amplitude of V_{in}?

SOLUTION

The VCO output frequency is

$$f = \frac{\text{pulses}}{\text{gate-pulse duration}} = \frac{560 \text{ pulses}}{0.1 \text{ s}} = 5600 \text{ Hz}$$

Therefore, the voltage is

$$V_{in} = \frac{f}{80} = \frac{5600 \text{ Hz}}{80} = 70 \text{ V}$$ ❏

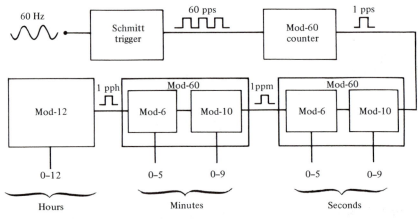

FIGURE 9-42   Digital clock block diagram

9-13.3 Digital Clocks

One of the most popular applications of counters is in digital clocks, that is, clocks that display the time of day. To be able to construct an accurate clock, an accurate frequency source is required. Digital clocks that operate from an ac power source can use the 60-Hz power frequency for this purpose. If the 60-Hz source is applied to a mod-60 counter, which effectively divides the frequency by 60, its output frequency is 1 Hz or 1 pulse per second, as shown in Fig. 9-42. This pulse is applied to another mod-60 counter, where it provides a ''seconds'' count and where it is again divided by 60 to provide 1 pulse per minute. This pulse, which provides a ''minutes'' count, is again divided by 60 to provide 1 pulse per hour and provides the ''hours'' count.

9-14
SUMMARY

Counters are very useful and versatile digital circuits with many commercial and industrial applications. The most basic counters are binary ripple counters, which are constructed by cascading flip-flops. When n flip-flops are cascaded, the counter has 2^n natural count states. Counters with n count states are referred to as mod-n counters. Binary ripple counters operate in an asynchronous mode. The propagation delay associated with this mode of operation limits the rate at which asynchronous counters can be clocked. The effects of propagation delay can be substantially reduced by using parallel counters that operate in a synchronous mode.

The state diagram for a counter can be altered by the use of feedback, which means to apply the output signal of certain flip-flops to clear previous flip-flops. However, if the state diagram is different from the natural-count state diagram, it always contains fewer states.

Higher-modulus counters can be constructed by cascading more easily configured lower-modulus counters. The modulus of the new counter is the product of the individual moduli. Counters can also be configured so that they count toward zero and are therefore called down counters. Circuits that count up can be combined with one that counts down to give us an up/down counter.

PROBLEMS

1. How many flip-flops are required to construct the following counters?
 (a) Mod-3
 (b) Mod-6
 (c) Mod-11
 (d) Mod-24

2. What is the highest decimal number to which a six-stage binary ripple counter can count?

3. How many flip-flops are required to construct a mod-128 flip-flop?

4. If a mod-50 counter is to be constructed, how many count states must be skipped?

5. If a 2.4-MHz clock signal is applied to a six-stage binary ripple counter, what will the frequency at the output of the sixth stage be?

6. How many flip-flops must be cascaded to scale a 25.6-kHz signal to 400 Hz at the output?

7. If 135 clock pulses are applied to a five-stage binary ripple counter which is initially reset to 00000, what is the last count state?

8. How many distinct count stages does a seven-stage binary ripple counter have?

9. Draw the timing diagram for the following binary ripple counter.

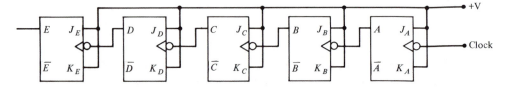

10. What is the modulus of the counter that generated the following set of waveforms?

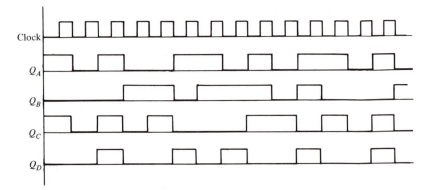

11. Tabulate the count states for the counter in Problem 10.

12. Draw the logic diagram for a mod-6 asynchronous counter whose count states are from 000 to 101.

13. Draw the decoding gates that are required to decode the mod-6 counter of Problem 12.

14. Determine the modulus of the mod-N counter in the following circuit if $t = 8.1925$ ms.

15. If each flip-flop of a mod-32 binary ripple counter has a 16-ns propagation delay, what is the maximum clock frequency that can be applied to the counter?

16. Design a counter to generate the set of waveforms shown with Problem 10, using the principles illustrated in Example 9-8.

17. Which four flip-flop outputs in the following counter should be applied to a four-input decoding AND gate so that the gate output is HIGH after six clock pulses?

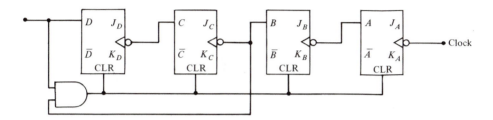

18. Four 7490 decade counters are cascaded as shown in Fig. 9-40. If each 7490 counts in straight binary count from 0000 to 1001, determine the reading of the four-stage counter after a single 0.20-s gate pulse if the clock frequency is 1645 Hz.

19. Draw the logic diagram for a mod-6 down counter with count states from 101 to 000. The flip-flops should trigger on the trailing edge of the clock pulses.

20. Determine the frequency of the unknown signal if the count shown by the ON LEDs was stored during one gate pulse.

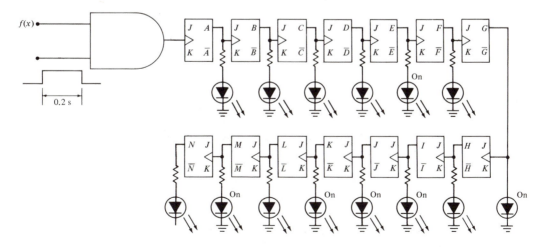

21. If after one gate pulse the indicating LEDs are ON as shown in the figure on p. 276, what is the frequency of generator 1?

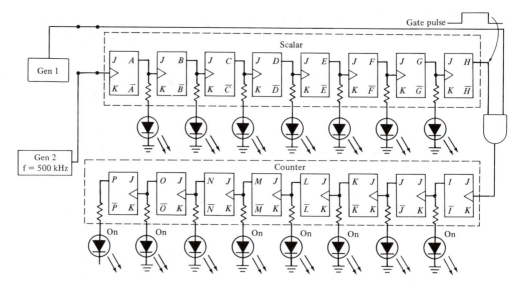

22. The following figure shows a counter being used to generate a control waveform. Determine the duration of the control waveform.

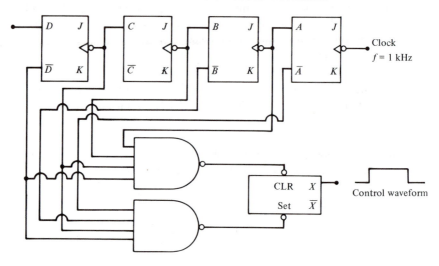

REFERENCES

Bartee, Thomas C., *Digital Computer Fundamentals, 4th ed.* New York: McGraw-Hill Book Company, 1977.

Levine, Morris E., *Digital Theory and Practice Using Integrated Circuits.* Englewood Cliffs, N.J.: Prentice-Hall, Inc., 1978.

The Logic Databook. Santa Clara, Calif.: National Semiconductor Corp., 1982.

Malvino, Albert P., and Donald P. Leach, *Digital Principles and Applications,* 3rd ed. New York: McGraw-Hill Book Company, 1981.

The TTL Logic Data Manual. Sunnyvale, Calif.: Signetics Corp., 1982.

Shift Registers and Shift Register Counters

INSTRUCTIONAL OBJECTIVES

In this chapter you are introduced to shift registers and shift register counters. After completing the chapter, you should be able to:

1. Describe the operation of a shift register.
2. List the four modes of operation of shift registers.
3. Draw the timing diagram associated with shifting a binary number into a serial shift register.
4. Compute the delay that may be introduced by a shift register in a data transfer operation.
5. Describe the operation of a bidirectional shift register.
6. Describe how a shift register is modified to make a ring counter.
7. Describe how a shift register is modified to make a Johnson counter.
8. Analyze the operation of a ring counter of a Johnson counter, which incorporates a self-correcting circuit, in going from an illegal state into its valid state diagram.
9. Determine the modules of a ring counter.
10. Describe the necessary decoding circuitry for any Johnson counter.
11. Describe why Johnson counters are normally even-moduli counters.
12. Describe the difference between static and dynamic shift registers.

SELF-EVALUATION QUESTIONS

The following questions relate to the material presented in this chapter. Read the questions prior to studying the chapter and, as you read through the material, watch for answers to the questions. After completing the chapter, return to this section and evaluate your comprehension of the material by answering the questions again.

1. What are the four modes of operation of shift registers?
2. How can multiplication and division of binary numbers by 2 be accomplished using shift registers?
3. What is the expression for computing the delay associated with transferring data through a serial shift register?

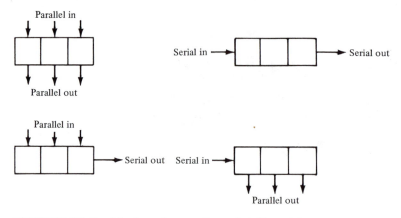

FIGURE 10-1   Modes of operation of shift registers

4.   How many clock pulses are required to shift the binary number 101101 into a six-stage serial shift register?
5.   How does a ring counter differ from a shift register?
6.   How does a ring counter differ from a Johnson counter?
7.   Why are Johnson counters normally even-moduli counters?
8.   What is another name for a Johnson counter?

10-1
INTRODUCTION

Shift registers are constructed by interconnecting flip-flops so that each flip-flop ''shifts'' its logic bit to the adjacent flip-flop when a clock pulse occurs. Such circuits find many applications related to *transfer* of momentary *storage* of data in digital systems.

There are two methods of entering or extracting data when using shift registers. The first method involves shifting the data into or out of the register one bit at a time in a series fashion. Such registers are therefore called *serial shift registers*. The second method involves shifting all bits in a binary number into the register at the same time. Registers that operate in this mode are called *parallel shift registers*. By combining the serial and parallel modes, we obtain the four methods of shift register operation shown in Fig. 10-1. In addition to data storage and transfer applications, shift registers are easily modified to implement several types of counters with some very attractive characteristics.

10-2
SERIAL SHIFT REGISTERS

An n-bit serial shift register is constructed by cascading n flip-flops; that is, there must be one flip-flop for each binary digit to be stored. For example, a shift register made up of five flip-flops in cascade is required to store the binary number 10101 momentarily; therefore, its *storage capacity* is 5 bits.

Figure 10-2 shows a serial shift register implemented with *JK* flip-flops. Binary

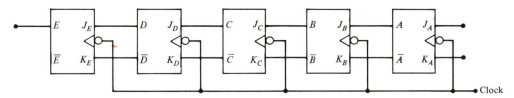

FIGURE 10-2   Serial shift register

numbers are shifted into this type of register one bit at a time by applying *control waveforms*, which change in synchronization with the clock, to the J and K inputs of the first flip-flop. The level of these control waveforms determines whether a binary 1 or 0 is shifted into the first flip-flop by a clock pulse. Binary numbers are shifted into a serial shift register bit by bit, starting with the most significant bit, by applying the following rules:

1. To store a binary 1 in flip-flop A, set J_A HIGH and K_A LOW and allow the clock to go through *one* cycle.
2. To store a binary 0 in flip-flop A, set J_A LOW and K_A HIGH and allow the clock to go through *one* cycle.
3. Any time that a binary 1 is stored in flip-flop A, it will be shifted into flip-flop B on the next clock pulse.
4. Any time that a binary 0 is stored in flip-flop A, flip-flop B will be reset LOW on the next clock pulse.
5. A binary 1 or 0 stored in any flip-flop of a serial shift register will be shifted into the adjacent "downstream" flip-flop on the next clock pulse.

Figure 10-3 shows the timing diagram that is generated by shifting binary number 10110 into the 5-bit shift register shown in Fig. 10-2. The most significant bit, a binary 1, is the first bit entered into the register. According to the rules above, this is accomplished by setting J_A HIGH and K_A LOW and applying *one* clock pulse. As can be seen in Fig. 10-3, a binary 1 is indeed stored in flip-flop A after one clock pulse; that is, A is HIGH. The next most significant bit is a binary 0, which is stored in flip-flop A by setting J_A LOW, K_A HIGH and applying one clock pulse. By again observing Fig. 10-3, we can see that on the second clock pulse, the binary 0 is entered into flip-flop A and the binary 1 that was in flip-flop A is now in flip-flop B. It was shifted into flip-flop B according to rule 3. On the third clock pulse, a binary 1 is entered into flip-flop A, the 0 that was in A is shifted into B and the 1 that was in B is shifted into C. By applying two more clock pulses, the binary 1 in flip-flop C will be shifted into E with all other bits shifted two positions as well. We have now shifted binary 10110 into the serial shift register with the most significant bit in flip-flop E and the least significant bit in flip-flop A. If we apply another clock pulse, each bit will be shifted and we will lose the bit in flip-flop E; therefore, we can see that we must have one flip-flop for each bit in the binary number to be stored and that we must apply exactly the same number of clock pulses as there are bits to be stored.

It should be pointed out that the order in which bits are shifted into serial shift registers is made somewhat arbitrarily; that is, numbers may be shifted into the register starting with either the most significant or the least significant bit. However, once a choice has been made in this regard, one must be consistent in the design of the total system.

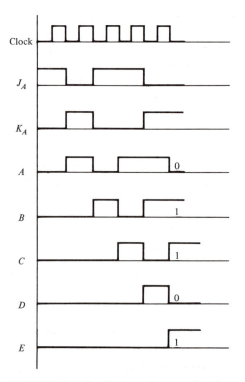

FIGURE 10-3   Timing diagram for the shift register of Fig. 10-2

10-3
SERIAL-IN, SERIAL-OUT SHIFT REGISTERS

Once data are entered into a shift register, as discussed in the preceding section, we must be able to retrieve the data. One way of doing so is by shifting it out, one bit at a time, in a serial fashion. Registers in which data are both entered and extracted in this manner are called *serial-in, serial-out shift registers*. A 4-bit serial-in, serial-out register and its associated timing diagram are shown in Fig. 10-4. As can be seen, during the first four clock pulses, binary number 1101 is shifted into the register and on the next four clock pulses the number is shifted out of the register. By observing the timing diagram, we can see that serial-in, serial-out shift registers can be used to delay data transfer between circuits where the delay equals the time of $n - 1$ clock pulses, where n is the number of flip-flops in the register.

EXAMPLE 10-1 ━━━━━━━━━━━━━━━━━━━━━━━━━━━━━━━━━━━━

How much delay is associated with a 5-bit serial-in, serial-out register if the clock frequency is 200 kHz?

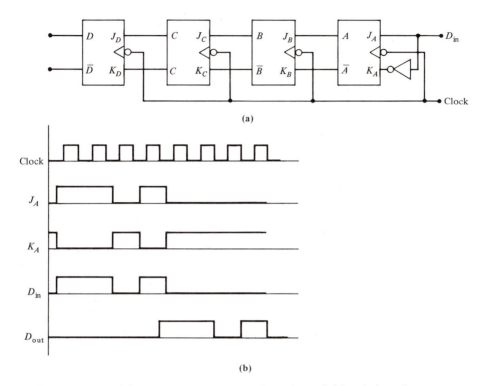

FIGURE 10-4 (a) Serial-in, serial-out shift register; (b) its timing diagram

SOLUTION

The time per clock pulse is

$$t = \frac{1}{f} = \frac{1}{200 \text{ kHz}} = 5 \text{ } \mu s$$

The total time delay is, therefore,

$$t_D = \left(\frac{\text{time}}{\text{pulse}}\right)(n - 1 \text{ clock pulses})$$

$$= (5 \text{ } \mu s)(4) = 20 \text{ } \mu s \qquad \square$$

10-4
PARALLEL-IN, SERIAL-OUT SHIFT REGISTERS

In applications where operating speed is a major consideration, parallel-in and/or parallel-out shift registers should be used, whereas if cost is a major consideration, serial-in and/or serial-out registers are a better choice. In many applications registers that combine parallel and serial operation are required. For example, we often transmit data from high-speed digital computers long distances as telemetry signals or via telephone lines. Because of the need for high-speed operation within the computer, parallel registers are generally used; however, to transmit data via a single telephone line, a serial mode of operation is required. By using a parallel-

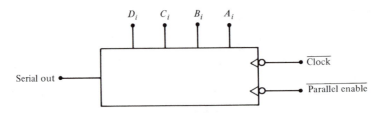

FIGURE 10-5   Parallel-in, serial-out register

in, serial-out register as shown in Fig. 10-5, we are able to satisfy the operating speed requirements of the computer while holding down costs by using a single line for long-distance data transmission.

The timing diagram generated by entering the binary number 1101 into the register of Fig. 10-5 in a parallel mode and shifting the number out of the register in a serial mode is shown in Fig. 10-6. Note that the levels representing the binary number to be entered ($D_iC_iB_iA_i$) are first set to the required levels (either HIGH or LOW); then the parallel-enable line (PE) is set HIGH. On the next clock pulse (in this case, clock pulse 2), the 4 bits are entered into the register. Since the most significant bit is in flip-flop D, the data-out line (D_{out}) is now HIGH. Prior to the next clock pulse the parallel-enable is reset LOW; therefore, the next three clock pulses serially shift the data onto the data-out line. Although all 4 bits of data are on the data-out line after the three clock pulses, one more clock pulse is required if the register is to be cleared.

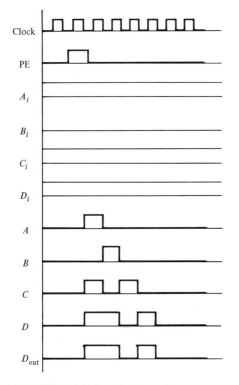

FIGURE 10-6   Timing diagram generated by the register in Fig. 10-5

10-5
IC SHIFT REGISTER CHIPS

Many configurations of shift registers are commercially available in the form of IC chips. One very popular chip is the 7491A 8-bit register. This device, which is classified as an MSI circuit, is a serial-in, serial-out shift register. The internal circuitry and the pin connections for the 7491A are shown in Fig. 10-7. As can be seen, the register is implemented with RS flip-flops and has two serial inputs labeled A and B. The data shifted into the register are the AND function of logic signals A and B; that is,

$$D_{in} = AB$$

The data inputs are positive-edge triggered with data shifted in and out in a serial fashion. The 7491A has no reset capability; therefore, initialization requires the shifting in of at least 8 bits of known data. Once the register is fully loaded, the Q output follows the inputs delayed by seven clock pulses. The complemented output from the last stage ($\overline{Q}$) is also available for simpler decoding applications.

Another very widely used shift register chip is the 7494, which contains the circuitry for both serial-in, serial-out and parallel-in, serial-out operation. The pin configuration and internal circuitry for this IC are shown in Fig. 10-8. As can be seen, the 7494 is a 4-bit register. To facilitate the parallel transfer of data from two sources, there are two sets of parallel inputs. Parallel inputs $D_1C_1B_1A_1$ are loaded by the logic level of control input PE_1, while inputs $D_2C_2B_2A_2$ are loaded by the logic level of input PE_2. When operated as a parallel-in, serial-out register, external connections to the 7494 are as shown in Fig. 10-9. Enable inputs PE_1 and PE_2 are normally set to logic 1 to load data into the register. However, in this circuit input PE_2 is connected to logic 0; therefore, we can see that parallel inputs $D_2C_2B_2A_2$ are not in use. To load the data from inputs $D_1C_1B_1A_1$, the register is first cleared by applying a logic 1 to load the register in a parallel fashion. Once loaded, data are shifted out of the register in serial form at output terminal S_0.

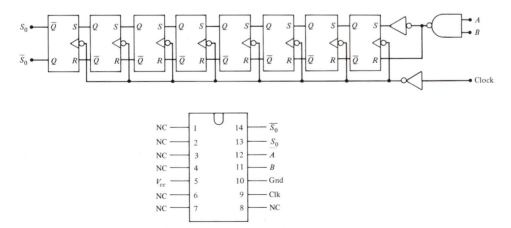

FIGURE 10-7 Logic circuit and pin configuration for the 7491A 8-bit register

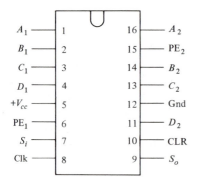

FIGURE 10-8 Logic diagram and pin configuration for the 7494 4-bit register

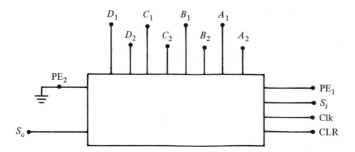

FIGURE 10-9 7494 parallel-in, serial-out connections

EXAMPLE 10-2

A 7494 register is wired to operate in its parallel-in, serial-out mode. The binary number 1011 is to be entered into the register on the $D_1C_1B_1A_1$ lines and then shifted out in serial fashion. Draw timing diagram for the device.

SOLUTION

Inputs $D_1C_1B_1A_1$ are set to the proper levels to enter binary 1011 into the register. The register is cleared by applying a logic 1 to the CLR input; then enable input PE_1 is set to logic 1 to enter the data. Since Q_D is set to logic 1, the output S_0 is at logic 1. Three clock pulses shift the data so that S_0 is HIGH due to the least significant bit; however, the bit is still in flip-flop D. One more clock pulse is required to clear the register of the number 1011. The timing diagram is shown in Fig. 10-10. ❏

10-6
BIDIRECTIONAL SHIFT REGISTERS

In each of the registers discussed to this point, operation in a serial mode has involved shifting data from flip-flop A toward flip-flop D. A circuit that shifts data in this direction is called a *shift-left register*. Registers that permit data to be

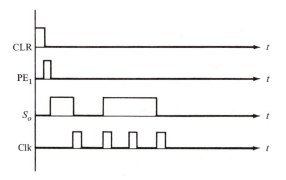

FIGURE 10-10 Timing diagram for Example 10-2

shifted to the right as well as to the left are very useful in digital circuits used for performing arithmetic operations. Such registers are called *bidirectional* or *shift-right, shift-left registers*. The block diagram of a register that operates in this manner is shown in Fig. 10-11. The level of the signal on the *mode* terminal selects shift-right or shift-left operation. For shift-left operation, the mode line is set HIGH and data on the serial input terminal are shifted from A_0 to D_0 as clock pulses are applied. For shift-right operation, the mode line is set LOW. Data are shifted from D_0 toward A_0 as clock pulses are applied. The logic diagram for a four-stage bidirectional shift register is shown in Fig. 10-12. Such registers are

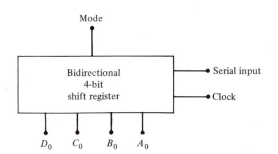

FIGURE 10-11 Bidirectional shift register block diagram

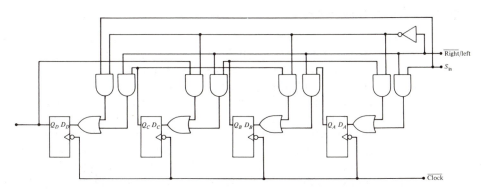

FIGURE 10-12 Logic diagram for a four-stage bidirectional register

widely used to perform arithmetic operations. For example, shifting data one position to the left is equivalent to multiplying by 2, whereas shifting data one position to the right is equivalent to dividing by 2. These operations are illustrated in the following examples.

EXAMPLE 10-3 ██

A four-stage bidirectional register contains the binary number 0011, which is equivalent to 3_{10}. What are the contents of the register if the data are shifted left one position?

SOLUTION

The contents of the register become 0110_2, which is equivalent to 6_{10}; thus the original contents have been effectively multiplied by 2.   ❏

EXAMPLE 10-4 ██

A four-stage bidirectional register contains the binary number 1010, which is equivalent to decimal 10. What are the contents of the register if the data are shifted right one position?

SOLUTION

The contents of the register become 0101_2, which is equivalent to 5_{10}; thus the original contents have been effectively divided by 2.   ❏

10-7
SHIFT REGISTER COUNTERS

Shift registers are very useful in implementing several special counter circuits. Two of the most widely used shift register counters are ring counters and Johnson counters, both of which are discussed in some detail in the following paragraphs.

10-7.1   Ring Counters

Ring counters are so named because of the feedback paths from the last flip-flop back to the first flip-flop, which permits data to circulate through the closed loop or *ring*. The feedback paths are shown in Fig. 10-13 and, as can be seen, are from Q_9 to J_0 and from $\overline{Q}_9$ to K_0. Ring counters utilize one flip-flop for each count state in their count sequence; hence they appear to be rather inefficient in their use of flip-flops. However, ring counters frequently require no decoding gates, which usually compensates for the cost of the extra flip-flops.

Typically, all the flip-flops in a ring counter except one are *reset* while the remaining flip-flop is *set* HIGH. As clock pulses are applied, the single 1 is circulated through the counter, thus generating the timing diagram shown in Fig. 10-14. This same timing diagram was obtained with the decade counter discussed

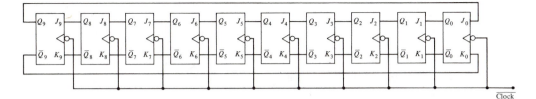

FIGURE 10-13 *Ring counter*

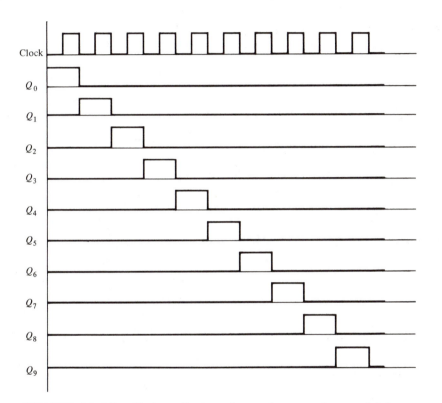

FIGURE 10-14 Timing diagram for a ring counter containing a single 1

in Section 9-9; however, as was mentioned, decoding gates were required with that counter. Ring counters that circulate a single 1 are essentially self-decoding. By observing the timing diagram of Fig. 10-14, we can readily see that a 10-stage ring counter functions as a decade counter without the need for decoding gates.

To this point we have considered the waveform generated by circulating a single 1 through a ring counter without regard to how we ensure that only one flip-flop is initially set HIGH. This can be accomplished by *setting* the desired flip-flop and *clearing* each of the other flip-flops, as shown in Fig. 10-15. If noise, or some other factor, causes the counter to be loaded with a count state other than the desired count state, it will cycle through an invalid state diagram. Ring counters have $2^N - N$ invalid states. If a counter starts in an invalid state, it will cycle through an invalid state diagram and will never count properly until steered into a valid state by an external influence. The state diagram for the four-stage ring

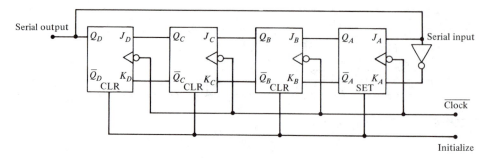

FIGURE 10-15   Setting a single 1 in a ring counter

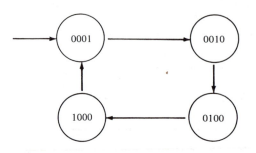

FIGURE 10-16   State diagram for a four-stage ring counter

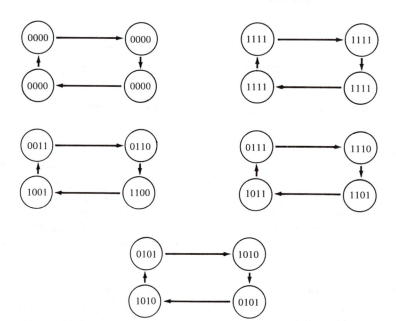

FIGURE 10-17   Invalid state diagrams for a four-stage ring counter

counter of Fig. 10-15 is shown in Fig. 10-16. A four-stage ring counter also has five invalid state diagrams that the counter may cycle through. These are shown in Fig. 10-17. By connecting a self-correcting logic circuit to a ring counter, it will be steered into its valid state diagram. One possible self-correcting circuit for a four-stage ring counter is shown in Fig. 10-18. When the self-correcting circuit becomes part of the ring counter, it no longer simply circulates the single binary 1 through the loop as described earlier. Nonetheless, the state diagram shown in Fig. 10-15 is still valid. The complete four-stage ring counter, including the self-correcting circuit, is shown in Fig. 10-19.

The operation of the circuit is best understood by the truth table shown in Table 10-1. The "present states" represents the 2^4 count states for any four-stage counter. The "present control levels" tabulates the logic levels on the control inputs due to the outputs at Q_A, Q_B, Q_C, and Q_D, while the "next states" tabulates the count states after each clock pulse due to the control levels at the time of the clock pulse. By observing the "present states" and "next states," we can draw the state diagram shown in Fig. 10-20. This shows the valid state diagram as well as the invalid states through which the counter progresses, due to the operation of the self-correcting circuit, to enter the valid count-state diagram. As can be seen, the counter cycles through a maximum of four count states before entering the valid state diagram.

Biquinary Ring Counter. Figure 10-21 shows a five-stage ring counter followed by a T flip-flop. The circuit is sometimes called a biquinary counter because the five count states associated with the ring counter are multiplied by 2 by the T flip-flop. This gives us 10 count states; hence the circuit is a mod-10, or decade,

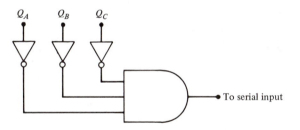

FIGURE 10-18 Self-correcting circuit

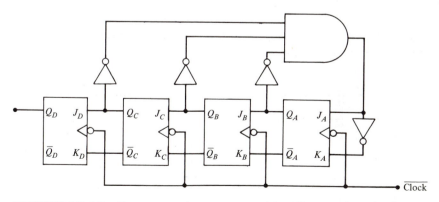

FIGURE 10-19 Four-stage ring counter with self-correcting circuit

TABLE 10-1. / Truth Table for the Self-Correcting Four-Stage Ring Counter in Fig. 10-19

Present States				Present Control Levels				Next States			
Q_D	Q_C	Q_B	Q_A	J_D	J_C	J_B	J_A	Q'_D	Q'_C	Q'_B	Q'_A
0	0	0	0	0	0	0	1	0	0	0	1
0	0	0	1	0	0	1	0	0	0	1	0
0	0	1	0	0	1	0	0	0	1	0	0
0	0	1	1	0	1	1	0	0	1	1	0
0	1	0	0	1	0	0	0	1	0	0	0
0	1	0	1	1	0	1	0	1	0	1	0
0	1	1	0	1	1	0	0	1	1	0	0
0	1	1	1	1	1	1	0	1	1	1	0
1	0	0	0	0	0	0	1	0	0	0	1
1	0	0	1	0	0	1	0	0	0	1	0
1	0	1	0	0	1	0	0	0	1	0	0
1	0	1	1	0	1	1	0	0	1	1	0
1	1	0	0	1	0	0	0	1	0	0	0
1	1	0	1	1	0	1	0	1	0	1	0
1	1	1	0	1	1	0	0	1	1	0	0
1	1	1	1	1	0	0	0	1	0	0	0

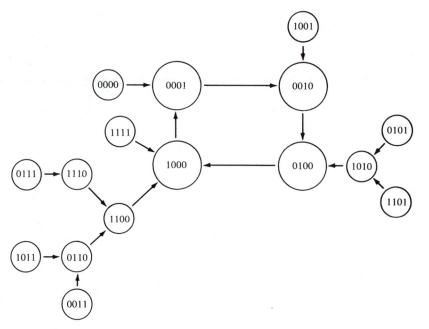

FIGURE 10-20 Count state routes into the valid state diagram for a four-stage ring counter

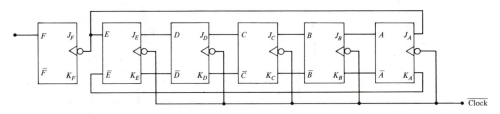

FIGURE 10-21 Biquinary ring counter

290

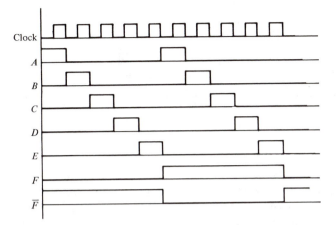

FIGURE 10-22 Timing diagram for the counter in Fig. 10-21

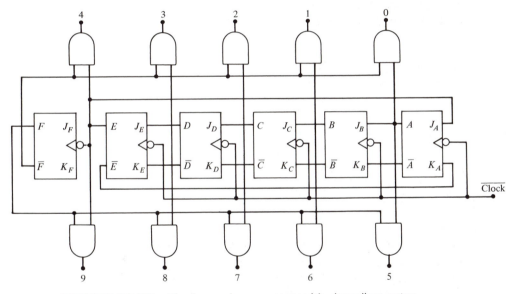

FIGURE 10-23 Biquinary ring counter with decoding gates

counter. The timing diagram for the circuit is shown in Fig. 10-22. From it we can see that the count states are not as simply defined as with the 10-stage ring counter; therefore, decoding gates are required to decode the count states. The complete circuit for the biquinary ring counter, including decoding gates, is shown in Fig. 10-23.

10-7.2 Johnson Counters

Another widely used shift register counter is the Johnson counter, which is shown in Fig. 10-24. Inspection of the logic diagram shows that the only difference between the ring counter of the preceding section and the Johnson counter is the

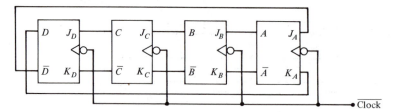

FIGURE 10-24   *Johnson counter*

points of application of the feedback. Feedback in the Johnson counter is inter-changed, or twisted, with respect to ring counters; therefore Johnson counters are sometimes referred to as *twisted-ring counters*. The circuit is also called a shift counter. The truth table for the four-stage Johnson counter shown in Fig. 10-24 is tabulated in Table 10-2.

**TABLE 10-2. / Truth Table
for Johnson Counter in
Fig. 10-24**

Clock Pulses	Q_D	Q_C	Q_B	Q_A
0	0	0	0	0
1	0	0	0	1
2	0	0	1	1
3	0	1	1	1
4	1	1	1	1
5	1	1	1	0
6	1	1	0	0
7	1	0	0	0
8	0	0	0	0

By observing the count states we see that on the clock pulse a logic 1 is set into flip-flop *A* of the counter. Each subsequent clock pulse enters another logic 1 until the register is full. The next clock pulse resets flip-flop *A* and subsequent pulses reset each flip-flop, one at a time, until the register contains all binary 0's. Additional clock pulses cause the count sequence to repeat, as can be observed by the timing diagram shown in Fig. 10-25, which is for the Johnson counter in Fig. 10-24.

Johnson counters have $2N$ valid count states and $2^N - 2N$ disallowed states, where N is the number of flip-flops. Johnson counters are easily decoded by use of the decoding AND gates shown in Fig. 10-26.

Since Johnson counters have $2N$ count states, they always have an *even* number of count states unless the circuit is modified. An odd-number-length state diagram can be obtained by using feedback from the complemented output of the next-to-the-last flip-flop to the *J* input of the first flip-flop, as shown in Fig. 10-27.

Self-Starting, Self-Correcting Johnson Counter. As with ring counters dis-cussed in the preceding section, Johnson counters may encounter an invalid count state at turn-on or due to noise or some other cause. By incorporating a self-

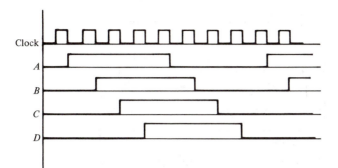

FIGURE 10-25 Timing diagram for a Johnson counter

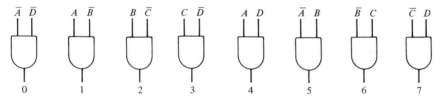

FIGURE 10-26 Decoding gates for a Johnson counter

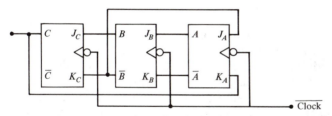

FIGURE 10-27 Johnson counter with odd number of count states

correcting circuit as shown in Fig. 10-28 on p. 294, the counter is directed into the desired state diagram.

Johnson Counter Implemented with the 7495 Shift Register. The 7495 IC shift register can be externally wired as shown in Fig. 10-29 to function as a Johnson counter. The inverter is needed because the Johnson counter requires the complement of the output to be fed back to the input; however, since $\overline{Q}$ outputs are not available with 7495 chips, the Q output must be complemented. The register is initially reset by applying a logic 1 pulse to the Clock-left/Mode terminal.

10-8
STATIC AND DYNAMIC REGISTERS

Shift registers are often categorized according to their circuit technologies as *static* or *dynamic*. Registers in the 54/7400 TTL family use flip-flops as their storage element and are categorized as *static* registers. There are two classifications of

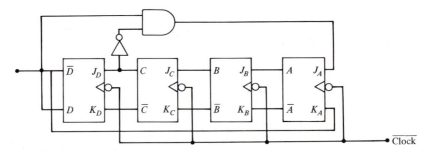

FIGURE 10-28 Self-starting, self-correcting Johnson counter

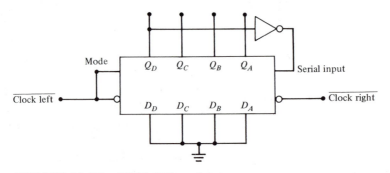

FIGURE 10-29 7495 shift register

static registers, bipolar and MOS. Registers that use TTL devices are classified as bipolar static registers, while registers that use MOSFETs are classified as MOS static registers. NMOS, PMOS, and CMOS technologies are used in MOS static registers; however, CMOS is the most widely used.

Static shift registers are very attractive because they can store data indefinitely without requiring continual clocking. An example of IC static shift register using CMOS technology is shown in Fig. 10-30. The figure shows a 4031, which is a 64-stage static shift register. The register has two input terminals, labeled *Data in* and *Recirculate in,* as well as a Mode control input. Data at the Data in terminal are shifted into the register on each positive transition of the clock if the Mode control is HIGH. Data at the Recirculate in are shifted into the register on each positive transition of the clock if the Mode control is LOW.

Dynamic shift registers are registers that utilize MOS technology. The basic storage element in dynamic registers is the inherent gate capacitance of a MOS-

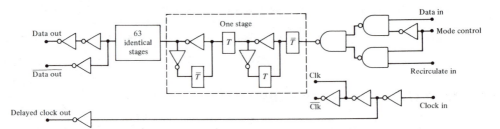

FIGURE 10-30 4031 64-stage CMOS static shift register (Courtesy of RCA Solid State Division)

FET. Since capacitors cannot retain their charge indefinitely, they must be re-charged periodically by clocking the device at a rate sufficient to maintain charge. Consequently, dynamic shift registers have a *minimum clock frequency*.

An example of an IC dynamic register that uses CMOS technology is shown in Fig. 10-31. This 200-stage dynamic shift register is a 4062A, which has a minimum single-phase clock frequency of 10 kHz over the rated temperature range of the device.

10-9
APPLICATIONS

Applications for shift registers and counters are numerous and interesting. Applications cover a wide range of topics, including generation of control waveforms, introduction or measurement of time delay, CRT refresh memory, serial-to-parallel or parallel-to-serial data conversion, and as pseudo-random-number generators. Three of these applications are discussed in the following paragraphs.

10-9.1 Control Waveform Generators

An irregular but periodic waveform, or pulse train, can be produced with a ring counter. For example, suppose that we wish to produce the waveform shown in Fig. 10-32. As can be seen, the pulse train represents an irregular sequence of binary 1's and 0's; however, it is periodic in that three identical cycles are shown. Each cycle includes a combination of nine 1's and 0's; therefore, a nine-stage ring counter, as shown in Fig. 10-33, is required. We can enter the initial conditions with the *initialize* input. Once these levels are set, we can cycle the control waveform through the ring counter, thus repeatedly generating the desired waveform.

10-9.2 Recirculating Shift Registers

Recirculating shift registers differ from ring counters in that the bit pattern stored in the flip-flops of the recirculating shift register is entered as data via the data-

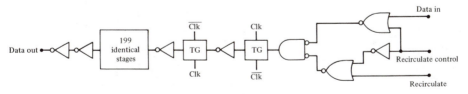

FIGURE 10-31 4062A 200-stage CMOS dynamic shift register (Courtesy of RCA Solid State Division)

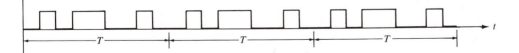

FIGURE 10-32 Periodic control waveform with period of T

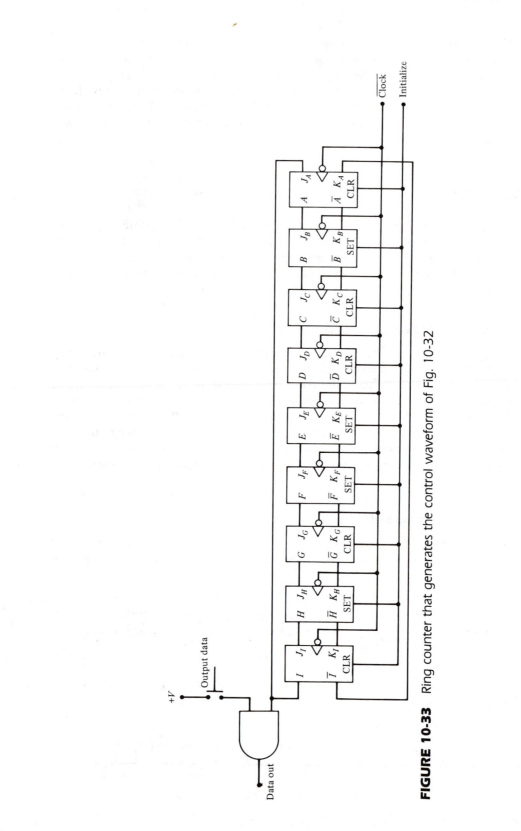

FIGURE 10-33　Ring counter that generates the control waveform of Fig. 10-32

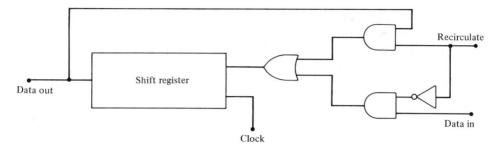

FIGURE 10-34 *Recirculating shift register*

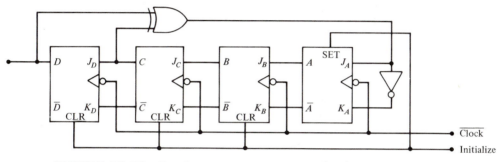

FIGURE 10-35 *Random number generator circuit*

input terminal, whereas the bit pattern in a ring counter is the result of setting or resetting the flip-flops. Recirculating shift registers, or circulating memories, are useful as temporary memories in applications such as refreshing waveforms displayed on CRT screens.

A basic logic diagram for a recirculating shift register is shown in Fig. 10-34. When the Recirculate input is LOW, data are shifted into or out of the shift register by clock pulses. When the Recirculate input is HIGH, the contents of the shift register are continually recirculated through the register.

10-9.3 Pseudo-Random-Number Generators

Another interesting application of shift registers is as pseudo-random-number generators. In this application the numbers generated are random; however, the sequence of random numbers repeats after the circuit progresses through all its count states. The number of count states is equal to

$$2^n - 1$$

where n is the number of stages in the register. A random-number generator circuit is shown in Fig. 10-35. The circuit is very similar to an ordinary shift register except that Exclusive-OR gates are used in the feedback network. The circuit must be initialized. The count states generated by the pseudo-random number shown in Fig. 10-35 are tabulated in Table 10-3.

TABLE 10-3. / Count States of Pseudo-Random-Number Generator of Fig. 10-35

Q_D	Q_C	Q_B	Q_A	Count
0	0	0	1	1
0	0	1	0	2
0	1	0	0	4
1	0	0	1	9
0	0	1	1	3
0	1	1	0	6
1	1	0	1	13
1	0	1	0	10
0	1	0	1	5
1	0	1	1	11
0	1	1	1	7
1	1	1	1	15
1	1	1	0	14
1	1	0	0	12
1	0	0	0	8
0	0	0	1	1

Pseudo-random-number generators find applications in cyptography, games, and other statistical systems, as well as in analog and digital test systems for generating timing pulses.

10-10 SUMMARY

Shift registers are implemented by cascading flip-flops and operating them in a synchronous mode. Such circuits are very useful for transfer or momentary storage of data and can be found with a wide variety of storage capacities and operational capabilities.

Shift registers are categorized according to the manner in which data are entered or extracted from the register as serial-in, serial-out; serial-in, parallel-out; parallel-in, serial-out; or parallel-in, parallel-out devices. Registers are also classified by direction of shift capability as right-shift, left-shift, or right/left-shift devices.

Shift registers can be connected to function as synchronous counters by the use of feedback. The primary types of shift-register counters are ring counters and Johnson counters. Both types of counters are easily constructed and easily decoded and find many interesting and varied applications.

PROBLEMS

1. Draw the circuit diagram of a five-stage serial shift register using negative-edge-triggered *JK* flip-flops.

2. Draw the timing diagram for the shift register in Problem 1 if the binary number 10110 is loaded into it starting with the most significant bit.

3. How much delay is associated with a 12-stage serial shift register if the clock frequency is 400 kHz?

4. If it is desired to delay a pulse by 40 μs, how many stages are required in a serial shift register clocked at 250 kHz?

5. How long will it take to load a 124-bit serial shift register if the clock is operating at 2 MHz?

6. A 256-bit serial shift register is used as a delay line. If the clock is operating at 2.5 MHz, how much delay is introduced?

7. A six-stage bidirectional register contains the binary number 000101. How many shift pulses, and in what direction, are needed for the register contents to equal 40_{10}?

8. An eight-stage shift register contains the binary number 01100000. How many shift pulses, and in what direction, are needed to divide the register contents by 32_{10}? What will the register contents be after applying the necessary shift pulses?

9. Draw the timing diagram for the following ring counter through eight clock pulses. Flip-flop A is initially SET and all other flip-flops are RESET.

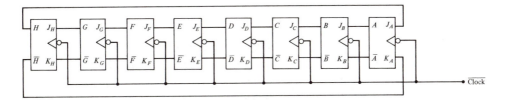

10. If the frequency of the clock in Problem 9 is 200 kHz, what is the period of the waveform at Q_H?

11. If the ring counter shown below is to be used to generate the following control waveform, indicate which flip-flops must be set and which must be reset.

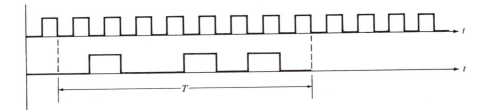

12. How many flip-flops are needed to implement a mod-12 counter in a Johnson counter configuration?

13. Determine the contents of a 12-stage bidirectional register after each clock pulse for the right/left control waveform shown if a logic 1 causes data to be shifted left and a logic 0 causes data to be shifted right. Assume that the binary equivalent of 84_{10} is initially stored in the register with the rightmost flip-flop containing the least significant bit.

14. A mod-10 Johnson counter is to be implemented with 7476 *JK* flip-flops and 7408 quad two-input AND gates. Draw a block diagram for the counter showing pin-to-pin connections.

15. If 7476 *JK* flip-flops cost 40 cents and 7408 quad two-input AND gates cost 25 cents, which is cheaper to construct: a mod-10 Johnson counter with decoding gates or a 10-stage ring counter that functions as a decade counter?

16. List the count sequence for the following pseudo-random-number generator if the register initially contains binary number 001.

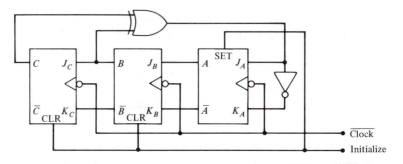

REFERENCES

CMOS Databook. Santa Clara, Calif.: National Semiconductor Corp., 1981.

COS/MOS Integrated Circuits. New York: RCA Corp., 1980.

Levine, Morris E., *Digital Theory and Practice Using Integrated Circuits,* Englewood Cliffs, N.J.: Prentice-Hall, Inc., 1978.

Triebel, Walter A., *Integrated Digital Electronics*. Englewood Cliffs, N.J.: Prentice-Hall, Inc., 1979.

Williams, Gerald E., *Digital Technology*. Chicago: Science Research Associates, Inc., 1982.

Numeric and Alphanumeric Codes

INSTRUCTIONAL OBJECTIVES

In this chapter you are introduced to several numeric and alphanumeric codes that are commonly used in digital systems. After completing the chapter, you should be able to:

1. List the most commonly used code in digital systems.
2. Encode decimal numbers in BCD.
3. Encode decimal numbers in excess-3 code.
4. Perform arithmetic operations with BCD numbers.
5. Perform arithmetic operations with numbers in excess-3 code.
6. List an important feature of the excess-3 code.
7. Describe the terms *weighted* and *unweighted codes*.
8. Convert binary numbers to Gray code.
9. Convert Gray code numbers to straight binary.
10. List an important application of the Gray code and make calculations involving this application.
11. List two important alphanumeric codes and their primary applications.
12. Describe the meaning of the term *parity*.
13. List two types of parity.
14. Describe the function of parity bits.
15. Given a set of data and the necessary parity bits, locate any error in the data.

SELF-EVALUATION QUESTIONS

The following questions deal with the material presented in this chapter. Read the questions prior to studying the chapter and, as you read through the material, watch for answers to the questions. After completing the chapter, return to this section and evaluate your comprehension of the material by answering the questions again.

1. What is the most commonly used code in digital systems?
2. What is a weighted code?
3. What is an unweighted code?
4. List a self-complementing code.

5. Which code is sometimes referred to as a unit-distance code?
6. What does the term *alphanumeric* mean?
7. What is the most widely used alphanumeric code?
8. What does the term *parity* mean?

11-1
INTRODUCTION

Even though calculators, computers, adding machines, and other computational systems operate as binary systems, information is rarely supplied to them in binary form. Numerical data, other than binary, alphabetical characters, and mathematical symbols must be converted to a format consisting of only 1's and 0's, called a *bit pattern*. These bit patterns are often in the form of a *code*. There are a number of codes used with digital systems, each with unique features. Some codes are very suitable when performing arithmetic operations. Others offer advantages when alphabetical characters are involved or for error detection. In this chapter we discuss some of the codes that are widely used in digital systems.

11-2
BINARY-CODED-DECIMAL (BCD) NUMBERS

In Chapter 6 we investigated the use of octal and hexadecimal numbers, our primary interest being to reduce the number of digits in data handled directly by an operator. Conversion between binary and octal or hexadecimal numbers was seen to be straightforward; however, the problem of unfamiliarity with the number systems still remained. As has been stated and restated, we human beings work best with decimal numbers. Therefore, *binary*-coded-decimal numbers offer a very feasible approach to entering information into digital systems.

There are many binary-coded-decimal systems, most of which use 4 binary bits to represent a decimal digit. With 4 binary bits, there are 2^4 or 16 possible combinations of 1's and 0's. Of the 16, ten are required to represent the 10 decimal digits when developing a BCD code, which leaves six excess, or redundant, states. As a result of these redundant states many BCD codes have been developed; however, we shall discuss only a few in this chapter.

11-3
THE 8421 BCD CODE

The most widely used number code is the 8421 BCD code shown in Table 11-1. Because of its predominance, when we refer to BCD we mean the 8421 code unless specifically stated otherwise. The 8421 BCD code is a weighted code where the designation "8421" indicates the binary weights (2^3 2^2 2^1 2^0) of the 4 bits.

The designation "binary-coded decimal" means that *each* decimal digit is represented by the corresponding 4-bit binary code shown in Table 11-1. Because of this binary–decimal relationship, the 8421 code is sometimes referred to as a *mixed-base code* in that it is binary within each 4-bit group but decimal from

TABLE 11-1. / Four-Bit Binary Codes

Decimal	8421 BCD
0	0000
1	0001
2	0010
3	0011
4	0100
5	0101
6	0110
7	0111
8	1000
9	1001
Invalid combinations for 8421 BCD code	1010
	1011
	1100
	1101
	1110
	1111

group to group. Using this code provides us with a very direct means of conversion from the decimal to the binary system, as shown in the following examples.

EXAMPLE 11-1

Encode the following decimal numbers in 8421 BCD.

(a) 25   (b) 74   (c) 361

SOLUTION

Each decimal digit is replaced with the corresponding 4-bit binary group from Table 11-1.

(a)   2   5   decimal
    0010   0101   8421 BCD

(b)   7   4   decimal
    0111   0100   8421 BCD

(c)   3   6   1   decimal
    0011   0110   0001   8421 BCD   ❏

EXAMPLE 11-2

Decode the following 8421 BCD numbers.

(a) 1001   0110   0010
(b) 0101   1000   0100
(c) 0100   1001   0011   0001

SOLUTION

(a) 1001   0110   0010   8421 BCD
       9   6   2   decimal

(b) 0101   1000   0100   8421 BCD
       5   8   4   decimal

(c) 0100   1001   0011   0001   8421 BCD
       4   9   3   1   decimal   ❏

11-4
8421 BCD ADDITION

The ease of conversion between decimal numbers and 8421 BCD numbers makes the code very attractive from this vantage point; however, it is less attractive with regard to addition, in that the rules for ordinary binary addition do not apply. The reason is that when we add various binary numbers, the sum may equal one of the invalid combinations shown in Table 11-1. When this occurs, the following correction is required.

> If the result of an 8421 BCD addition is greater than 1001, or if a carry is generated, add 0110 to the result to take us beyond the six invalid combinations.

The following examples demonstrate how 8421 BCD addition is performed.

EXAMPLE 11-3

Add the following 8421 numbers.

(a) 0011 + 0110
(b) 1001 + 0101
(c) 1000 + 1001
(d) 0100 0101 + 0001 0110
(e) 1001 0101 + 0111 0110

SOLUTION

(a)   0011
           0110
           ‾‾‾‾
           1001

Since 1001 is a valid code group, no correction is required; therefore, the result is as shown.

(b)   1001
           0101
           ‾‾‾‾
           1110   an invalid group; therefore, add 0110
           0110
           ‾‾‾‾
0001   0100   8421 BCD result
   1   4   decimal result

```
(c)        1000
        +  1001
0001       0001    carry a 1 into the tens column
        +  0110    add 0110 due to the carry
0001       0111    8421 BCD result
   1          7    decimal result

(d)        0100   0101
           0001   0110
           0101   1011    invalid group for LSB digit
                  0110    add 0110
           0001   0001    add carry to MSB digit
           0110   0001    8421 BCD result
              6      1    decimal result

(e)        1001   0101
           0111   0110
0001       0000   1011    add 0110 to both groups because of
           0110   0110    carry and invalid group
           0110
           0001   0001    sum of LSB column, including carry
0001       0111   0001    8421 BCD result
   1          7      1    decimal result                          ❏
```

11-5
THE EXCESS-3 BCD CODE

The excess-3 code is another important BCD code. This *unweighted* code is so named because we add 3_{10}, or 11_2, to each 8421 BCD group to obtain the excess-3 code groups. As with all 4-bit BCD codes, the excess-3 code uses only 10 of the possible 16 states listed in Table 11-2. When converting numbers from decimal

**TABLE 11-2. / Excess-3
BCD Code**

Decimal	Excess-3 BCD
0	0011
1	0100
2	0101
3	0110
4	0111
5	1000
6	1001
7	1010
8	1011
9	1100
	1101
	1110
Invalid	1111
combinations	0000
for excess-3	0001
BCD code	0010

to an excess-3 number, a decimal 3 is added to each decimal digit prior to encoding as an excess-3 number, as shown in the following examples.

EXAMPLE 11-4

Encode the following decimal numbers in excess-3 BCD.

(a) 24   (b) 39   (c) 167

SOLUTION

(a)   2   4
     $+\underline{3}$ $+\underline{3}$
       5   7

0 1 0 1   0 1 1 1 $= 24_{10}$

The same result can be obtained directly from Table 6-4.

(b)   3   9
     $+\underline{3}$ $+\underline{3}$
       6   12

0 1 1 0   1 1 0 0 $= 39_{10}$

If adding 3 to either digit generates a carry, do *not* add it to the higher-order digit of the original decimal number. In other words, the sum of 9 plus 3 remains intact as 12.

(c)   1   6   7
     $+\underline{3}$ $+\underline{3}$ $+\underline{3}$
       4   9   10

0 1 0 0   1 0 0 1   1 0 1 0 $= 167_{10}$ ❏

The most attractive feature of the excess-3 code is that it is a *self-complementing* code and is therefore very useful in complementary arithmetic. A self-complementing code is defined as one for which the code for the 9's complement of a decimal number is the 1's complement of the code for the decimal number. For example, the excess-3 code for 587_{10} is given as

$$587_{10} = 1\ 0\ 0\ 0\quad 1\ 0\ 1\ 1\quad 1\ 0\ 1\ 0_{\text{excess-3}}$$

and the 9's complement of 587_{10}, obtained by subtracting each digit from the highest digit in the number set, which is 9, is

      9   9   9
    $-\underline{5}$ $-\underline{8}$ $-\underline{7}$
      4   1   2

The excess-3 code for 412_{10} is given as

$$412_{10} = 0\ 1\ 1\ 1\quad 0\ 1\ 0\ 0\quad 0\ 1\ 0\ 1_{\text{excess-3}}$$

which is the 1's complement of the excess-3 code for 587_{10}.

11-6
EXCESS-3 ARITHMETIC

Excess-3 addition basically follows the rules for ordinary binary addition. However, beyond these basic rules, two additional rules must be applied:

1. When there is no carry from a 4-bit group, subtract 3(0011) from that group to obtain the excess-3 code for the digit.

2. When there is a carry from a 4-bit group, add 3(0011) to that group to obtain the excess-3 code for the digit and add 3(0011) to any new column generated by the last carry.

When we add two numbers, each of which is in excess of the binary numbers involved by 3, we obtain a result that exceeds the expected sum by 6. Therefore, subtracting 3, according to rule 1, leaves us in excess by 3; it is thus in excess-3 code. A carry is an indication that we are in the invalid states for excess-3 code. Therefore, we add 3, according to rule 2, to obtain a valid excess-3 code group. The following examples illustrate excess-3 addition.

EXAMPLE 11-5 ▬▬▬▬▬▬▬▬▬▬▬▬▬▬▬▬▬▬▬▬▬▬▬▬

Express each of the following decimal numbers in excess-3 code and perform the addition indicated.

(a) 6 + 3   (b) 14 + 32   (c) 325 + 174

SOLUTION

	Decimal			Excess-3	
(a)	6			1001	
	+ 3			+ 0110	
	9			1111	no carry
				− 11	subtract 3
				1100	sum in excess-3

	Decimal		Excess-3		
(b)	14		0100	0111	
	+ 32		+ 0110	0101	
	46		1010	1100	no carry
			− 11	− 11	subtract 3
			0111	1001	sum in excess-3

	Decimal				
(c)	315	0110	0100	1000	
	+ 174	+ 0100	1010	0111	
	489	1010	1110	1111	no carry
		− 11	− 11	− 11	subtract 3
		0111	1011	1100	sum in excess-3 ❏

EXAMPLE 11-6 ━━━━━━━━━━━━━━━━━━━━━━━━━━━━━━━━━━━━━

Express each of the following decimal numbers in excess-3 code and perform the addition indicated.

(a) 8 + 4   (b) 26 + 15   (c) 97 + 78

SOLUTION

```
         Decimal              Excess-3
(a)        8                    1011
         + 4                  + 0111
         ───                  ──────
          12         0001      0010    carry is generated
                             + 11      + 11    add 3 to both bit groups
                             ────      ────
                             0100      0101    sum in excess-3

(b)       26         0101      1001
        + 15       + 0100      1000
        ───        ──────      ────
          41         1001      0001
                     0001              carry is generated
                     ────
                     1010    + 11      add 3 due to carry
                     − 11              subtract 3, no carry
                     ────      ────
                     0111      0100    sum in excess-3

(c)       97         1100      1010
        + 78       + 1010      1011
        ───        ──────      ────
         175  0001   0110      0101    carry generated   (left code group)
                     0001              carry generated   (right code group)
                     ────
                     0111
             + 11   + 11      + 11     add 3 to each bit group
             ────   ────      ────
             0100   1010      1000     sum in excess-3              ❏
```

Subtraction in excess-3 can be done using the rules for ordinary binary subtraction or by using 1's-complement addition. In either case, the difference will be straight binary; therefore, 3 (0011) must be added to the difference to put it in excess-3 code, as illustrated in the following examples.

EXAMPLE 11-7 ━━━━━━━━━━━━━━━━━━━━━━━━━━━━━━━━━━━━━

Express each of the following decimal numbers in excess-3 code and perform the indicated subtraction.

(a) 7 − 2   (b) 28 − 16   (c) 286 − 163

SOLUTION

```
       Decimal   Excess-3
(a)       7      1010
        − 2      0101
        ───      ────
          5       101    straight binary
                + 11     add 3
                 ────
                 1000    difference in excess-3
```

```
(b)     28       0101   1011
      − 16     − 0100   1001
               0001   0010   straight binary
               + 11   + 11   add 3
               0100   0101   difference in excess-3
```

```
(c)    286      0101   1011   1001
      − 163   − 0100   1001   0110
       123      0001   0010   0011   straight binary
               + 11   + 11   + 11   add 3
               0100   0101   0110   difference in
                                    excess-3                    ❏
```

Binary subtraction can be done using 1's-complement addition with excess-3 code groups, as shown in the following examples.

EXAMPLE 11-8

Express each of the following decimal numbers in excess-3 code and perform the indicated subtraction by use of 1's-complement addition.

(a) $8 - 1$ (b) $5 - 4$ (c) $28 - 16$

SOLUTION

```
        Decimal    Excess-3 Codes      1's-Complement Addition
(a)        8          1001                 1001
         − 1        − 0100               + 1011
           7                              10110   add 1's complement
                                              1   end-around carry
                                           0111   straight binary
                                           + 11   add 3
                                           1010   difference in excess-3
```

```
(b)        5          1000                 1000
         − 4        − 0111                 1000
           1                              10000   add 1's complement
                                            + 1   end-around carry
                                           0001   straight binary
                                           + 11   add 3
                                           0100   difference in excess-3
```

```
(c)       28     0101   1011      0101     1011
        − 16   − 0100   1001    + 1011     0110
          12                     10000    10001   add 1's complement
                                   + 1        1   end-around carry
                                  0001     0010   straight binary
                                  + 11     + 11   add 3
                                  0100     0101   difference in excess-3    ❏
```

11-7
THE GRAY CODE

The Gray code is another unweighted BCD code. This code is not well suited to arithmetic operations but is widely used in input/output devices as well as in other peripheral equipment on computers. The Gray code is shown in Table 11-3 together with the corresponding binary numbers. We can see by observing each number of the Gray code that each number differs from the preceding number by a single binary bit. For example, in going from decimal 10 to 11, where the corresponding Gray code numbers are 1111 and 1110, we can see that only the low-order bit changed. Because of this characteristic, the Gray code is sometimes referred to as a *unit-distance code*. Such codes are very useful in instrumentation applications such as *shaft encoders* to measure displacement and *linear encoders* to measure linear displacement.

Two angular shaft encoders are shown in Fig. 11-1. The encoder shown in Fig. 11-1a is using a straight binary code to measure angular displacement. Since

**TABLE 11-3. / Gray Code
from 0_{10} to 15_{10}**

Decimal	Gray	Binary
0	0000	0000
1	0001	0001
2	0011	0010
3	0010	0011
4	0110	0100
5	0111	0101
6	0101	0110
7	0100	0111
8	1100	1000
9	1101	1001
10	1111	1010
11	1110	1011
12	1010	1100
13	1011	1101
14	1001	1110
15	1000	1111

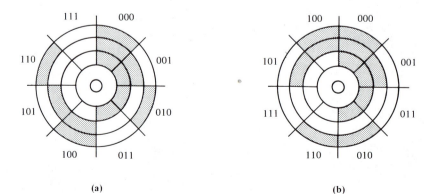

(a)   (b)

FIGURE 11-1   Angular shaft encoders for measuring angular displacement: (a) binary; (b) Gray

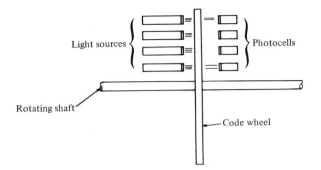

FIGURE 11-2 Optomechanical system for displaying angular displacement in Gray code

several bits may change at the same time, very large measurement errors may occur. The encoder shown in Fig. 11-1b, which is using the Gray code, is much less prone to error, since only one bit changes in going from any Gray code group to the adjacent code group. Figure 11-2 shows a typical optomechanical system, which incorporates a shaft encoder and an optical code reader. The light and dark areas on the code wheel correspond to binary 1's and 0's, respectively.

Each concentric ring represents a binary weighted digit; therefore, increasing the number of concentric rings increases the *resolution* of the encoder. The resolution is given as

$$\text{Res} = \frac{360°}{2^n}$$

where Res is the resolution, which is the shaft rotation in degrees corresponding to a change of one binary digit in the Gray code and n is the number of concentric rings on the shaft encoder.

EXAMPLE 11-9 ━━━━━━━━━━━━━━━━━━━━━━━━━━━━━━━━━━━━━━━

How many concentric rings are required to obtain a resolution of at least 1°?

SOLUTION

The expression for resolution states that

$$\text{Res} = \frac{360°}{2^n}$$

For our problem

$$1° = \frac{360°}{2^n}$$

Rewriting the expression gives us

$$2^n = \frac{360°}{1°} = 360$$

To solve for n, we must take the natural logarithm of both sides of the expression, which is written as

$$\ln 2^n = \ln 360 \quad \text{or} \quad n \ln 2 = \ln 360$$

Evaluating the natural logarithms gives us

$$n(0.693) = 5.89$$

Solving for n yields

$$n = \frac{5.89}{0.693} = 8.45$$

Therefore, the shaft encoder must contain nine concentric rings to provide a resolution of at least $1°$. ❏

11-8
BINARY–GRAY-CODE CONVERSION

To facilitate interfacing shaft encoders that use the Gray code and microprocessors or computers that recognize binary numbers, it is often necessary to convert Gray code numbers to binary, or vice versa.

11-8.1   Gray-Code-to-Binary Conversion

The conversion from Gray code to binary makes use of *modulo-2 addition*, which obeys the rules of binary addition but discards any carry. The symbol for modulo-2 addition is $\oplus$ and the rules for this type of addition are

$$0 \oplus 0 = 0$$
$$0 \oplus 1 = 1$$
$$1 \oplus 0 = 1$$
$$1 \oplus 1 = 0 \quad \text{(carry is discarded)}$$

Using modulo-2 addition and the following procedure permits us to readily convert Gray code numbers to binary.

1. Write the Gray code number.

2. The most significant bit (MSB) for the binary number is the same as the MSB of the Gray code number.

3. Add, by the rules of modulo-2 addition, the MSB of the binary number to the second MSB of the Gray code number to obtain the second MSB of the binary number.

4. Repeat step 3 for each remaining bit of the Gray code using the next MSB of the binary number just determined.

The following example illustrates the use of these rules to convert from Gray code to an equivalent binary number.

EXAMPLE 11-10 ━━━━━━━━━━━━━━━━━━━━━━━━━━

Convert the Gray code number 11011010 to an equivalent binary number.

SOLUTION

Write the Gray code number and perform modulo-2 addition.

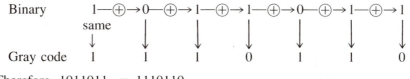

Therefore, $11011010_{Gray} = 10010011_2$. ❏

11-8.2 Binary-to-Gray-Code Conversion

Binary numbers are converted to Gray code by applying essentially the same procedure as above in reverse, as indicated by the following rules:

1. Write the binary number.

2. The MSB of the Gray code number is the same as the MSB of the binary number.

3. Add by the rules of modulo-2 addition the MSB of binary number to the second MSB of the binary number to obtain the second MSB of the Gray code number.

4. Repeat step 3 for the remaining bits of the binary number.

The following example illustrates the use of these rules to express a binary number in Gray code.

EXAMPLE 11-11 ━━━━━━━━━━━━━━━━━━━━━━━━━━

Express the binary number 10011011 in Gray code.

Solution

Write the binary number and perform modulo-2 addition.

Binary 1—$\oplus$→0—$\oplus$→1—$\oplus$→1—$\oplus$→0—$\oplus$→1—$\oplus$→1

     same ↓ ↓ ↓ ↓ ↓ ↓

     ↓

Gray code 1 1 1 0 1 1 0

Therefore, $10011011_2 = 1110110_{Gray}$. ❏

11-9
OTHER FOUR-BIT BCD CODES

There are many other 4-bit codes besides those discussed to this point. Some of the more common weighted codes are shown in Table 11-4. Each code has par-

TABLE 11-4. / Weighted Four-Bit BCD Codes

Decimal	2421	3321	4221	5211	5311	5421	6311	7421
0	0000	0000	0000	0000	0000	0000	0000	0000
1	0001	0001	0001	0001	0001	0001	0001	0001
2	0010	0010	0010	0011	0011	0010	0011	0010
3	0011	0011	0011	0101	0100	0011	0100	0011
4	0100	0101	1000	0111	0101	0100	0101	0100
5	1011	1010	0111	1000	1000	1000	0111	0101
6	1100	1100	1100	1001	1001	1001	1000	0110
7	1101	1101	1101	1011	1011	1010	1001	1000
8	1110	1110	1110	1101	1100	1011	1011	1001
9	1111	1111	1111	1111	1101	1100	1100	1010

ticular applications for which it is best suited. As with the codes discussed previously, decimal numbers larger than single-digit numbers are encoded one digit at a time.

11-10
FIVE-BIT CODES

Although single decimal digits can be encoded with 4-bit codes, a fifth bit allows us to code numbers more easily as well as providing a means for error detection. Table 11-5 shows some of the more common 5-bit codes.

The 51111 code is a weighted, self-complementing code. Its most attractive characteristic is the ease by which it can be decoded by the use of basic logic gates. The 63210 code is a weighted code, except for the code group corresponding to decimal 0. For other code groups, except 0, the weights are 6, 3, 2, 1, 0, reading from left to right. The most significant characteristic of this code is the fact that each code group has exactly two binary 1's, which is a very useful feature for error detection. One of the primary applications for the 63210 code is for mass storage of digital data on magnetic drums or tape.

The 2-of-5 code is an unweighted code that finds applications in the telephone and communications industries. As with the 63210 code, there are exactly two binary 1's in each code group, which, again, is useful for error detection.

TABLE 11-5. / Common Five-Bit Codes

Decimal	51111	63210	2-of-5	Shift-Counter
0	00000	00110	00011	00000
1	00001	00011	00101	00001
2	00011	00101	00110	00011
3	00111	01001	01001	00111
4	01111	01010	01010	01111
5	10000	01100	01100	11111
6	11000	10001	10001	11110
7	11100	10010	10010	11100
8	11110	10100	10100	11000
9	11111	11000	11000	10000

The shift-counter code, or Johnson code as it is sometimes called, is an unweighted code that finds applications in electronic counters. One of its most attractive characteristics, as was so with the 51111 code, is the ease of decoding with basic logic gates, which was discussed in Chapter 10.

11-11
ALPHANUMERIC CODES

In addition to the binary codes discussed to this point that are used to represent decimal digits 0 through 9, there are several binary codes used to represent both alphabetical and numerical characters. These codes are called *alphanumeric codes* and are used primarily with input and output (I/O) equipment. The two most widely used alphanumeric codes are ASCII (American Standard Code for Information Interchange) and EBCDIC (Extended Binary-Coded Decimal Interchange Code).

11-11.1 ASCII Code

ASCII is a widely accepted code that allows standardization of interface hardware between computers and input/output devices, such as keyboards, teletypewriters, printers, and monitors. The code is a 7-bit code that permits both lowercase and uppercase alphabetical characters, as well as special characters such as mathematical symbols, and more than 30 command and control statements to be coded.

Table 11-6 shows the ASCII code. The code format is

$$X_6 \ X_5 \ X_4 \quad X_3 \ X_2 \ X_1 \ X_0$$

where the 4 low-order bits correspond to the row and the 3 high-order bits correspond to the column in which the symbol appears. For example, the ASCII code for the letter A is 100 0001 in binary or 41 in hexadecimal. The meanings of the code abbreviations in Table 11-6 are listed in Table 11-7.

When using ASCII, decimal digits are represented by the 8421 BCD code preceded by 011; therefore, computational systems need to save only the 4 low-order bits to retain the decimal value. Since many computers and calculators use either 8- or 16-bit word length, the 4-bit ASCII representation of decimal digits can be entered very compactly as single 8- or 16-bit words.

11-11.2 EBCDIC Code

Another frequently encountered alphanumeric code is EBCDIC. This code is very similar to ASCII except that it is an 8-bit code. With 8 bits there are 2^8 or 256 code combinations. Both lowercase and uppercase letters can be represented as well as numerous symbols and commands and decimal numbers. These decimal numbers are again represented by the 8421 BCD code preceded by 1111, as shown in Table 11-8, which is a partial EBCDIC code table. The meanings of the EBCDIC machine codes are listed in Table 11-9 on p. 318.

TABLE 11-6. / ASCII Code

MSB / LSB	Binary / Hex	000 / 0	001 / 1	010 / 2	011 / 3	100 / 4	101 / 5	110 / 6	111 / 7
0000	0	NUL	DEL	SP	0	@	P	.	P
0001	1	SOH	DC1	!	1	A	Q	a	q
0010	2	STX	DC2	"	2	B	R	b	r
0011	3	ETX	DC3	#	3	C	S	c	s
0100	4	EOT	DC4	$	4	D	T	d	t
0101	5	END	NAK	%	5	E	U	e	u
0110	6	ACK	SYN	&	6	F	V	f	v
0111	7	BEL	ETB	'	7	G	W	g	w
1000	8	BS	CAN	(	8	H	X	h	x
1001	9	HT	EM	)	9	I	Y	i	y
1010	A	LF	SUB	*	:	J	Z	j	z
1011	B	VT	ESC	+	;	K	[	k	{
1100	C	FF	FS	,	<	L	/	l	l
1101	D	CR	GS	–	=	M	]	m	}
1110	E	SO	RS	.	>	N	Λ	n	~
1111	F	SI	US	/	?	O	–	o	DEL

TABLE 11-7. / Tabulation of ASCII Machine Codes and Their Meanings

ACK	Acknowledge	FS	Form separator
BEL	Bell	GS	Group separator
BS	Backspace	HT	Horizontal tab
CAN	Cancel	LF	Line feed
CR	Carriage return	NAK	Negative acknowledge
DC1–DC4	Direct control	NUL	Null
DEL	Delete idle	RS	Record separator
DLE	Data link escape	SI	Shift in
EM	End of medium	SO	Shift out
ENQ	Enquiry	SOH	Start of heading
EOT	End of transmission	STX	Start text
ESC	Escape	SUB	Substitute
ETB	End of transmission block	SYN	Synchronous idle
ETX	End text	US	Unit separator
FF	Form feed	VT	Vertical tab

TABLE 11-8. / EBCDIC Code

MSB →		00				01				10				11			
LSB ↓ Binary	Hex	00	01	10	11	00	01	10	11	00	01	10	11	00	01	10	11
Hex →		0	1	2	3	4	5	6	7	8	9	A	B	C	D	E	F
0000	0	NUL	DEL	DS		SP	&	—									0
0001	1	SOH	DC1	SOS						a	j			A	J		1
0010	2	STX	DC2	FS	SYN					b	k	s		B	K	S	2
0011	3	ETX	DC3							c	l	t		C	L	T	3
0100	4	PF	RES	BYP	PN					d	m	u		D	M	U	4
0101	5	HY	NL	LF	RE					e	n	v		E	N	V	5
0110	6	LC	BS	EOB	UC					f	o	w		F	O	W	6
0111	7	DEL	IL	PRE	EOT					g	p	x		G	P	X	7
1000	8		CAN							h	q	y		H	Q	Y	8
1001	9	RLF	EM	SM						i	r	z		I	R	Z	9
1010	A	SMM	CC				!		:								
1011	B	VT				.	$		#								
1100	C	FF	IFS		DC4		*	%	@								
1101	D	CR	IGS	ENQ	NAK	(	)	_	'								
1110	E	SO	IRS	ACK		+	;		=								
1111	F	SI	IUS	BEL	SUB			?	"								

TABLE 11-9. / Tabulation of EBCDIC Machine Codes and Their Meanings

ACK	Acknowledge	IRS	Interchange record separator
BEL	Bell	IUS	Interchange unit separator
BS	Backspace	LC	Lowercase
BYP	Bypass	LF	Line feed
CAN	Cancel	NAK	Negative acknowledgement
CC	Cursor control	NL	New line
CR	Carriage return	NUL	Null
DC1	Device control 1	PF	Punch off
DC2	Device control 2	PN	Punch on
DC3	Device control 3	PRE	Prefix
DC4	Device control 4	RES	Restore
DEL	Delete	RLF	Reverse line feed
DLE	Data link escape	RS	Reader stop
DS	Digit select	SI	Shift in
EM	End of medium	SM	Set mode
ENQ	Enquiry	SMM	Start of manual message
EOB	End of block	SO	Shift out
EOT	End of transmission	SOH	Start of heading
ETX	End of text	SOS	Start of significance
FF	Form feed	SP	Space
FS	Field separation	STX	Start of text
HT	Horizontal tab	SUB	Substitute
IFS	Interchange file separator	SYN	Synchronous idle
IGS	Interchange group separator	UC	Uppercase
IL	Idle	VT	Vertical tabulation

11-12
ERROR-DETECTING CODES

Because of the two-state nature of digital information, there is less chance of error when digital data are transmitted than when transmitting analog data, but the need for error checking still exists. There are many possible causes of error, including power-line transients, intermittent power failure, electrical noise, or marginally operating equipment, which may cause a single bit to change states.

When using binary codes, there is *some* built-in protection against errors due to the six invalid combinations associated with 4-bit codes. For example, if when transmitting binary 1001, noise causes the second MSB to change from 0 to 1, the number is received as 1101, which is a valid binary number. Therefore, we have no way of knowing that an error has occurred. On the other hand, if the numbers are in 8421 BCD, the system quickly detects the error, since 1101 is an invalid code group. However, the error would have gone undetected if either of the 1's in the number 1001 had changed, since either change would result in a valid 8421 BCD number. Therefore, we can see that the six redundant states

TABLE 11-10. / Even and Odd Parity Bits

Even Parity		Odd Parity	
Data	Parity Bit	Data	Parity Bit
0 1 0 0 1 0	0	1 1 0 1 0 1	1
1 1 0 1 0 0	1	0 0 1 0 1 1	0
0 1 1 1 1 0	0	1 0 1 1 0 1	1
0 1 0 1 1 0	1	1 0 1 0 0 1	1 error
1 0 1 1 0 0	0 error	0 1 1 1 1 1	0
1 0 0 1 1 1	0	1 0 1 0 1 1	1

TABLE 11-11. / Row and Column Parity Bits

Data	Parity Bit Rows
0 1 0 1 0 1	0
1 0 1 0 1 1	1
0 1 0 0 1 1	0
0 1 ①1 0 0	1 ← error
1 0 0 0 1 1	0
1 1 1 0 0 1	1
1 0 1 1 1 1	0
1 1 0 0 1 1	Parity bits columns

↑
error

enable us to detect *some* but not all errors. To improve the reliability of transmitted data, codes with *parity* are often used. Parity is the addition of a bit to a binary word to provide increased confidence in the integrity of transmitted data.

Coded data can utilize one of two types of parity, which are called *even parity* and *odd parity*. Data transmitted using *even* parity must have an even number of 1's in each code group, including the parity bit. On the other hand, data transmitted using *odd* parity must have an odd number of 1's in each code group, including the parity bit. The use of parity bits for error detection is illustrated in the following data. Odd parity is generally preferred because it is not possible to transmit a code group with all 0's when using it. The addition of the parity bit, as shown in Table 11-10, serves only to indicate that an error is present. There is no indication of which bit is incorrect. A technique that identifies the incorrect bit makes use of parity bits for both the rows and the columns, as shown in Table 11-11. Odd parity is used here, but even parity could just as well have been used. Since there should be an odd number of 1's, including the parity bits, in all rows and columns, we can first identify the row and column in which an error exists. We can then identify the particular bit in error. This is the bit where the row and column that contain the error intersect.

11-13
THE PARITY GENERATOR

When a code group such as BCD is to be transmitted with parity, a parity generator such as the one shown in Fig. 11-3 is required. The logic circuit shown will

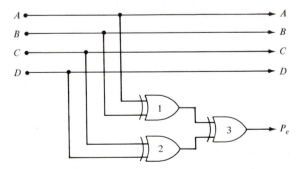

FIGURE 11-3 Logic circuit to generate even parity bit

**TABLE 11-12. / Truth
Table for the Parity
Generator of Fig. 11-3**

Inputs	Outputs
$ABCD$	$ABCDP_e$
0000	00000
0001	00011
0010	00101
0011	00110
0100	01001
0101	01010
0110	01100
0111	01111
1000	10001
1001	10010

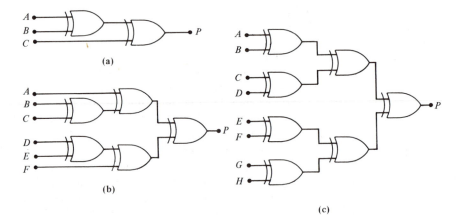

FIGURE 11-4   Parity bit generators for (a) 3-bit, (b) 6-bit, and
(c) 8-bit code groups

generate an even parity bit for a 4-bit code group. As can be seen, the circuit has
four inputs and five outputs. Four of the outputs are the same as the inputs and
the fifth output is the parity bit. The truth table for the circuit is shown in Table
11-12. By tracing the signals through the circuit we can see that a logic 1 appears
at the P_e terminal whenever an odd number of the inputs are a logic 1. The odd
number of HIGH inputs plus the parity bit produced even parity. Odd parity can
be achieved by following Exclusive-OR gate 3 with an inverter. Circuits for
generating the parity bit for code groups containing 3, 6, and 8 bits are shown in
Fig. 11-4.

11-14
PARITY-CHECKING CIRCUIT

When digital data that include a parity bit are transmitted between digital systems,
the integrity of the data can be determined by using a parity-checking circuit.
Figure 11-5 shows a parity-checking circuit used to check the integrity of trans-

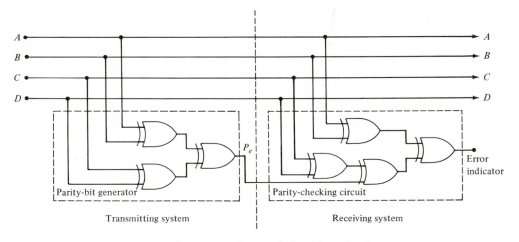

FIGURE 11-5 Parity generating and checking circuit

mitted data when they reach their destination. As can be seen, at the receiver the transmitted data are combined with the parity bit to check for the existence of a single error. If a single bit is in error, the output of the parity-checking circuit goes HIGH. If two errors are present, correct parity is maintained; therefore, no error signal is generated by the parity-checking circuit. Such circuits are useful only where the probabilities of double errors are very, very low.

11-15
SUMMARY

Codes are widely used in digital systems to permit numerical data other than binary, alphabetical characters, and mathematical symbols to be expressed as a combination of 1's and 0's. The most widely used code is the BCD code, which is a compromise between binary and decimal numbers.

Some codes offer advantages when dealing with numerical data, whereas others are more suitable when dealing with alphabetical characters or symbols. To ensure the integrity of transmitted data, a parity bit is transmitted with the data to provide error detection.

PROBLEMS

1. Encode the following decimal numbers in 8421 BCD.

(a) 12   (b) 37
(c) 89   (d) 136
(e) 168   (f) 243

2. Decode the following 8421 BCD numbers to decimal.

(a) 0110   (b) 1001
(c) 01000110   (d) 10010010
(e) 01011001   (f) 100101010001

3. Add the following BCD numbers.

(a) 0010 + 0011   (b) 0001 + 0110
(c) 0110 + 0111   (d) 1001 + 0101
(e) 01010100 + 00110101   (f) 01010111 + 10000101

4. Encode the following decimal numbers in excess-3 code.

(a) 7   (b) 24
(c) 49   (d) 136
(e) 321   (f) 1234

5. Decode the following excess-3 code groups to decimal.

(a) 0101   (b) 0111
(c) 01101010   (d) 00110100
(e) 010010010101   (f) 101010110110

6. Add the following excess-3 numbers.

(a) 0110 + 1010   (b) 1010 + 0010
(c) 1001 + 0111   (d) 10010101 + 01111001
(e) 10110011 + 01101010   (f) 10101001 + 10101001

7. Subtract the following excess-3 numbers.

(a) 1011 − 1001   (b) 1010 − 0111
(c) 10000101 − 01100101   (d) 01101010 − 00110101
(e) 10101011 − 01100100   (f) 100101110101 − 010100110100

8. Express the following binary numbers in Gray code.

(a) 10101   (b) 110011
(c) 1011010   (d) 11101001
(e) 100011101   (f) 1011010101

9. Convert the following Gray code numbers in binary.

(a) 11101   (b) 101011
(c) 1100110   (d) 11101001
(e) 101110101   (f) 1110101001

10. A shaft encoder that is encoded in Gray code has eight concentric rings. What is the resolution of the encoder?

11. How many concentric rings are required on a shaft encoder to obtain a resolution of at least $2°$?

12. Code the following in ASCII.

(a) ADD 15   (b) PAGE 7
(c) CODED DATA   (d) BINARY
(e) DIGITAL   (f) ALPHANUMERIC

13. Code the following in EBCDIC.

(a) ECL
(b) DIP
(c) CMOS

14. Add the necessary parity bit to each of the following words so that each word is shown with even parity.

(a) 101101   (b) 10001110
(c) 10101101   (d) 11101010
(e) 1011100101   (f) 110001011011

15. Locate the error in the following data, which are being transmitted with odd parity.

Data	Row Parity Bit
101010	0
110101	1
011100	1
011111	0
011101	1
110110	Column parity bit

16. Draw the logic diagram for a parity generator for a 5-bit code.

REFERENCES

Floyd, Thomas L., *Digital Fundamentals*. Columbus, Ohio: Charles E. Merrill Publishing Company, 1982.

Kershaw, John D., *Digital Electronics and Logic Systems*. North Scituate, Mass.: Breton Publishers, 1983.

Nashelsky, Louis, *Introduction to Digital Technology,* 3rd ed. New York: John Wiley & Sons, Inc., 1983.

Santoni, Andy, ''IEEE-488 Compatible Instruments,'' *EDN*, Nov. 5, 1979, pp. 91–96.

Encoding, Decoding, and Multiplexing

INSTRUCTIONAL OBJECTIVES

In this chapter you are introduced to digital logic circuitry associated with encoding, decoding, and multiplexing. After completing the chapter, you should be able to:

1. Describe what encoders are and where they are used.
2. Describe the purpose of the priority function in encoders.
3. Describe what decoders are and where they are used.
4. Describe what multiplexers are and how they work.
5. Describe what code converters are and how they work.
6. Describe what demultiplexers are and how they work.
7. Design a basic encoder, decoder, or converter circuit.
8. Show how multiplexers can be used to implement Boolean functions directly from a truth table.

SELF-EVALUATION QUESTIONS

The following questions relate to the material presented in this chapter. Read the questions prior to studying the chapter and, as you read through the material, watch for answers to the questions. After completing the chapter, return to this section and evaluate your comprehension of the material by answering the questions again.

1. What is the primary function of an encoder?
2. What purpose does the priority function serve in priority encoders?
3. What two types of priority encoders are commercially available in IC form?
4. What types of logic gates are employed in encoders?
5. What is the primary function of a decoder?
6. What is the primary function of a code converter?
7. How many output lines does a multiplexer have?
8. By what other name are multiplexers known?
9. How many input lines does a demultiplexer have?
10. What does TDM stand for, and how do such circuits work?

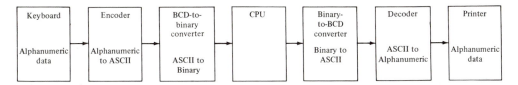

FIGURE 12-1 *Block diagram of the data conversion steps in a typical digital computational system*

12-1
INTRODUCTION

Data that are transmitted between digital systems or subsystems are usually transmitted in code form. In this chapter we deal with code-handling circuitry, which handles the various codes that were introduced in Chapter 11. To be able to provide a code representation of alphanumeric data that enter digital systems via a keyboard or other input device, a digital circuit that transforms to input to the proper code is required. This digital circuit is called an *encoder*. As a general rule, *encoding* means to express decimal numbers, alphabetical characters, or special symbols as a combination of binary digits in coded form.

To convert the coded binary digits back to decimal numbers, letters, and special symbols, the encoding process is reversed. The reverse process is called *decoding* and the digital circuit is called a *decoder*.

It is also frequently necessary to convert from one code to another, such as from BCD to excess-3 code. The digital circuit required to do this is called a *code converter*. The block diagram shown in Fig. 12-1 illustrates the process of encoding, decoding, and converting signals between the input and output of a computational system.

12-2
ENCODERS

Encoders are circuits that convert a single input into an equivalent binary output. For example, a decimal keyboard must provide a binary output that corresponds to the particular key pressed. The encoder converts the signal generated each time a key is pressed into its binary equivalent. If only one key is pressed at a time, the encoding can be accomplished with gates. However, if two or more inputs are permitted to be active at the same time, a priority must be established. A circuit that operates in this manner is called a *priority encoder* and is discussed in the next section.

One of the earliest encoder circuits, but one that is still fairly widely used, is constructed with diode OR gates. A basic diode OR-gate encoder that will encode decimal numbers 1 through 5 into binary is shown in Fig. 12-2. As shown, when the switch is set to the position corresponding to the desired decimal value, the equivalent binary outputs go HIGH.

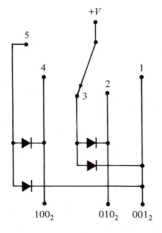

FIGURE 12-2 Diode OR-gate encoder for decimal numbers 1 through 5

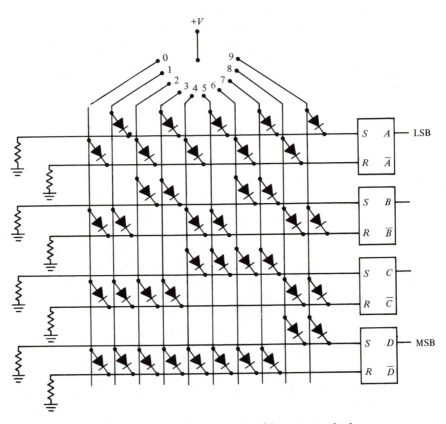

FIGURE 12-3 Decimal encoding matrix with storage devices

12-2.1 Decimal-to-Binary Encoders

Diode encoders are often constructed on a printed-circuit board in a rectangular matrix form, as shown in Fig. 12-3. The circuit shown encodes decimal numbers 0 through 9 into binary and stores the binary number in *RS* flip-flops. The circuit is called a *decimal-to-binary encoder*.

As might be expected, it is possible to reduce the number of diodes used in the encoder in Fig. 12-3 by combining input functions as shown in Fig. 12-4. The circuit requires four OR gates, as shown in Fig. 12-4a. Gate 1 produces the least significant binary bit, which is a binary 1 for each odd decimal number. Gate 2 produces the next least significant bit, which is HIGH for decimal numbers 2, 3, 6, and 7. Gate 3 produces the next-higher-order bit and is HIGH for decimal numbers 4, 5, 6, and 7. Gate 4 produces the most significant bit and is HIGH for decimal numbers 8 and 9.

Since the circuit shown in Fig. 12-4 encodes only the decimal number 0 through 9, the binary output is in BCD code form; therefore, the circuit is sometimes referred to as a *decimal-to-BCD encoder*.

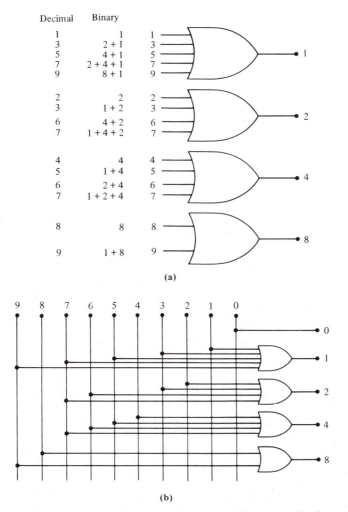

(a)

(b)

FIGURE 12-4 (a) Encoder OR gates; (b) encoder circuit

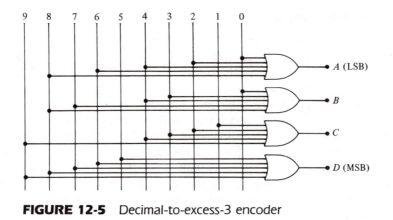

FIGURE 12-5   Decimal-to-excess-3 encoder

12-2.2   Decimal-to-Excess-3 Encoders

The basic approach discussed for decimal-to-binary encoding can be used to encode decimal numbers in excess-3 code, as shown in Fig. 12-5.

12-2.3   Priority Encoders

In systems where two or more inputs may go HIGH simultaneously, a priority must be established as to which input the system will respond. A logic circuit that produces an output signal in accordance with an established priority, in the event that inputs go HIGH simultaneously, is called a *priority encoder*. The priority feature is particularly useful in microcomputers.

Priority may be established on the basis of the magnitude of decimal numbers appearing at the inputs of the encoder, or on the basis of many other considerations, such as on the basis of the job titles in a company or by virtue of the relative magnitude of physical parameters, such as temperature or pressure. Consider the block diagram of a decimal-to-BCD encoder shown in Fig. 12-6 as well as the relationship between decimal and BCD numbers shown in Table 12-1. We can see by observing the table that output A is HIGH whenever decimal numbers 1, 3, 5, 7, and 9 are being encoded and that A is LOW whenever decimal numbers 2, 4, 6, and 8 are being encoded.

The following statements regarding the priority of decimal inputs to produce an output at A can be made.

1. A should be HIGH if decimal input 1 is HIGH and decimal inputs 2, 4, 6, and 8 are LOW.
2. A should be HIGH if decimal input 3 is HIGH and decimal inputs 4, 6, and 8 are LOW.
3. A should be HIGH if decimal input 5 is HIGH and decimal inputs 6 and 8 are LOW.
4. A should be HIGH if decimal input 7 is HIGH and decimal input 8 is LOW.
5. A should be HIGH if decimal input 9 is HIGH.

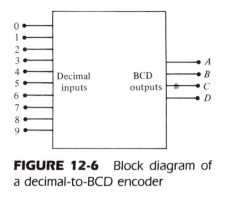

FIGURE 12-6 Block diagram of a decimal-to-BCD encoder

TABLE 12-1. / Decimal and BCD Numbers

Decimal Digit	BCD DCBA
0	0000
1	0001
2	0010
3	0011
4	0100
5	0101
6	0110
7	0111
8	1000
9	1001

The logic circuitry required to produce the A output is shown in Fig. 12-7.

By observing Table 12-1 and applying the same line of reasoning as above, we can write the following statements with regard to output B.

1. B should be HIGH if decimal input 2 is HIGH and decimal inputs 4, 5, 8, and 9 are LOW.
2. B should be HIGH if decimal input 3 is HIGH and decimal inputs 4, 5, 8, and 9 are LOW.
3. B should be HIGH if decimal input 6 is HIGH and decimal inputs 8 and 9 are LOW.
4. B should be HIGH if decimal input 7 is HIGH and decimal inputs 8 and 9 are LOW.

The logic circuitry required to produce the B output is shown in Fig. 12-8.

The same reasoning process as applied above can be used to develop the logic

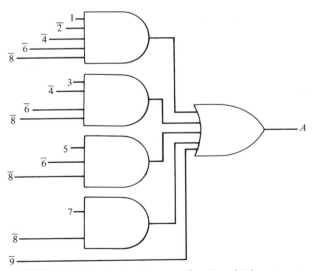

FIGURE 12-7 Logic diagram for the A-bit output of a decimal-to-BCD priority encoder

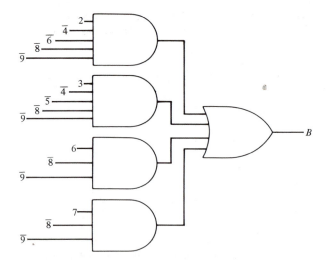

FIGURE 12-8 Logic diagram for *B*-bit output of a decimal-to-BCD priority encoder

circuitry for the *C* output. The following statements relating to the priority associated with bit *C* can be made.

1. *C* should be HIGH if decimal input 4 is HIGH and decimal inputs 8 and 9 are LOW.
2. *C* should be HIGH if decimal input 5 is HIGH and decimal inputs 8 and 9 are LOW.
3. *C* should be HIGH if decimal input 6 is HIGH and decimal inputs 8 and 9 are LOW.
4. *C* should be HIGH if decimal input 7 is HIGH and decimal inputs 8 and 9 are LOW.

The logic circuitry required to produce the *C* output is shown in Fig. 12-9.

Finally, for the *D* output the following single statement applies.

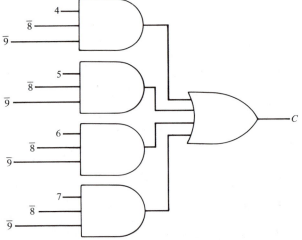

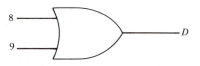

FIGURE 12-9 Logic diagram for the *C*-bit output of a decimal-to-BCD priority encoder.

FIGURE 12-10 Logic diagram for the *D*-bit output of a decimal-to-BCD priority encoder

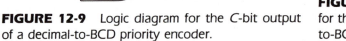

1. *D* should be HIGH if decimal input 8 is HIGH or if decimal input 9 is HIGH.

The logic circuitry required for the *D*-bit output is shown in Fig. 12-10. The logic circuit for the complete decimal-to-BCD encoder is shown in Fig. 12-11.

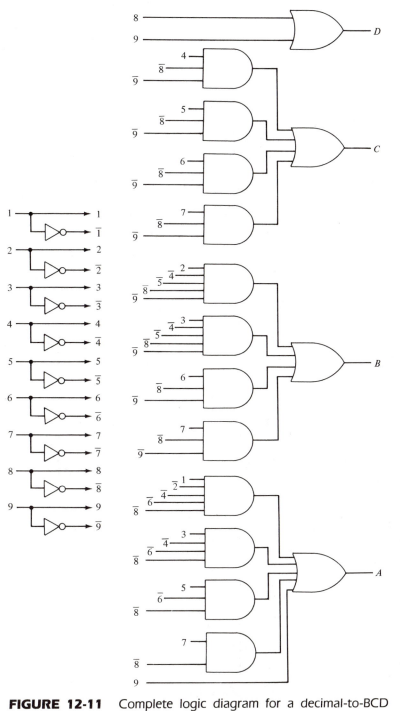

FIGURE 12-11 Complete logic diagram for a decimal-to-BCD priority encoder

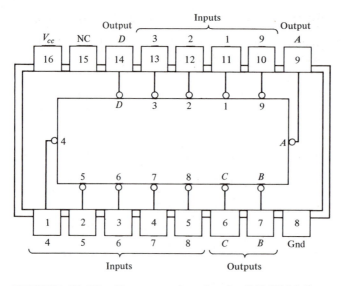

FIGURE 12-12 Pin connections for the 74147 10-line-decimal-to-4-line-BCD encoder

TABLE 12-2. / 74147 Truth Table

Inputs									Outputs			
1	2	3	4	5	6	7	8	9	D	C	B	A
H	H	H	H	H	H	H	H	H	H	H	H	H
×	×	×	×	×	×	×	×	L	L	H	H	L
×	×	×	×	×	×	×	L	H	L	H	H	H
×	×	×	×	×	×	L	H	H	H	L	L	L
×	×	×	×	×	L	H	H	H	H	L	L	H
×	×	×	×	L	H	H	H	H	H	L	H	L
×	×	×	L	H	H	H	H	H	H	L	H	H
×	×	L	H	H	H	H	H	H	H	H	L	L
×	L	H	H	H	H	H	H	H	H	H	L	H
L	H	H	H	H	H	H	H	H	H	H	H	L

H, high logic level; L, low logic level; ×, irrelevant.

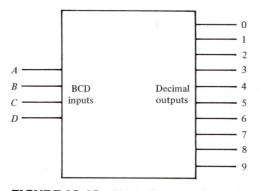

FIGURE 12-13 Block diagram of a BCD-to-decimal decoder

12-2.4 IC PRIORITY ENCODERS

In general, two types of priority encoders are available in IC form: 8-line-to-3-line and 10-line-to-4-line encoders. These are categorized as MSI devices and are available from a number of manufacturers. Priority encoders available in the 7400 TTL series are the 74147 and the 74148, which are 10-line-to-4-line and 8-line-to-3-line encoders, respectively.

The pin connections for the 74147 10-line-decimal-to-4-line-BCD encoder are shown in Fig. 12-12. Table 12-2 shows the truth table for the device. As can be seen, data inputs and outputs are active at the LOW logic level; thus the outputs at A, B, C, and D correspond to the highest-order LOW input. The highest priority is assigned to $\bar{9}$; $\bar{1}$ has the lowest priority. The implied decimal zero condition requires no input, as zero is encoded when all nine input data lines are HIGH.

12-3
DECODERS

Decoders are circuits that convert some type of code into a recognizable number or character. They are generally used to convert various codes to decimal numbers or to letters of the alphabet. In the following paragraphs several commonly used decoders are discussed.

12-3.1 BCD-to-Decimal Decoders

Decoders that produce a decimal output which corresponds to a 4-bit BCD input are called *BCD-to-decimal decoders*. The block diagram for this very popular type of decoder, which is also called a *4-line-to-10-line decoder,* is shown in Fig. 12-13.

By referring to Table 12-1 we can write Boolean algebra expressions for each decimal digit. For example, the Boolean expression to produce a HIGH at the decimal 0 output is

$$\text{output } 0 = \overline{A}\,\overline{B}\,\overline{C}\,\overline{D}$$

This can be implemented with a four-input AND gate. By continuing to use Table 12-1, we can implement the complete decoder circuit shown in Fig. 12-14.

The relationships shown schematically in Fig. 12-14 are summarized in Table 12-3.

TABLE 12-3. / BCD-to-Decimal Decoding Functions

Decimal Digit	BCD D C B A	Minterms
0	0 0 0 0	$\bar{D}\ \bar{C}\ \bar{B}\ \bar{A}$
1	0 0 0 1	$\bar{D}\ \bar{C}\ \bar{B}\ A$
2	0 0 1 0	$\bar{D}\ \bar{C}\ B\ \bar{A}$
3	0 0 1 1	$\bar{D}\ \bar{C}\ B\ A$
4	0 1 0 0	$\bar{D}\ C\ \bar{B}\ \bar{A}$
5	0 1 0 1	$\bar{D}\ C\ \bar{B}\ A$
6	0 1 1 0	$\bar{D}\ C\ B\ \bar{A}$
7	0 1 1 1	$\bar{D}\ C\ B\ A$
8	1 0 0 0	$D\ \bar{C}\ \bar{B}\ \bar{A}$
9	1 0 0 1	$D\ \bar{C}\ \bar{B}\ A$

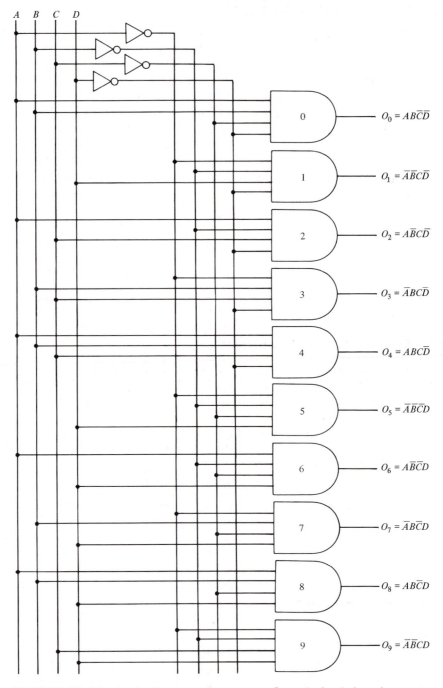

FIGURE 12-14   Logic diagram of an excess-3-to-decimal decoder

12-3.2   Excess-3-to-Decimal Decoders

Since the excess-3 code is a 4-bit code, the excess-3-to-decimal decoder is also categorized as a 4-line-to-10-line decoder. Table 12-4 shows the relationship between decimal digits and excess-3 code groups as well as the Boolean terms from which the logic circuitry for the decoder can be implemented.

TABLE 12-4. / Decimal and Excess-3 Code Numbers and Minterms

Decimal Digit	Excess-3 D C B A	Minterms
0	0 0 1 1	$\bar{D}\ \bar{C}\ B\ A$
1	0 1 0 0	$\bar{D}\ C\ \bar{B}\ \bar{A}$
2	0 1 0 1	$\bar{D}\ C\ \bar{B}\ A$
3	0 1 1 0	$\bar{D}\ C\ B\ \bar{A}$
4	0 1 1 1	$\bar{D}\ C\ B\ A$
5	1 0 0 0	$D\ \bar{C}\ \bar{B}\ \bar{A}$
6	1 0 0 1	$D\ \bar{C}\ \bar{B}\ A$
7	1 0 1 0	$D\ \bar{C}\ B\ \bar{A}$
8	1 0 1 1	$D\ \bar{C}\ B\ A$
9	1 1 0 0	$D\ C\ \bar{B}\ \bar{A}$

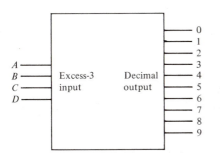

FIGURE 12-15 Block diagram of an excess-3-to-decimal decoder

The block diagram for an excess-3-to-decimal decoder is shown in Fig. 12-15. The complete logic diagram is shown in Fig. 12-16.

12-3.3 BCD-to-Seven-Segment Decoders

The BCD-to-seven-segment decoder accepts a BCD number at its inputs and produces the output signals that are required to provide a digital readout. The readout is displayed as a decimal number by a device called a *seven-segment display,* which is shown in Fig. 12-17. As can be seen, the device consists of seven light-emitting segments. By applying the correct voltages to certain combinations of the segments, each of the 10 decimal digits can be displayed. Figure 12-18 shows the segments that are generally used to display each decimal digit.

There are several methods of implementing seven-segment displays, the most popular being light-emitting diodes (LEDs) and liquid-crystal displays (LCDs). In either case the basic operation is the same and is shown in Fig. 12-19. As shown, each segment can be illuminated with either a HIGH or a LOW voltage.

By observing the segments that are used to display each decimal number, as shown in Fig. 12-18, we can tabulate what we observe pictorially. This is done in Table 12-5. Next, we need to tabulate the digits that utilize each segment.

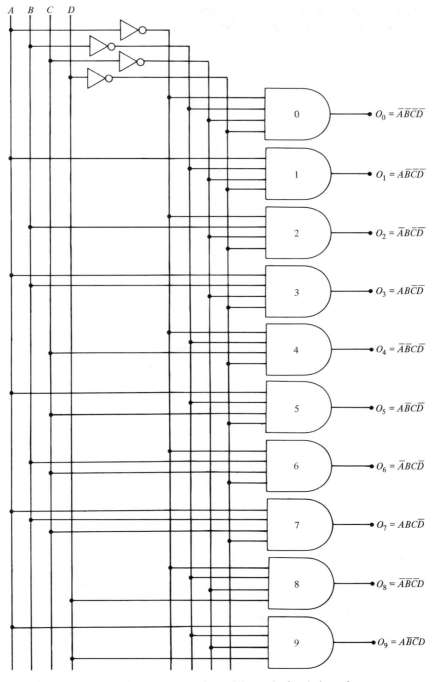

FIGURE 12-16 Logic diagram of a BCD-to-decimal decoder

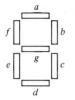

FIGURE 12-17 Seven-segment display showing position of segments

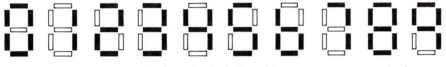

FIGURE 12-18 Display of decimal digits with a seven-segment device

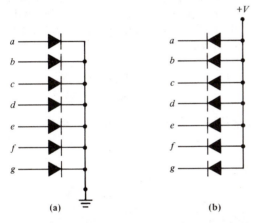

FIGURE 12-19 Schematic diagram of seven-segment LED devices: (a) activated with HIGH inputs (common cathode type); (b) activated with LOW inputs (common anode type)

TABLE 12-5. / Segment Utilization for Seven-Segment Display

Digit	Segments Used
0	a, b, c, d, e, f
1	b, c
2	a, b, g, e, d
3	a, b, c, d, g
4	b, c, f, g
5	a, c, d, f, g
6	c, d, e, f, g
7	a, b, c
8	a, b, c, d, e, f, g
9	a, b, c, f, g

From Table 12-5 we can see that segment a is utilized when displaying decimal digits 0, 2, 3, 5, 7, 8, 9. By using the OR operator, we can write

$$a = 0 + 2 + 3 + 5 + 7 + 8 + 9$$
$$b = 1 + 2 + 3 + 4 + 7 + 8 + 9 + 0$$
$$c = 1 + 3 + 4 + 5 + 6 + 7 + 8 + 9 + 0$$
$$d = 0 + 2 + 3 + 5 + 6 + 8$$
$$e = 0 + 2 + 6 + 8$$
$$f = 0 + 4 + 5 + 6 + 8 + 9$$
$$g = 2 + 3 + 4 + 5 + 6 + 8 + 9$$

If each of these decimal numbers is represented in BCD code, each of the expressions above can be rewritten as Boolean algebra equations. Table 12-6 tabulates the Boolean terms for each BCD code group. By replacing the decimal numbers in the OR expressions above with the corresponding minterms from Table 12-6, we obtain the following Boolean expressions:

$$a = \overline{A}\,\overline{B}\,\overline{C}\,\overline{D} + \overline{A}B\overline{C}\,\overline{D} + AB\overline{C}\,\overline{D} + A\overline{B}\,C\overline{D} + ABC\overline{D} + \overline{A}\,\overline{B}\overline{C}D + A\overline{B}\,\overline{C}D$$

$$b = \overline{A}\,\overline{B}\,\overline{C}\,\overline{D} + \overline{A}\,B\overline{C}\,\overline{D} + A\overline{B}\,\overline{C}\,\overline{D} + AB\overline{C}\,\overline{D} + \overline{A}\,B\overline{C}D + ABC\overline{D} + \overline{A}\,\overline{B}\,\overline{C}D + A\overline{B}\,\overline{C}D$$

$$c = \overline{A}\,\overline{B}\,\overline{C}\,\overline{D} + A\overline{B}\,\overline{C}\,\overline{D} + AB\overline{C}\,\overline{D} + \overline{A}\,\overline{B}C\overline{D} + A\overline{B}C\overline{D} + \overline{A}BC\overline{D} + ABC\overline{D} + \overline{A}\,\overline{B}\,\overline{C}D + A\overline{B}\,\overline{C}D$$

$$d = \overline{A}\,\overline{B}\,\overline{C}\,\overline{D} + \overline{A}\,B\overline{C}\,\overline{D} + AB\overline{C}\,\overline{D} + A\overline{B}C\overline{D} + \overline{A}BC\overline{D} + \overline{A}\,\overline{B}\,\overline{C}D$$

$$e = \overline{A}\,\overline{B}\,\overline{C}\,\overline{D} + \overline{A}B\overline{C}\,\overline{D} + \overline{A}BC\overline{D} + \overline{A}\,\overline{B}\,\overline{C}D$$

$$f = \overline{A}\,\overline{B}\,\overline{C}\,\overline{D} + A\overline{B}\,\overline{C}\,\overline{D} + AB\overline{C}\,\overline{D} + \overline{A}BC\overline{D} + \overline{A}\,\overline{B}\,\overline{C}D + A\overline{B}\,\overline{C}D$$

$$g = \overline{A}B\overline{C}\,\overline{D} + AB\overline{C}\,\overline{D} + A\overline{B}C\overline{D} + \overline{A}\,\overline{B}C\overline{D} + \overline{A}BC\overline{D} + ABC\overline{D} + \overline{A}\,\overline{B}\,\overline{C}D$$

By using these Boolean expressions, we can implement the logic circuitry for the BCD-to-seven-segment decoder. For example, segment a is illuminated by constructing the logic circuitry that is necessary to perform the Boolean expression associated with segment a. The circuit required for segment a is shown in Fig. 12-20.

TABLE 12-6. / Minterms That Correspond to BCD Numbers

Decimal	D C B A	Minterms
0	0 0 0 0	$\overline{D}\ \overline{C}\ \overline{B}\ \overline{A}$
1	0 0 0 1	$\overline{D}\ \overline{C}\ \overline{B}\ A$
2	0 0 1 0	$\overline{D}\ \overline{C}\ B\ \overline{A}$
3	0 0 1 1	$\overline{D}\ \overline{C}\ B\ A$
4	0 1 0 0	$\overline{D}\ C\ \overline{B}\ \overline{A}$
5	0 1 0 1	$\overline{D}\ C\ \overline{B}\ A$
6	0 1 1 0	$\overline{D}\ C\ B\ \overline{A}$
7	0 1 1 1	$\overline{D}\ C\ B\ A$
8	1 0 0 0	$D\ \overline{C}\ \overline{B}\ \overline{A}$
9	1 0 0 1	$D\ \overline{C}\ \overline{B}\ A$

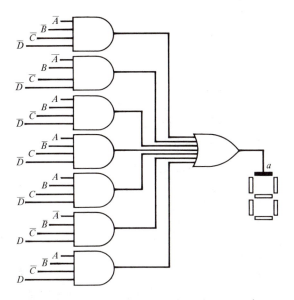

FIGURE 12-20 Decoding logic to activate segment *a* of a seven-segment device

The logic diagram for the complete BCD-to-seven-segment decoder is shown in Fig. 12-21 on p. 340.

12-4
INTEGRATED-CIRCUIT BCD-TO-SEVEN-SEGMENT DECODER/DRIVERS

One of the more popular BCD-to-seven-segment decoder/drivers is the 9317. This TTL/MSI device accepts four inputs in BCD code form and decodes the four inputs to provide the output required to drive a seven-segment display. The logic diagram for the 9317 is shown in Fig. 12-22. Because of the driver circuitry the device can be used to directly drive seven-segment displays that use incandescent lamps as well as light-emitting diodes.

The 9317 also provides the necessary circuitry for automatically blanking the leading and/or trailing zeros in a multidigit number. For example, in a system that has eight-digit readout capabilities, the number 052.0700 would be displayed as 52.07. This number is more easily read, since it is in the form in which we would normally write the number. Blanking leading and trailing zeros is achieved by use of the RBO and RBI inputs shown in Fig. 12-23. Leading zero suppression is achieved by connecting the ripple blanking output (RBO) of one decoder to the ripple blanking input (RBI) of the next-lower stage. The most significant digit position is always blanked if a 0 code appears on its input *and* the ripple blanking input is HIGH. The next most significant bit is blanked if a 0 code appears at its BCD inputs and the most significant bit is 0 as indicated by a HIGH on its ripple blanking output terminal. The blanking output of each decoder is connected to the blanking input of the decoder for the next-lower-order stage. A HIGH on the RBO terminal of any decoder indicates that a 0 code on its BCD inputs and all more significant digits are also 0.

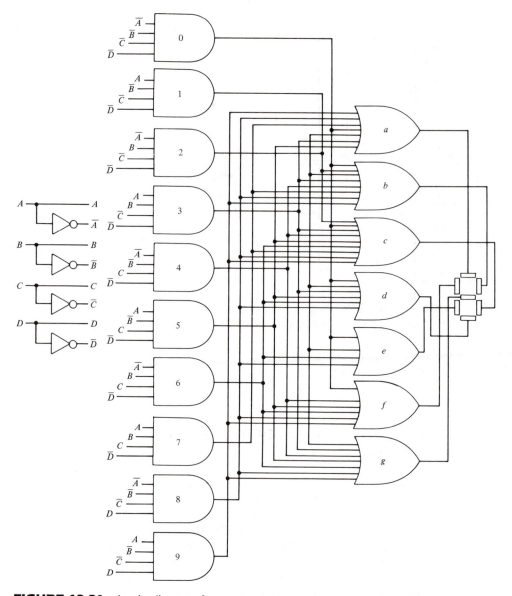

FIGURE 12-21   Logic diagram for a complete seven-segment decoder

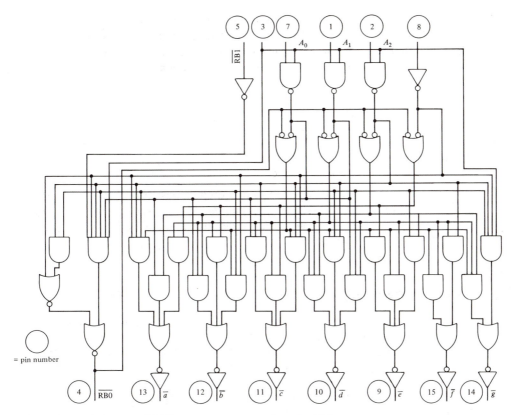

FIGURE 12-22 TTL 9317 seven-segment decoder/driver

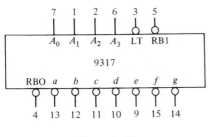

V_{CC} = pin 16
Gnd = pin 8

FIGURE 12-23 Pin connections for the 9317 seven-segment decoder/driver

For digits to the right of the decimal point, the least significant digit is blanked if a 0 code appears at its BCD inputs and its RBI input is HIGH. This same procedure blanks all 0's to the right of the least significant nonzero bit. The RBI input on the decoder for the least significant bit is connected to a positive voltage or left open.

12-5
CODE CONVERTERS

Circuits that translate decimal numbers to a digital code or translate a digital code to decimal numbers are called *encoders* and *decoders*, respectively. Sometimes it is desirable to convert from one binary code to another. Circuits used to accomplish this are called *converters*.

12-5.1   BCD-to-Excess-3 Converters

Conversion from BCD to excess-3 code is quite straightforward. BCD code groups are converted to excess-3 code by adding 3 (0011_2) to the BCD code for each digit. This conversion is readily accomplished by using a 4-bit parallel full adder with one set of inputs permanently wired to fixed voltages that correspond to 0011_2. The other set of input terminals accept the BCD code that is to be converted, and the output is the excess-3 code equivalent of the input. A logic diagram for BCD-to-excess-3 conversion is shown in Fig. 12-24. Since there can never be a carry into the fifth column, the carry output of the full adder for the most significant bit is not used.

12-5.2   Binary-to-Gray-Code Converters

Recall that the Gray code is an unweighted code in which only one digit changes in going from one number to the next. To facilitate our discussion of binary-to-Gray-code conversion, and vice versa, the Gray code is listed in Table 12-7.

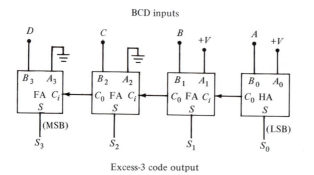

FIGURE 12-24   BCD-to-excess-3 converter

TABLE 12-7. / Comparison of Binary Numbers and Gray Code

Binary	Gray Code
0000	0000
0001	0001
0010	0011
0011	0010
0100	0110
0101	0111
0110	0101
0111	0100
1000	1100
1001	1101
1010	1111
1011	1110
1100	1010
1101	1011
1110	1001
1111	1000

Binary numbers are converted to Gray code as follows:

1. The most significant bit of the binary number becomes the leftmost bit of the Gray code combination.
2. The second bit from the left in the Gray code equals the sum, in modulo-2 addition, of the two most significant bits of the binary number.
3. The third bit from the left in the Gray code equals the sum, in modulo-2 addition, of the second and third most significant bits of the binary number, and so on.

Modulo-2 addition is performed with Exclusive-OR gates. A converter circuit that converts binary numbers to Gray code is shown in Fig. 12-25.

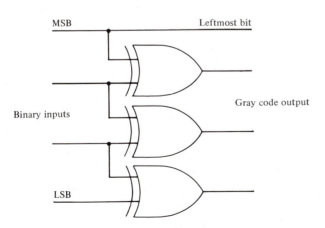

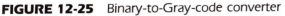

FIGURE 12-25 Binary-to-Gray-code converter

12-5.3 Gray-Code-to-Binary Converters

Numbers represented in Gray code can be converted to binary as follows:

1. The most significant bit in binary equals the leftmost Gray code bit.
2. The second binary bit is the modulo-2 sum of the two leftmost Gray code bits.
3. The third binary bit is the modulo-2 sum of the second binary bit and the third Gray code bit.
4. All other binary bits are equal to the modulo-2 sum of the corresponding Gray code bit and the adjacent (next more significant) binary bit.

A converter circuit that converts Gray code to binary is shown in Fig. 12-26.

12-6
MULTIPLEXERS

Multiplexers, or *data selectors*, are logic circuits with many input lines but only one output line. Such circuits are designed so that digital information on each input line is routed onto the single output line; hence the name "data selector." The routing of the data on a particular input line to the output is controlled by SELECT inputs. The process of selecting 1 out of N input lines and transmitting the data to a single output line is called *multiplexing*. For example, suppose that a computer receives data in serial form from several different sources, such as card readers, teletypewriters, the keyboard of CRT terminals, various measuring instruments, or from sensing devices such as temperature sensors. The purpose of the multiplexer is to select only one source at a time in a system such as the one shown in Fig. 12-27.

A basic two-input logic circuit that is capable of multiplexing is shown in Fig. 12-28. The logic level applied to the S input determines which AND gate is enabled and, therefore, which input signal appears at the output. The Boolean expression for the output is

$$X = AS + B\overline{S}$$

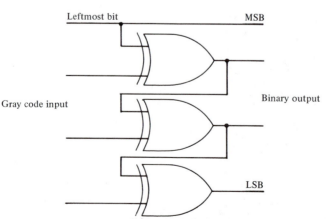

Leftmost bit   MSB

Gray code input   Binary output

LSB

FIGURE 12-26 Gray-code-to-binary converter

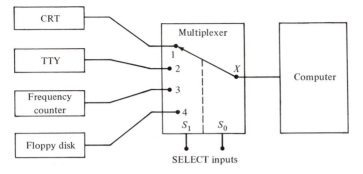

FIGURE 12-27 Selection of one input at a time by multiplexing

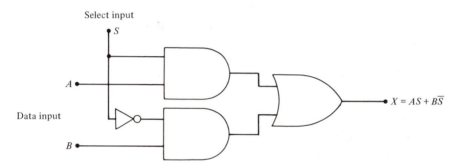

FIGURE 12-28 Logic diagram of a two-channel multiplexer

From this expression we can see that when S is HIGH, $\overline{S}$ is LOW; therefore, the output expression can be written as

$$X = A \cdot 1 + B \cdot 0$$

Also, when S is LOW, $\overline{S}$ is HIGH, so the output expression can be written as

$$X = A \cdot 0 + B \cdot 1$$

Therefore, we can see that when $S = 1$, the signal at input A appears at the output, whereas when $S = 0$, the signal at input B appears at the output.

The technique described above can be used to implement a four-input, or four-channel, multiplexer such as the one shown in Fig. 12-29. The Boolean expression for the circuit is

$$X = AS_0S_1 + B\overline{S}_0S_1 + CS_0\overline{S}_1 + D\overline{S}_0\overline{S}_1$$

From this expression we can develop a truth table (Table 12-8).

We have not considered the concept of time sharing when multiplexing data. As a way of doing so, suppose that we wish to send messages from three different sources along a single transmission line in a data communications system. The messages are to be multiplexed and transmitted serially with the start and com-

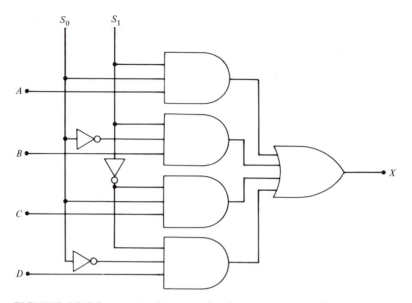

FIGURE 12-29   Logic diagram of a four-channel multiplexer

TABLE 12-8. / Truth Table for Multiplexer in Fig. 12-29

Input S_1 S_0	Output
0 0	$X = D$
0 1	$X = C$
1 0	$X = B$
1 1	$X = A$

pletion of the transmission essentially simultaneous for all three messages, as far as a human observer can tell. For example, the three messages OK, HI, and LO from multiplexer channels 1, 2, and 3 are to be multiplexed and transmitted serially. This is done by transmitting the first letter of each message serially, followed by the serial transmission of the second letter of the message. The output of the multiplexer is a bit stream of letters "OHLKIO." This type of multiplexing is called *time-division multiplexing* (TDM) because it divides the time during which each message is transmitted into time intervals.

12-7
DEMULTIPLEXERS

Demultiplexers perform the reverse operation that multiplexers perform in that they take a single input signal and distribute it to several output lines. A 1-line-to-4-line demultiplexer circuit is shown in Fig. 12-30. As can be seen, the input

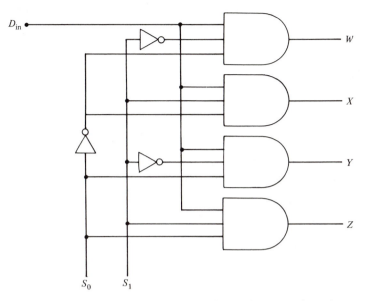

FIGURE 12-30 Logic diagram of a 1-line-to-4-line de-multiplexer

data line is connected to each AND gate. The two SELECT lines permit only one gate at a time to be enabled, in accordance with the following expressions:

$$W = D_{in}\bar{S}_1\bar{S}_0$$

$$X = D_{in}S_1\bar{S}_0$$

$$Y = D_{in}\bar{S}_1S_0$$

$$Z = D_{in}S_1S_0$$

12-8
APPLICATIONS

There are many applications for the different types of logic circuits that have been discussed in this chapter. For example, between the keys and the digital circuitry of calculators there must be an encoder to convert the decimal inputs into binary.

EXAMPLE 12-1

Show how diode OR gates can be used to implement an active-HIGH keyboard encoder for the decimal digits 0 through 9 shown in Fig. 12-31.

SOLUTION

The calculator keyboard code is tabulated in Table 12-9, which shows a straight BCD code except for the decimal 0 digit. The pushbutton circuit for an active-

7
8
9
4
5
6
1
2
3
0

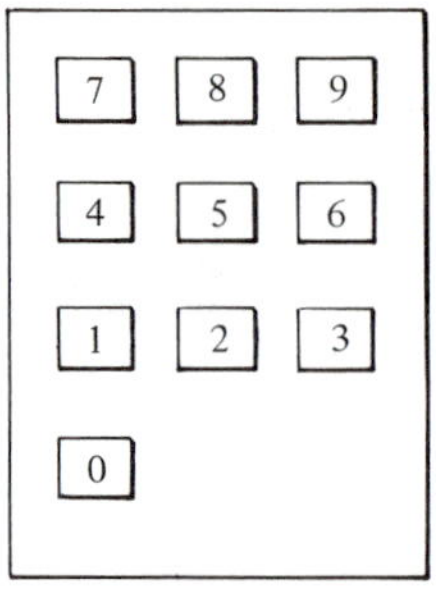

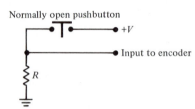

FIGURE 12-31 Decimal calculator keyboard

Normally open pushbutton
+V
Input to encoder
R

FIGURE 12-32 Circuit to provide active-HIGH input to encoder

TABLE 12-9. / Calculator Keyboard Code

Decimal Digit Input	Binary-Coded Output $W\ X\ Y\ Z$
1	0 0 0 1
2	0 0 1 0
3	0 0 1 1
4	0 1 0 0
5	0 1 0 1
6	0 1 1 0
7	0 1 1 1
8	1 0 0 0
9	1 0 0 1
0	1 0 1 0

HIGH output is shown in Fig. 12-32. By observing Table 12-9 we can write the following Boolean expressions that show which keys cause a HIGH to exist at each output:

$$W = 8 + 9 + 0$$

$$X = 4 + 5 + 6 + 7$$

$$Y = 2 + 3 + 6 + 7 + 0$$

$$Z = 1 + 3 + 5 + 7 + 9$$

Using the expressions above, we can implement the keyboard shown in Fig. 12-33.

Multiplexers also find many applications, including data or frequency selection,

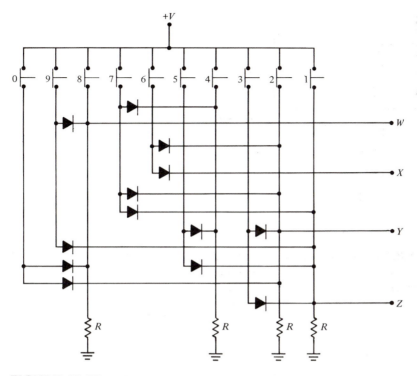

FIGURE 12-33 Diode logic keyboard encoder

parallel-to-serial conversion, logic function implementation directly from a truth table, and time-sharing functions related to digital readout circuits. The following example illustrates how a multiplexer can be used directly to implement a truth table.

EXAMPLE 12-2

Show how to implement the truth table in Fig. 12-34 with the eight-channel multiplexer shown.

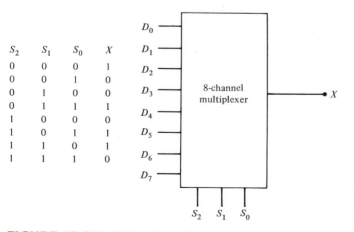

S_2	S_1	S_0	X
0	0	0	1
0	0	1	0
0	1	0	0
0	1	1	1
1	0	0	0
1	0	1	1
1	1	0	1
1	1	1	0

FIGURE 12-34 Eight-channel multiplexer

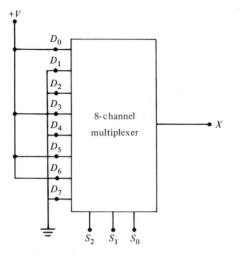

FIGURE 12-35 Eight-channel multiplexer showing external connections required to implement the truth table shown in Fig. 12-34

SOLUTION

To implement the truth table with the multiplexer shown, we must connect each data input line to the logic level that corresponds to the binary number in the X column. When the SELECT lines are set to any combination shown, the output corresponds to the value of X for that combination. The input connections are as shown in Fig. 12-35. ❏

12-9
SUMMARY

The circuits that have been discussed in this chapter are logic circuits constructed to perform a particular function. Encoders made up of logic gates convert data from a single input into an equivalent multiple output binary or binary-coded number. Encoders are generally available in 8-line-to-3-line or 10-line-to-4-line form. Most commercially available encoders include a priority feature that outputs to the higher priority input when two inputs are activated simultaneously.

Decoders are circuits that convert binary-coded inputs into a more readily recognized number or character. The most commonly used decoders are 4-line to 1-of-10-line, 4-line to 1-of-16-line decoders, and 4-line to seven-segment.

Converters are logic circuits that are used to convert one binary code to another. The most commonly used converters are for converting binary numbers to Gray code, or vice versa.

Multiplexers are the logic circuit equivalent of rotary selector switches in that they have many input lines and a single output line. The primary applications of the multiplexers are for data-routing purposes and for implementing logic circuits in minterm form directly from truth tables.

Demultiplexers can also be viewed as the logic circuit equivalent of rotary selector switches; however, unlike multiplexers, which have many inputs and a single output, demultiplexers have a single input and many outputs.

PROBLEMS

1. Draw the logic diagram using diode logic for an octal-to-BCD encoder.

2. Draw the logic diagram using AND and OR gates for an octal-to-BCD encoder.

3. Design a priority encoder that will permit computer access on a company's computer according to the following priorities:

Priority	Person or Department
1	Company president
2	Design engineering
3	Sales and service
4	Manufacturing

4. A police department wishes to install a priority encoder for their incoming telephone calls according to the following priorities:

Priority	Source of Calls
1	911 emergency calls
2	Emergency telephones along highways
3	Company security office telephones
4	Homes

Design the system.

5. The following decoding gates are for BCD-to-decimal decoding. If the output of each gate is HIGH, what BCD code is applied to the inputs? The least significant bit is applied to terminal A. To which gate is an invalid BCD code being applied?

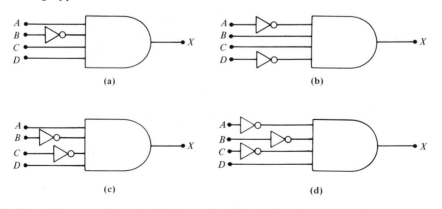

6. Draw the decoding circuit for each of the following code groups if a HIGH output is required.

(a) 1010   (b) 1100   (c) 10101
(d) 11001   (e) 100011   (f) 101001

7. Draw the logic diagram for a decoding circuit that will indicate the presence of the following BCD code groups by producing a HIGH output: 1001, 0110, 0010, 0101. The output is LOW for all other BCD numbers.

8. Repeat Problem 5 if the decoding gates are for excess-3-to-decimal decoding.

9. Draw the logic circuit to activate segment "f" of a seven-segment device.

10. Draw the converter circuit that is required to convert the binary number 101101 to Gray code, then determine the Gray code for the given binary number.

11. Draw the converter circuit that is required to convert Gray code 100110 to binary, then determine the binary number that corresponds to the given Gray code.

12. Draw the logic diagram for a 3-line-to-8-line decoder with active HIGH inputs and outputs.

13. The circuit shown is used for parallel-to-serial conversion of data. Draw the serial pulse train in relationship to the clock pulse train that corresponds to the data stored in the 8-bit register when they are placed on the output line in serial form.

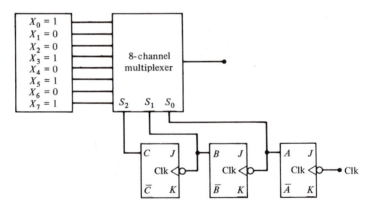

14. Draw the logic diagram for the four-channel multiplexer shown below in block diagram form and determine the output state for the conditions shown.

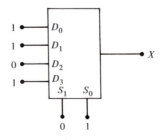

15. The three messages ONE, TWO, and OFF are to be transmitted by time-division multiplexing with the first letter of each word, starting with ONE.

This is followed by the second letter and then the third letter. If each letter is transmitted in ASCII code, show the resulting output pulse train.

16. Show the block diagram and external connections that are necessary to implement the logic function indicated by the following truth table.

C	B	A	X
0	0	0	1
0	0	1	0
0	1	0	1
0	1	1	1
1	0	0	0
1	0	1	1
1	1	0	1
1	1	1	0

REFERENCES

Blakeslee, Thomas R., *Digital Design with Standard MSI and LSI,* 2nd ed. New York: John Wiley & Sons, Inc., 1979.

The Logic Databook. Santa Clara, Calif.: National Semiconductor Corp., 1982.

Sandige, Richard S., *Digital Concepts Using Standard Integrated Circuits.* New York: McGraw-Hill Book Company, 1978.

The TTL Data Book. Dallas, Tex.: Texas Instruments, Inc., 1976.

13

Arithmetic Circuits with Registers

INSTRUCTIONAL OBJECTIVES

In this chapter we combine the arithmetic circuits that were discussed in Chapter 7 and the registers from Chapter 10. After completing the chapter, you should be able to:

1. Describe the function of the various registers that are used in conjunction with arithmetic circuits.
2. Describe what the term *settling time* means.
3. Describe what is meant by *overflow* as it relates to arithmetic circuits.
4. List the signed binary operations for which overflow conditions are possible.
5. Describe the operation of serial adders.
6. Describe the operation of binary multipliers.
7. Show that shifting the contents of a shift register one position to the left is equivalent to multiplication by binary 2.
8. Show that shifting the contents of a register one position to the right is equivalent to division by binary 2.
9. Describe the operation of an arithmetic-logic unit.

SELF-EVALUATION QUESTIONS

The following questions relate to the material presented in this chapter. Read the questions prior to studying the chapter and, as you read through the material, watch for answers to the questions. After completing the chapter, return to this section and evaluate your comprehension of the material by answering the questions again.

1. What arithmetic operations make possible an overflow condition?
2. List an advantage and a disadvantage of serial adders compared with parallel adders.
3. Where is the final sum obtained with a parallel adder temporarily stored?
4. If the contents of a shift register are shifted 2 bits to the left, what arithmetic operation is this equivalent to?
5. What does the term *settling time* mean?
6. What is the purpose of the MODE input on arithmetic-logic units?
7. When performing multiplication, what value is added to the contents of the accumulator if the bit in the 2^0 position of the multiplier is a binary 1?

13-1
INTRODUCTION

Arithmetic circuits and registers were discussed separately in earlier chapters; however, since these types of circuits are often used together in computational systems, in this chapter we discuss systems that contain both arithmetic circuits and registers.

Arithmetic circuits, such as adder and subtracter circuits, form the basis for a unit that is capable of performing many arithmetic operations, called an *arithmetic-logic unit* (ALU). The ALU works in conjunction with several operating registers, including an *accumulator* and a *data register*, which temporarily store the results of the arithmetic operations.

The chapter begins with a discussion of basic adders used in conjunction with registers for temporary storage of a sum and concludes with a discussion of ALUs.

13-2
PARALLEL FULL ADDERS WITH REGISTERS

Computational systems such as calculators and computers generally add two binary numbers, consisting of several binary bits, at a time. The numbers to be added are placed in the adder via registers, where they are temporarily stored, as shown in Fig. 13-1. The registers are made up of flip-flops. The output of each flip-flop is set HIGH or LOW depending on the binary bit stored in it.

The numbers to be added are parallel-loaded into registers A and B by applying a "load" pulse. Since the register outputs are applied to the inputs of the full adders, the numbers are added with the sum present at the S_0 through S_4 inputs

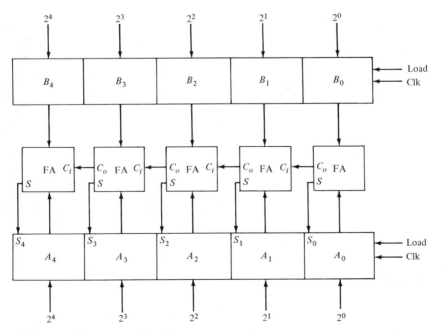

FIGURE 13-1 Parallel full adder with registers

of the accumulator (register A). The bits of the two numbers are, in effect, all added simultaneously. However, since some finite time is required for any carries generated to propagate from one adder to the next, the add-cycle time is considered to be two clock periods. This total time required for the addition to be complete is called the *settling time*. For example, if each full adder of an 8-bit parallel binary adder has a 0.1-μs delay, the settling time is approximately 0.8 μs.

The next step is to store the sum back in the accumulator and enter another number in register B and repeat the cycle until as many numbers as are desired are added and the final sum is stored in the accumulator. In a computer the final sum would probably be transferred to the main memory for storage, which would free the adder circuit for additional arithmetic operations.

13-3
COMBINED ADDER/SUBTRACTER CIRCUITS
WITH REGISTERS

Many modern computers use the 2's-complement system, which was discussed in Chapter 6 to represent negative numbers and to perform subtraction using adder circuits. To extend the capabilities of computers to include arithmetic operations with negative numbers, it is common practice to include one more bit in binary numbers than is necessary to represent a particular number. The additional bit is the leftmost bit and is used exclusively to designate the sign of a number and is therefore called the *sign bit*. By convention, and for other reasons that will be discussed later, a 0 in the sign-bit position indicates that the number is positive, whereas a 1 in the sign-bit position means that the number is negative. Both positive and negative numbers, or both, can be added and stored temporarily using the circuit shown in Fig. 13-2.

The addition/subtraction function is controlled by the MODE switch. When the MODE switch is set to ADD, the output of the true-complement circuit is identical to the contents of register B and is added to the number in the accumulator. When the MODE switch is set to SUBTRACT, the output of the true-complement circuit is the 1's complement of the number in register B. This is added to the carry-in to give us the 2's complement of a number in register B. This number is then added to the number in the accumulator. The result, which is equal to the accumulator contents minus the register B contents, is then stored in the accumulator.

The adder/subtracter circuit shown in Fig. 13-2 is the essence of a very widely used computer circuit because it provides a relatively straightforward method for adding and subtracting positive and negative numbers. As before, the final sum, or difference, would probably be transferred to the main memory of a computer for storage, thus freeing the adder/subtracter circuit for additional arithmetic operations.

13-4
OVERFLOW

When an arithmetic operation produces a result that is larger than the accumulator can accommodate, a condition called *overflow* exists. The limitations on the size of numbers that can be handled by an *n*-bit register are $2^{n-1} - 1$ positive numbers

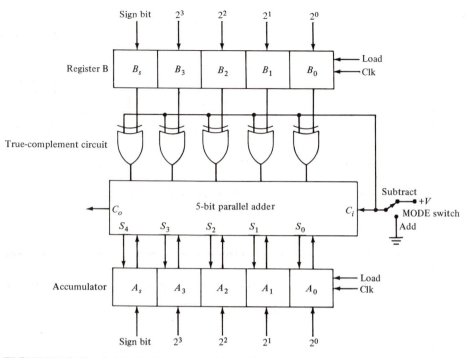

FIGURE 13-2 Added/subtracter with registers

and 2^{n-1} negative numbers. Thus an 8-bit register is restricted to numbers between $+127$ and -128; that is, the binary equivalent of decimal numbers outside this range causes an overflow condition to exist. Table 13-1 summarizes the arithmetic operations discussed to this point and indicates whether an overflow condition is possible.

A circuit that can be used to detect overflow in the parallel adder/subtracter of Fig. 13-2 is shown in Fig. 13-3. The output of the Exclusive-OR gate C will be LOW whenever A_s and B_s in Fig. 13-2 are equal. This causes the bottom input terminal of the AND gate to be HIGH. The output of Exclusive-OR gate D will be HIGH whenever B_s and S_4 are different. This will cause the middle input terminal to be HIGH. From the description above we can see that the function of Exclusive-OR gate C is to compare signs; thus we can aptly call it a *sign comparator*. Also, the function of Exclusive-OR gate D is to compare the magnitude of B_s and S_4; therefore, we will designate it a *magnitude comparator*. If the signs are alike and B_s and S_4 are different, an overflow condition exists; therefore, when

TABLE 13-1. / Summary of Signed Binary Operations and Overflow Conditions

Arithmetic Operation	Addition of Signed Numbers	2's-Complement Addition	Result and Possible Condition
$A + B$	$(A) + (B)$	$A + B$	Result always positive; overflow possible
$A - B$	$(A) + (-B)$	$A + B_{2'sC}$	Result either positive or negative; no overflow
$-A + B$	$(-A) + (B)$	$A_{2'sC} + B$	Result either positive or negative; no overflow
$-A - B$	$(-A) + (-B)$	$A_{2'sC} + B_{2'sC}$	Result always negative; overflow possible

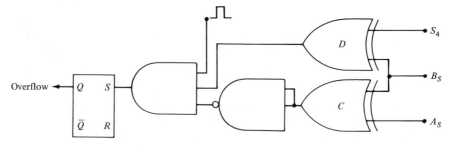

FIGURE 13-3 Overflow detectors

a CHECK pulse is applied to the upper input of the AND gate, the overflow flip-flop will be set.

EXAMPLE 13-1

Indicate which of the following arithmetic operations cause an overflow condition.

(a) $1001 + 0110$ (b) $1101 + 1010$ (c) $-1011 - 0011$

SOLUTION

(a) The sum of the numbers with their sign-bit designation is

$$
\begin{aligned}
\text{sign bits} \searrow & \\
[0]\,1001 &\leftarrow \text{contents of accumulator} \\
B_s \rightarrow [0]\,0110 &\leftarrow \text{contents of register } B \\
S_4 \rightarrow \ \ 01111 &
\end{aligned}
$$

The signs are alike, indicating a possible overflow condition; however, B_s and S_4 are also alike, which indicates that no overflow occurred.

(b) The sum of the numbers with their sign-bit designation is

$$
\begin{aligned}
\text{sign bits} \searrow & \\
[0]\,1101 &\leftarrow \text{contents of accumulator} \\
B_s \rightarrow [0]\,1010 &\leftarrow \text{contents of register B} \\
S_4 \rightarrow \ \ 10111 &
\end{aligned}
$$

The signs are alike indicating a possible overflow condition. Also, since B_s and S_4 are not equal, an overflow will occur.

(c) Since both numbers are negative, they must be expressed in 2's-complement form as

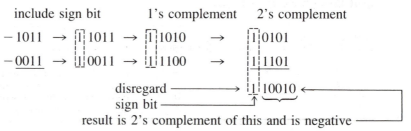

include sign bit   1's complement   2's complement

$-1011 \rightarrow [1]\,1011 \rightarrow [1]\,1010 \ \rightarrow \ [1]\,0101$

$-\underline{0011} \rightarrow [1]\,0011 \rightarrow [1]\,1100 \ \rightarrow \ [1]\,1101$

disregard $\longrightarrow [1]\,10010 \leftarrow$
sign bit $\longrightarrow$
result is 2's complement of this and is negative $\longrightarrow$

The 2's complement of 0010 is

$$0010 \rightarrow 1101 \quad \text{(1's complement)}$$
$$\underline{+\,1}$$
$$1110 \leftarrow \text{2's complement}$$

The register contents and the sum is

$$10101 \leftarrow \text{contents of accumulator}$$
$$B_s \rightarrow \underline{11101} \leftarrow \text{contents of register B}$$
$$S_4 \rightarrow 11110$$

❏

13-5
SERIAL ADDERS

Circuits that add two numbers on a bit-by-bit basis, rather than simultaneously as in a parallel adder, are called *serial adders*. The process by which numbers are added by serial adders is closely related to the way we add numbers on paper, that is, adding the least significant bits and adding any "carry" to the next-higher-order bits. The circuitry for serial adders is much simpler than for parallel adders; however, they perform addition at a much slower rate. The *add cycle* equals $n + 1$ clock periods where the $+1$ clock period is for the sign bit.

A block diagram for a 4-bit serial adder is shown in Fig. 13-4. The A and B registers, which in a serial adder are *shift registers*, are used to store the numbers to be added as well as to store the sum of the two numbers. In addition to the shift registers, serial adders contain a single full adder and single flip-flop in which any carry-out of the full adder is momentarily stored so that it can be added to the next-higher-order bits on the next clock pulse.

Observe that the Sum terminal of the full adder is connected to the input terminal of the flip-flop for the MSB of the number stored in shift register A (A_3 in this case). This transfers the sum of each pair of bits into A_3 so that the final sum of the numbers originally stored in shift registers A and B are stored in shift register A.

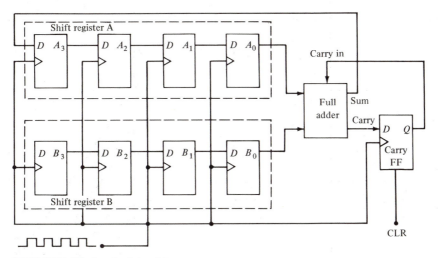

FIGURE 13-4 Serial adder

The way that numbers are added by a serial adder is perhaps best understood by a numerical example.

EXAMPLE 13-2 ▄▄▄▄▄▄▄▄▄▄▄▄▄▄▄▄▄▄▄▄▄▄▄▄▄▄▄▄▄▄▄▄▄▄▄▄

Show the steps involved in adding the 4-bit binary numbers listed in Table 13-2 with the serial adder in Fig. 13-4.

TABLE 13-2 / Register Contents for Example 13-2

Register A	Register B
$A_3A_2A_1A_0$	$B_3B_2B_1B_0$
0 1 1 0	0 1 1 1

SOLUTION

Table 13-3 summarizes the steps involved in the addition process. As shown in the table, the initial contents of shift registers A and B are the numbers to be added. Also, we shall assume that the carry FF has been initially cleared as indicated.

As is shown, a Sum and a Carry signal for the LSBs and carry FF appear as soon as the registers are loaded, that is, "before the first clock pulse."

TABLE 13-3. / Summary of Steps Involved in Serial Addition

First clock pulse. On the first clock pulse, the full-adder Sum for the LSBs is transferred to flip-flop A_3 and the full-adder Carry sets the carry flip-flop and is added to the next higher-order bits, thus giving us a full-adder Sum of 0 and Carry of 1.

Second clock pulse. On the second clock pulse, the full-adder Sum of A_1, B_1, and the carry is transferred to flip-flop A_3 at the same time the existing value in A_3 is shifted into A_2. Also, the full-adder Carry is added to the next higher-order bits (A_2 and B_2), giving us a full-adder Sum of 1 and Carry of 1.

Third clock pulse. On the third clock pulse, the 1 in A_2 is shifted in A_1, and 0 in A_3 is shifted into A_2, and the full-adder Sum is transferred to A_3. Also, the full-adder Carry is added to the MSBs, producing a Sum of 1 and a Carry of 0.

Fourth clock pulse. On the fourth clock pulse, the 1 in A_1 is shifted into A_0, the 0 in A_2 is shifted into A_1, the 1 in A_3 is shifted into A_2, and the full-adder Sum is transferred to A_3. Thus, at the completion of the fourth clock pulse, the contents of register A is 1101, which is, of course, the sum of the original contents of the A and B registers. ❑

13-5.1 Serial Adders with Three Registers

Rather than transfer the bit sums back into register A, and ultimately store the final sum there, it is sometimes advantageous to use a third register in serial adder circuits, as shown in Fig. 13-5. The circuit shown requires an additional register. However, the numbers to be added are more easily entered into registers A and B than with the circuit in Fig. 13-4.

13-6
BCD ADDERS

In some digital systems arithmetic operations are performed with binary-coded-decimal (BCD) numbers rather than as straight binary arithmetic. Recall from Chapter 11 that in the BCD code each decimal digit is expressed as an equivalent 4-bit binary code group. The rules of BCD addition presented in Chapter 11 are summarized below to refresh your memory:

1. Add the BCD code groups for each decimal digit using ordinary binary arithmetic.

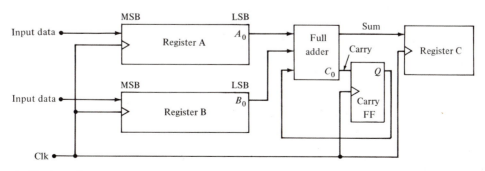

FIGURE 13-5 Serial adder with three registers

2. If the sum is 1001 (decimal 9) or less, the sum is valid BCD code group; therefore, no correction is required.
3. If the sum is greater than 1001, a correction factor of 0110 must be added to obtain a valid BCD code group. The addition will generate a carry that must be added to the next-higher-order code group.

If a BCD adder is to be able to carry out the arithmetic steps above, it must be able to do the following:

1. Add two 4-bit BCD code groups according to the rules of ordinary binary addition.
2. Determine if the sum is greater than binary 1001.
3. If the sum is greater than 1001, add 0110 to the sum and generate a carry that must be added to the next-higher-order code group.

A circuit that is capable of carrying out the requirements for BCD addition is shown in Fig. 13-6. This circuit is capable of adding two single-digit decimal numbers expressed in BCD code form. If multiple-digit decimal numbers in BCD code form are to be added, the circuitry shown in Fig. 13-6 is simply paralleled, as shown in the following example.

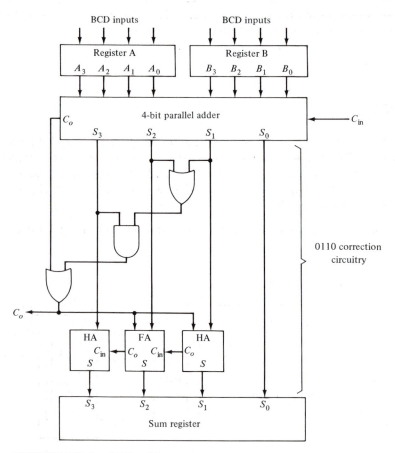

FIGURE 13-6 BCD adder

EXAMPLE 13-3 ━━━━━━━━━━━━━━━━━━━━━━━━━━━━━━━━━━

Show that the parallel BCD adder shown in Fig. 13-7 produces the same results as the following BCD addition.

```
     Decimal       BCD
        53         0101      0011
      + 47         0100      0111
      ─────        ─────     ─────
       100         1001      1010   invalid code group
                       1 ←         0110   add 6
                               ─────
invalid code group   1010      10000
       add 6         0110         │
                   ─────          │
                    10000         ↓
       ↓                          
      0001         0000      0000
```

SOLUTION

Since two-digit decimal numbers expressed in BCD form are to be added, the parallel BCD adder shown in Fig. 13-7 is required.   ❑

13-7
BINARY MULTIPLICATION

Most mini- and microcomputers perform multiplication as successive additions. As such, multiplication of two binary numbers can be performed in a manner very similar to the pencil-and-paper multiplication algorithm of two binary, or decimal, numbers: that is, the multiplicand is multiplied by each digit in the multiplier and the appropriately positional partial products are then added. In binary multiplication each binary digit is either a 0 or a 1; therefore, each partial product either equals the multiplicand or it equals zero. To illustrate, consider the following multiplication process:

```
        111        multiplicand
        101        multiplier
       ─────
        111   ⎫
        000   ⎬   partial products
        111   ⎭
      ───────
      100011       product
```

The first partial product is obtained by multiplying the multiplicand by the LSB of the multiplier. The second partial product is obtained by multiplying the multiplicand by the next-higher-order bit of the multiplier, and the third partial product is obtained by multiplying the multiplicand by the MSB of the multiplier.

There are several ways that circuits can be constructed to perform binary multiplication. We shall examine two approaches, one that makes use of registers in which the contents can be shifted to the right and one that uses logic gates and adders.

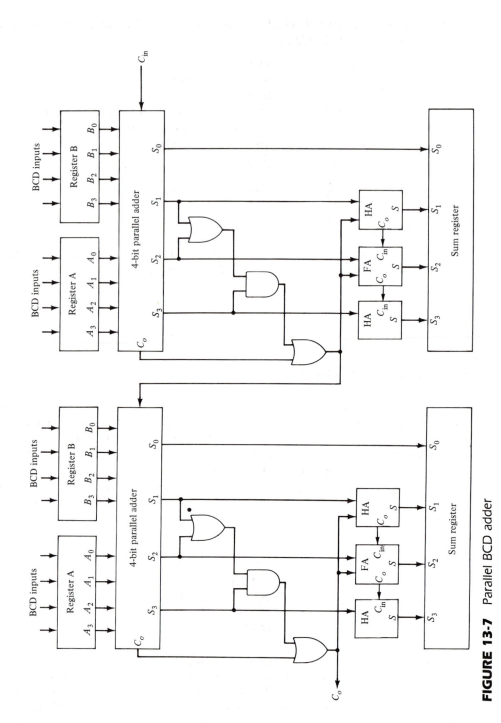

FIGURE 13-7   Parallel BCD adder

13-7.1 Binary Multipliers Using Registers

The multiplier circuit shown in Fig. 13-8 on p. 366 contains three registers, one in which the multiplicand is entered, and one where the final product appears. The multiplication process is carried out as follows:

1. Examine the sign bits. This may be done with the sign-bit comparator circuit discussed earlier in this chapter. If the signs are alike, the sign associated with the product will be positive, whereas unlike signs indicate a negative product. The sign bit for the product is stored until multiplication is complete.

2. Examine the 2^0 bit of the multiplier. If it is a 1, add the multiplicand to the contents of the accumulator, which should be 0 at this time. If the 2^0 bit is 0, no addition is required.

3. Shift the contents of the multiplier register one position to the right and the contents of the multiplicand register one position to the left and examine the bit that is now in the 2^0 position of the multiplier. If it is 1, add the number now in the multiplicand register to the contents of the accumulator.

4. Shift the contents of the multiplier register one position to the right and the contents of the multiplicand register one position to the left and examine the bit that is now in the 2^0 position of the multiplier. If it is 1, add the number now in the multiplicand register to the contents of the accumulator.

5. Continue shifting the multiplier to the right until the MSB is in the 2^0 position while shifting the multiplicand to the left. Add the number in the multiplicand register to the contents of the accumulator each time the 2^0 bit of the multiplier is 1.

6. Place the proper value in the sign-bit flip-flop of the accumulator. Multiplication is now complete.

111	multiplicand
101	multiplier
111	multiplicand $\times$ 2^0 bit of multiplier
0000	accumulator contents
0111	new accumulator contents
1110	multiplicand after shift left
10	multiplier after shift right
0000	multiplicand $\times$ 2^0 bit of multiplier
0111	accumulator contents
0111	accumulator contents unchanged because 2^0 bit is 0
11100	multiplicand after shift left
1	multiplier after shift right
11100	multiplicand $\times$ 2^0 bit of multiplier
0111	accumulator contents
100011	final product stored in accumulator

13-7.2 Binary Multipliers Using Logic Circuitry

Binary multiplication can be accomplished by using parallel AND gates to obtain partial products, which are then added to obtain the final product. The multiplier circuit shown in Fig. 13-9 performs multiplication in this manner. The circuit shown is a 2-bit by 4-bit multiplier, where $N_3N_2N_1N_0$ is the multiplicand, M_1M_0 is the multiplier, $A_3A_2A_1A_0$ and $B_3B_2B_1B_0$ are partial products, and $P_5P_4P_3P_2P_1P_0$ is the final product.

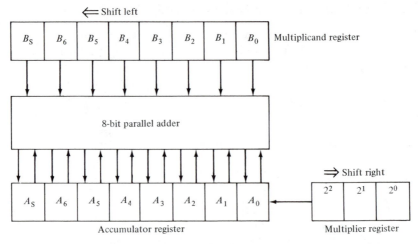

FIGURE 13-8   Binary multiplier circuit

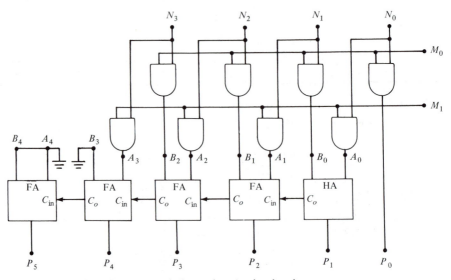

FIGURE 13-9   Binary multiplier using logic circuitry

EXAMPLE 13-4

Show the binary values for the partial products and the final product in Fig. 13-9 when the multiplicand in 1010 and the multiplier is 10.

SOLUTION

The binary values are shown in Fig. 13-10.   ❏

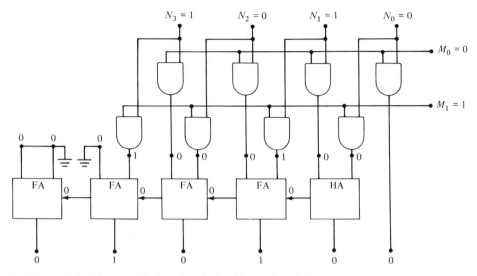

FIGURE 13-10 Multiplier circuit for Example 13-4

13-8
MULTIPLICATION AND DIVISION BY BINARY 2

When a binary number is multiplied by binary 2 (10), the product is the same as is obtained by shifting the contents of the multiplicand register one position to the left. Conversely, when a binary number is divided by binary 2, the quotient is the same as is obtained by shifting the contents of the dividend register one bit to the right. The following examples illustrate this procedure.

EXAMPLE 13-5

Show that shifting the contents of the multiplicand register shown below one position to the left is equivalent to multiplication by 2

$$0 \ 1 \ 0 \ 1 \qquad \text{multiplicand register}$$

SOLUTION

The present register contents, 0101_2, are equivalent to decimal 5. When shifted one bit to the left, the register contents become 1010_2, which is equivalent to decimal 10; therefore, the shift is equivalent to multiplication by 2. ❏

EXAMPLE 13-6

Show that shifting the contents of the following dividend register two positions to the right is equivalent to division by 4.

$$1 \ 1 \ 0 \ 0 \qquad \text{dividend register}$$

SOLUTION

The present register contents, 1100_2, are equivalent to decimal 12. When shifted one position to the right, the contents become 0110_2, the equivalent of decimal 6, corresponding to division by 2. When shifted one more position to the right, the contents become 0011_2, the equivalent of decimal 3, corresponding to another division by 2 or overall division by 4. ❑

13-9
THE ARITHMETIC-LOGIC UNIT

This section, in effect, ties together the earlier part of the chapter in that it combines the arithmetic circuits discussed into a single unit that performs arithmetic operations. This unit is called the *arithmetic-logic unit* (ALU). A simplified functional pin-out of the 74181, which is the basic arithmetic-logic unit in the 7400 TTL series, is shown in Fig. 13-11.

The 74181 accepts two 4-bit binary numbers, A and B, and a carry-in as *data inputs*. There are also five *control inputs*—a Mode input and four Select lines—which determine the *operation* performed on the inputs. The Mode input determines whether the output is an arithmetic or a logical function of the inputs, while the four Select lines select any 1 of 16 possible arithmetic or logic operations listed in Table 13-3.

The outputs of the 74181 include the 4-bit result, which appears at the F outputs; a Carry-out labeled C_{n+4}; an $A = B$ output; a Generate output; and a Propagate output. By noting that the $+$ sign signifies a logic OR operation and that the word "plus" means the arithmetic sum of the inputs, the outputs can be determined by carefully following the function table of Table 13-3.

As an example of the operation of the 74181, suppose that the Select inputs

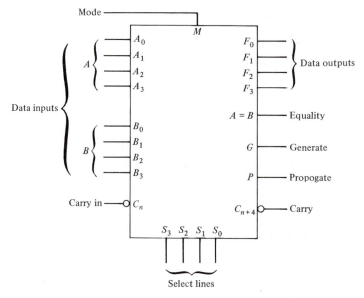

FIGURE 13-11 Pinout for the 74181 ALU with active-HIGH operands

TABLE 13-3. / Function Table for the 74181 ALU

$S_3 S_2 S_1 S_0$	Logic $M = 1$	M = 0 Arithmetic Operations	
		Arithmetic $C_2 = 1$ (No Carry)	Arithmetic $C_2 = 0$ (Carry)
0 0 0 0	$F = \overline{A}$	$F = A$	$F = A$ PLUS 1
0 0 0 1	$F = A + B$	$F = A + B$	$F = (A + B)$ PLUS 1
0 0 1 0	$F = AB$	$F = A + \overline{B}$	$F = (A + \overline{B})$ PLUS 1
0 0 1 1	$F = 0$	$F =$ MINUS 1 (2's comp.)	$F = 0$
0 1 0 0	$F = \overline{AB}$	$F = A$ PLUS $A\overline{B}$	$F = A$ PLUS $A\overline{B}$ PLUS 1
0 1 0 1	$F = \overline{B}$	$F = (A + B)$ PLUS $A\overline{B}$	$F = (A + B)$ PLUS $A\overline{B}$ PLUS 1
0 1 1 0	$F = A \oplus B$	$F = A$ MINUS B MINUS 1	$F = A$ MINUS B
0 1 1 1	$F = A\overline{B}$	$F = A\overline{B}$ MINUS 1	$F = A\overline{B}$
1 0 0 0	$F = \overline{A} + B$	$F = A$ PLUS AB	$F = A$ PLUS AB PLUS 1
1 0 0 1	$F = \overline{A \oplus B}$	$F = A$ PLUS B	$F = A$ PLUS B PLUS 1
1 0 1 0	$F = B$	$F = (A + \overline{B})$ PLUS AB	$F = (A + \overline{B})$ PLUS AB PLUS 1
1 0 1 1	$F = AB$	$F = AB$ MINUS 1	$F = AB$
1 1 0 0	$F = 1$	$F = A$ PLUS A[a]	$F = A$ PLUS A PLUS 1
1 1 0 1	$F = A + \overline{B}$	$F = (A + B)$ PLUS A	$F = (A + B)$ PLUS A PLUS 1
1 1 1 0	$F = A + B$	$F = (A + B)$ PLUS A	$F = (A + \overline{B})$ PLUS A PLUS 1
1 1 1 1	$F = A$	$F = A$ MINUS 1	$F = A$

[a]Each bit is shifted to the next more significant position.

$S_3 S_2 S_1 S_0$ are set to 0110 and the Mode input M is LOW. Since M is LOW the ALU is in the arithmetic mode and therefore selects the A MINUS B function. The number on the B inputs is subtracted from the number on the A inputs and the difference appears on the function (F) outputs. If the M input is set HIGH, the ALU is put into the logic mode and $A \oplus B$ is produced at the outputs. A basic block diagram of an ALU and the associated registers is shown in Fig. 13-12.

13-10
SUMMARY

In this chapter we have discussed the use of registers in conjunction with arithmetic circuits. In some circuits registers are an integral part of the arithmetic circuit,

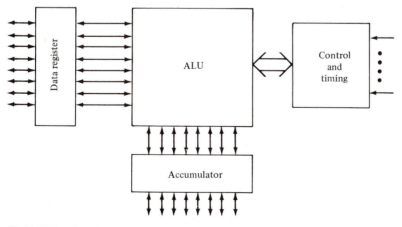

FIGURE 13-12 ALU and associated registers

whereas in others registers provide a place to store data momentarily. The data stored may be numbers on which arithmetic operations are to be performed, data representing a partially complete arithmetic operation, or the final result of an arithmetic operation.

Arithmetic-logic units are a combination of arithmetic circuits, which, when used in conjunction with registers, provides us with a system that is capable of performing both arithmetic and logic functions as well as providing a means of storing the associated data.

PROBLEMS

1. What is the approximate delay for each full adder of a 12-bit parallel binary adder if the settling time for the circuit is 1.5 μs?

2. Develop a table showing all possible input conditions for the overflow detector shown in Fig. 13-3, and state which cases produce an overflow when all entries in the table are positive.

3. Show the steps involved in adding the negative numbers $(-1001) + (-0110)$. The steps involved include 1's- and 2's-complement forms of the numbers.

4. Develop a table like Table 13-3 showing the steps involved in the serial addition of $1010 + 0011$.

5. What is the add-cycle time for an 8-bit serial adder that is driven by a 1-MHz clock?

6. Show the steps involved in adding the BCD equivalent of decimal numbers $57 + 46$ with the BCD adder shown in Fig. 13-7.

7. Show that the 0110 correction circuit adds 0000 to the BCD output of the 4-bit parallel adder in Fig. 13-6 for valid BCD 0111.

8. Show that the 0110 correction circuit adds 0110 to the BCD output of the 4-bit parallel adder in Fig. 13-6 for invalid BCD 1011.

9. Show what signals must be present on the data-input lines, the Select lines, and the Mode line of a 74181 ALU so that 5_{10} is subtracted from 12_{10}.

10. Show what signals must be present on the data-input lines, the Select lines, and the Mode line of a 74181 ALU so that the Exclusive-OR logic operation is performed.

REFERENCES

Deem, William, Kenneth Muchow, and Anthony Zeppa, *Digital Computer Circuits and Concepts,* 3rd ed. Reston, Va.: Reston Publishing Co., Inc., 1980.

Malvino, Albert P., and Donald P. Leach, *Digital Principles and Applications,* 3rd ed. New York: McGraw-Hill Book Company, 1981.

Tocci, Ronald J., *Digital Systems: Principles and Applications,* rev. ed. Englewood Cliffs, N.J.: Prentice-Hall, Inc., 1980.

The TTL Logic Data Manual. Sunnyvale, Calif.: Signetics Corp., 1982.

14

D/A and A/D Conversion

INSTRUCTIONAL OBJECTIVES

In this chapter you are introduced to digital-to-analog and analog-to-digital conversion techniques. After completing the chapter, you should be able to:

1. Explain the difference between analog and digital data.
2. Describe what is meant by *inverting* and *noninverting amplifiers*.
3. Describe how operational amplifiers can be used to perform mathematical operations such as summing, integration, and a differentiation.
4. Make calculations related to summing amplifiers.
5. Make calculations related to integrating amplifiers.
6. Describe how a binary-weighted ladder is used to accomplish D/A conversion.
7. Make calculations related to binary-weighted ladders.
8. List advantages of *R*-2*R* ladders over binary-weighted ladders for D/A conversion.
9. List five methods of analog-to-digital conversion.
10. Make calculations related to the voltage-to-frequency method of A/D conversion.
11. List an advantage and a disadvantage of simultaneous-type A/D converters.
12. Describe the operation of the various types of A/D converters discussed.
13. Define terms related to D/A and A/D converters.

SELF-EVALUATION QUESTIONS

The following questions relate to the material presented in this chapter. Read the questions prior to studying the chapter and, as you read through the material, watch for answers to the questions. After completing the chapter, return to this section and evaluate your comprehension of the material by answering the questions again.

1. What is the difference between analog and digital data?
2. Why are operational amplifiers so named?
3. Why are *R*-2*R* ladder networks preferred over binary-weighted ladders for D/A conversion?
4. Why is a resistor sometimes placed in parallel with the capacitor in an integrating op-amp circuit?
5. How is the weighted value assigned to each digital input voltage in a D/A converter?

6. Is there any difference between accuracy and resolution, and if so, what is the difference?
7. What does the term *monotonicity* mean?
8. What is an advantage and a disadvantage of simultaneous-type A/D converters?
9. What is an advantage of ramp-type A/D converters compared to simultaneous-type converters?
10. What does the term *quantum* mean with regard to A/D converters?
11. Describe the conversion cycle that successive approximation A/D converters go through.

14-1
INTRODUCTION

To this point we have dealt exclusively with digital signals and circuitry; however, the real world in which we live is an *analog* world. Analog parameters are continuous and have an infinite number of both positive and negative values. If we wish to use digital circuitry to transmit, measure, process, control, or store analog signals, we must be able to convert analog signals to a digital format, or vice versa.

The process of changing analog signals to equivalent digital signals is accomplished by using analog-to-digital (A/D) converters. Since the signal is then in a suitable form for digital circuitry, A/D converters are often referred to as *encoding devices*.

Digital-to-analog (D/A) conversion, which is considerably easier than A/D conversion, also finds many applications. For example, when digital circuitry is used in process control applications, digital signals may be converted to an analog signal for set-point adjustments.

The primary thrust of this chapter relates to A/D and D/A conversion techniques; however, since operational amplifiers play a very significant role in their operation, a brief discussion of operational amplifiers is in order.

14-2
OPERATIONAL AMPLIFIERS

Operational amplifiers, or *op-amps,* are one of the basic building blocks of D/A and A/D converters and as such must be discussed briefly. As their name implies, op-amps were originally designed to perform mathematical operations such as addition, subtraction, multiplication, division, differentiation, and integration. Interestingly enough, their primary applications were in *analog computers.*

Operational amplifiers are characterized by the following properties:

1. Very high open-loop gain.
2. Very high input impedance.
3. Broad bandwidth.
4. Very low output impedance.

Typical values for the parameters above, with the exception of the bandwidth, approach those of an ideal amplifier; however, in many op-amp applications considerable negative feedback is used, thus reducing gain and input impedance.

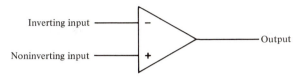

FIGURE 14-1 Op-amp symbol

The symbol for an op-amp is shown in Fig. 14-1. The plus $(+)$ and minus $(-)$ signs at the inputs indicate the phase relationship between each input and the output. The input with the minus sign is called the *inverting input* because there is a 180° phase inversion between a signal applied to this input terminal and the signal at the output. The input terminal marked with the plus sign is referred to as the *noninverting input* because a signal applied to it is in phase with the output signal.

In most applications, negative feedback is employed by feeding a portion of the op-amp output signal back to the input through a resistance network, as shown in Fig. 14-2. This provides us with a closed-loop system. With sufficient negative feedback, the closed-loop gain is independent of the open-loop gain and is determined by the values of resistors R_f and R_i. The *amplifier stage gain*, which is the ratio of output voltage to input voltage, is also equal to the ratio of the feedback resistor to the input resistor. This is written as

$$A = -\frac{V_o}{V_{in}} = \frac{R_f}{R_i} \qquad (14\text{-}1)$$

or

$$V_o = -V_{in}\frac{R_f}{R_i} \qquad (14\text{-}2)$$

The minus sign indicates phase inversion between the input and output.

EXAMPLE 14-1 ▬▬▬▬▬▬▬▬▬▬▬▬▬▬▬▬▬▬▬▬▬▬▬▬▬▬▬▬▬▬▬▬

Compute the output voltage for the circuit shown in Fig. 14-2 if $R_f = 7.5$ kΩ, $R_i = 1.5$ kΩ, and $V_{in} = 24$ mV.

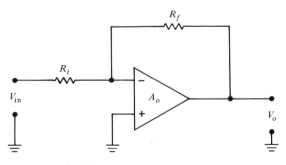

FIGURE 14-2 Inverting amplifier

SOLUTION

The output voltage is computed as

$$V_o = -V_{in}\frac{R_f}{R_i}$$

$$= -24 \text{ mV}\left(\frac{7.5 \text{ k}\Omega}{1.5 \text{ k}\Omega}\right)$$

$$= -(24 \text{ mV})(5) = -120 \text{ mV} \qquad \square$$

Based on the op-amp gain expression, one might conclude that a very large magnitude output voltage is attainable by simply using a large ratio of R_f to R_i. Theoretically, this is true, but in practice the output voltage cannot exceed the value of the supply voltage, which is generally from ± 10 V to ± 15 V.

14-2.1 Summing Amplifiers

Since our primary interest in op-amps relates to their use in D/A and A/D converters, there are a few specific op-amp circuits that are of major concern. One of these is the *summing amplifier* shown in Fig. 14-3. Equation 14-2 expresses the output voltage of the op-amp shown in Fig. 14-2 as the input voltage times the ratio of the feedback resistor to the input resistor. The same equation can be applied to each input voltage in Fig. 14-3 and is written as

$$V_{o(V_1)} = V_1\left(-\frac{R_f}{R_1}\right) \qquad V_{o(V_2)} = V_2\left(-\frac{R_f}{R_2}\right) \qquad V_{o(V_3)} = V_3\left(-\frac{R_f}{R_3}\right) \quad (14\text{-}3)$$

The total output voltage for the circuit is the algebraic sum of the portions of V_o due to each input as expressed in Eq. 14-3. The expressions in Eq. 14-3 can be combined into a single equation as follows:

$$V_o = -\left(V_1\frac{R_f}{R_1} + V_2\frac{R_f}{R_2} + V_3\frac{R_f}{R_3}\right) \quad (14\text{-}4)$$

This expression has significant implications regarding the conversion of binary numbers to their decimal equivalent. Recall from Chapter 6 that to convert a binary number to its decimal equivalent, we add the weight of the positions that

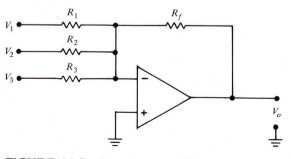

FIGURE 14-3   Summing amplifier

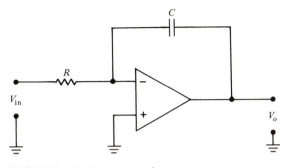

FIGURE 14-4   Integrating op-amp

contain a binary 1. For example, to convert binary 101 to its decimal equivalent, we write

$$101_2 = 1 \times 2^2 + 0 \times 2^1 + 1 \times 2^0$$
$$= \quad 4 \quad + \quad 0 \quad + \quad 1$$
$$= \quad 5_{10}$$

Since conversion from binary to decimal requires the addition of the weighted binary digits, we can readily accomplish this by use of the summing amplifier shown in Fig. 14-3.

14-4.2 Op-Amp Circuit for Integrating

The circuit shown in Fig. 14-4 is an op-amp with a capacitor as the feedback element. The output of this circuit is the integral of the input waveform. Logic signals are at either logic 0 or logic 1 and change from one to the other abruptly, as a step function. In the study of calculus one learns that the integral of a step function, written as

$$V_o = \frac{1}{RC} \int V_{in} \, dt \tag{14-5}$$

is a ramp function that can be written as

$$V_o = \frac{t}{RC} V_{in} \tag{14-6}$$

Due to the phase inversion associated with the op-amp when connected as an inverting amplifier, the ramp has a negative slope when the input charges from logic 0 to 1 and a positive slope when the input changes from logic 1 to 0. Figure 14-5 shows the input and output waveforms for Fig. 14-4.

Writing the gain equation, Eq. 14-1, for the circuit shown in Fig. 14-4 yields

$$\frac{V_o}{V_{in}} = -\frac{X_c}{R} \tag{14-7}$$

$$= -\frac{1}{2\pi f CR}$$

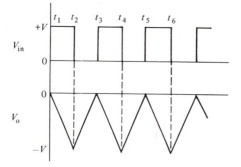

FIGURE 14-5   Input and output waveforms for an integrating op-amp

By observing Eq. 14-7, we can see that the gain increases as the frequency decreases and is theoretically infinite when $f = 0$ (direct current). The gain actually increases to the value of the open-loop gain. To deal with this large dc gain, a resistor is sometimes placed in parallel with the capacitor, as shown in Fig. 14-6.

By applying a square wave that is symmetrical to the horizontal axis (zero dc component), the output is also a symmetrical waveform about the horizontal axis, as shown in Fig. 14-7.

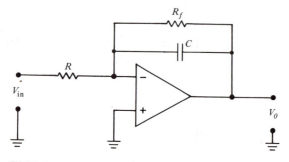

FIGURE 14-6   Integrating amplifier with R_f added to reduce dc gain

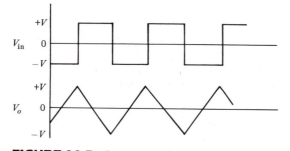

FIGURE 14-7   Symmetrical square-wave input and triangular output

14-2.3 Op-Amp Circuit for Differentiating

If the resistor and capacitor in the integrating circuit of Fig. 14-4 are interchanged as shown in Fig. 14-8, the output is the derivative of the input. The expression relating the input and output voltages is

$$V_o = RC \frac{d(V_{in})}{dt} \tag{14-8}$$

Differentiation is a mathematical operation that provides an indication of the rate at which a signal is changing. Thus, if the input signal applied to the circuit of Fig. 14-8 is a ramp voltage with a constant rate of change, the output will be a constant dc voltage that is proportional to the rate of change of the input. This is shown in Fig. 14-9.

By expressing Eq. 14-1 as a ratio of impedances, we can see that the gain of the op-amp differentiator is frequency dependent; that is, the gain increases with increasing frequency. At high frequencies the circuit is very susceptible to high-frequency noise. A solution to this problem is to add resistance in series with the input capacitor.

14-3
DIGITAL-TO-ANALOG CONVERSION

We begin our investigation of converting signals between the continuous-signal analog world and the discrete-step digital world with digital-to-analog (D/A) converters because it is somewhat easier and because D/A converters are also used as a subsystem in some analog-to-digital converters.

The basic problem in converting digital signals into equivalent analog signals

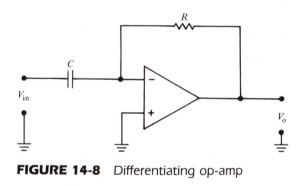

FIGURE 14-8 Differentiating op-amp

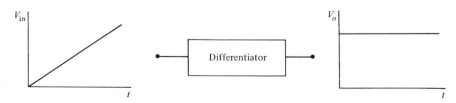

FIGURE 14-9 Differentiation of a ramp voltage

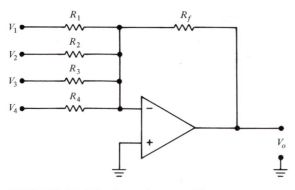

FIGURE 14-10 *Summing amplifier*

is changing the *n* digital voltage levels into one equivalent analog voltage. There are several ways to accomplish this conversion, some of which are discussed in the following paragraphs.

14-3.1
D/A CONVERSION USING BINARY-WEIGHTED LADDERS

The summing amplifier discussed in Section 14.2 can be viewed as a digital-to-analog conversion circuit. Recall that the output voltage for the summing amplifier shown in Fig. 14-10 is the sum of the input voltages, written as

$$V_o = - \left(\frac{R_f}{R_1} V_1 + \frac{R_f}{R_2} V_2 + \frac{R_f}{R_3} V_3 + \frac{R_f}{R_4} V_4 \right)$$

The weighted value assigned to each input voltage is the ratio of the feedback resistor R_f and the input resistor for that particular input voltage. For example, if the feedback resistor and all input resistors are equal in value, then the output voltage is written as

$$V_o = -(V_1 + V_2 + V_3 + V_4) \qquad (14\text{-}9)$$

However, if the input resistors are not equal in value, then the output voltage is the sum of the weighted value of the inputs. For example, if $R_1 = R_f$, $R_2 = R_f/2$, $R_3 = R_f/4$, and $R_4 = R_f/8$, the output voltage is given as

$$V_o = -(V_1 + 2V_2 + 4V_3 + 8V_4) \qquad (14\text{-}10)$$

Equation 14-10 means that each input will be multiplied by its weighting factor before being summed to become part of the output voltage. The weighting factor to convert digital voltages to equivalent analog voltages is related to the number of bits in the binary number. As a general expression, the weight of each bit in a 4-bit binary number is given as

$$(\text{LSB}) \quad 2^0 \text{ bit:} \quad R_f = 2^0 R_1 \quad \text{or} \quad R_1 = \frac{R_f}{2^0}$$

$$2^1 \text{ bit:} \quad R_f = 2^1 R_2 \quad \text{or} \quad R_2 = \frac{R_f}{2^1}$$

$$2^2 \text{ bit:} \quad R_f = 2^2 R_3 \quad \text{or} \quad R_3 = \frac{R_f}{2^2}$$

$$2^3 \text{ bit:} \quad R_f = 2^3 R_4 \quad \text{or} \quad R_4 = \frac{R_f}{2^3}$$

The circuit for the binary-weighted summing amplifier is shown in Fig. 14-11.

EXAMPLE 14-2 ■

Determine the output voltage of Fig. 14-11 if the following input voltages are applied:

$$V_1 = 1 \text{ V}$$

$$V_2 = 0 \text{ V}$$

$$V_3 = 1 \text{ V}$$

$$V_4 = 1 \text{ V}$$

SOLUTION

The output voltage is computed as

$$V_o = -[(1 \text{ V})(8) + (1 \text{ V})(4) + (0 \text{ V})(2) + (1 \text{ V})(1)]$$

$$= -(8 \text{ V} + 4 \text{ V} + 0 \text{ V} + 1 \text{ V})$$

$$= -13 \text{ V} \qquad\qquad ❑$$

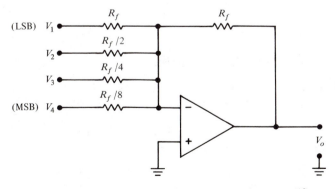

FIGURE 14-11   Binary-weighted summing amplifier

If we let the voltages V_1 through V_4 represent binary digits, then the summing amplifier shown in Fig. 14-11 can be viewed as a digital-to-analog converter. The voltages V_1 through V_4 are equivalent to binary 1101 or decimal 13. The largest binary number that we can represent with four bits is 1111, which is equivalent to decimal 15. Recall from Section 14.2 that the supply voltages for op-amps is generally in the range of 10 to 15 V. If we use the binary-weighted summing amplifier of Fig. 14-11, a binary input of 1111 will produce an analog output equal to

$$V_o = -[1(8) + 1(4) + 1(2) + 1(1)]$$

$$= -15 \text{ V}$$

Since the output voltage cannot exceed the value of the supply voltage, the output will be restricted to a value less than 15 if the supply voltage is less than 15 V. For example, if a supply voltage of 12 V is used, the output of the summing amplifier will be 12 V for binary inputs equivalent to decimal 12 or greater; therefore, our digital-to-analog conversion will be inaccurate for input values greater than 12.

One way to overcome the problem discussed in the preceding paragraph is to reduce the gain of the amplifier. This is most easily accomplished by reducing the value of the feedback resistor. For example, consider the situation just discussed, where the binary input is equivalent to decimal 15 and the maximum output of the op-amp is 12. This means that the output represents 80 percent of the input, or that

$$R_f = 0.8R \qquad (14\text{-}11)$$

This reduction factor can be applied to the gain expression, which gives us

$$V_o = -0.8(V_1 + 2V_2 + 4V_3 + 8V_4) \qquad (14\text{-}12)$$

Figure 14-12 summarizes graphically the weighted 4-bit summing amplifier, or the binary-weighted ladder D/A converter with a supply voltage of 15 V.

Binary-weighted ladder D/A converters have two disadvantages: the changing load that the binary source sees, and the need for several different values of precision resistors. Both of these disadvantages can be overcome by using R-$2R$ ladder D/A converters, which are discussed next.

14-3.2 D/A Conversion Using *R-2R* Ladders

Both of the disadvantages mentioned above with regard to binary-weighted ladder D/A converters can be overcome with R-$2R$ ladder D/A converters. The R-$2R$ ladder is a resistive network constructed using only two different values of resistors and whose analog output is a properly weighted sum of the digital inputs. Such a ladder, designed for a 4-bit binary input, is shown in Fig. 14-13.

We can analyze the resistive properties of the circuit by starting at node A, where we see a resistance of $2R$ looking either toward the terminating resistor or the LSB (2^0) input. The equivalent resistance for these two resistors is a single resistor whose value is R, shown in Fig. 14-14a on p. 382. Moving to node B,

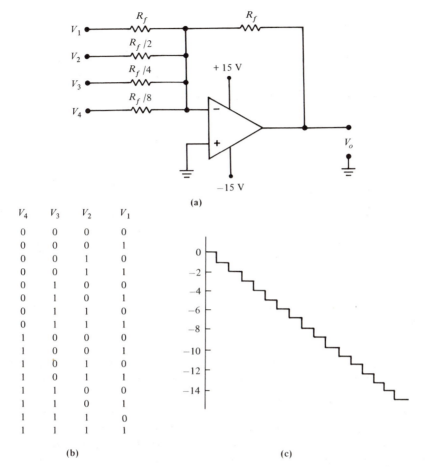

FIGURE 14-12 Four-bit D/A converter: (a) circuit diagram; (b) binary inputs; (c) waveforms

V_4	V_3	V_2	V_1
0	0	0	0
0	0	0	1
0	0	1	0
0	0	1	1
0	1	0	0
0	1	0	1
0	1	1	0
0	1	1	1
1	0	0	0
1	0	0	1
1	0	1	0
1	0	1	1
1	1	0	0
1	1	0	1
1	1	1	0
1	1	1	1

(b)   (c)

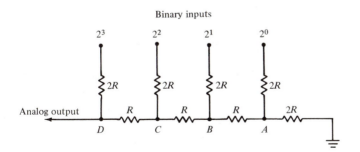

FIGURE 14-13 Four-bit R-2R ladder

we see a resistance of $2R$ looking toward either ground or toward the 2^1 input. The equivalent resistance to the right of node B is again a resistance of R, as shown in Fig. 14-14b. Moving to node C, we again see a resistance of $2R$ looking toward ground or toward the 2^2 input, which provides an equivalent resistance of R to the right of node C as shown in Fig. 14-14c. From the discussion above we

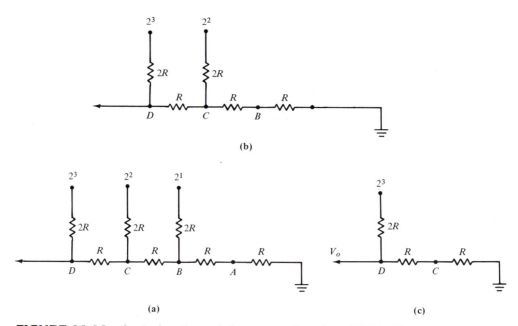

FIGURE 14-14   Analyzing the resistive properties of an R-2R ladder

can see that the total resistance from any node to ground or to the digital input is $2R$. This is true regardless of the number of digital inputs and regardless of whether the digital inputs are HIGH or LOW.

We can use ordinary circuit analysis techniques to determine the analog output voltage for the various binary inputs. For example, consider the circuit when the binary number 1000 is applied to the inputs. Since all inputs except the MSB are LOW, we can draw the ladder network as shown in Fig. 14-15. The equivalent resistance to the right of node D is $2R$; this means that the circuit in Fig. 14-15 can be replaced with the equivalent circuit shown in Fig. 14-16. Applying the voltage-divider equation to this equivalent circuit yields

$$V_o = V \frac{2R}{2R + 2R} = \frac{V}{2} \tag{14-13}$$

Thus, when the most significant of the digital inputs is logic 1, the analog output voltage is $V/2$.

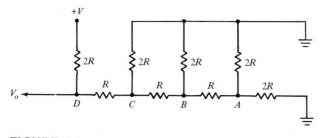

FIGURE 14-15   R-2R ladder with binary input of 1000

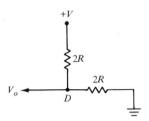

FIGURE 14-16 Equivalent resistance of *R-2R* ladder seen at node *D*

To determine the output voltage when the second most significant bit is HIGH, the ladder network can be drawn as shown in Fig. 14-17. The circuit to the right of node *C* can be replaced with an equivalent resistance of *2R*, as shown in Fig. 14-18. If we break the circuit at the dashed line in Fig 14-18 and replace the circuitry to the right of the dashed line with a Thévenin's equivalent circuit, the circuit in Fig. 14-18 can be redrawn as shown in Fig. 14-19. From Fig. 14-19 we can compute the analog output voltage when binary input 0100 is applied as

$$V_o = \frac{V}{2}\left(\frac{2R}{2R + 2R}\right) = \frac{V}{4} \tag{14-14}$$

By applying this same procedure to each digital input, we can obtain the corresponding analog output voltage for each. It should be pointed out that the MSB

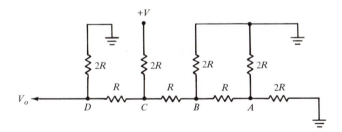

FIGURE 14-17 *R-2R* ladder with binary input of 0100

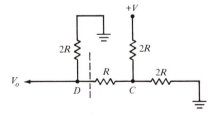

FIGURE 14-18 Applying Thévenin's theorem to *R-2R* ladder

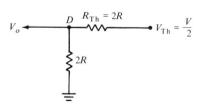

FIGURE 14-19 Voltage division at node *D* for a 4-bit *R-2R* ladder

contributes an analog output voltage equal to $V/2$ regardless of the number of bits in the digital input. The analog output voltages and the corresponding binary weights for a 10-bit R-$2R$ binary ladder are shown in Table 14-1.

The total output voltage is the sum of the contribution of each digital bit, written in the form of an equation as

$$V_o = \frac{V}{2} + \frac{V}{4} + \frac{V}{8} + \frac{V}{16} + \cdots + \frac{V}{2^n} \qquad (14\text{-}15)$$

where n is the number of bits in the digital input. By finding a common denominator for the terms in Eq. (14-15), the output voltage can be written as

$$V_o = \frac{V_0 2^0 + V_1 2^1 + V_2 2^2 + V_3 2^3 + \cdots + V_{n-1} 2^{n-1}}{2^n} \qquad (14\text{-}16)$$

where $V_0, V_1, V_2, V_3, \ldots, V_n$ are the digital input voltage levels.

EXAMPLE 14-3

Determine the output voltage for a 5-bit R–$2R$ ladder when the digital input is 10101 if 0 V corresponds to logic 0 and 5 V corresponds to logic 1.

SOLUTION

By using Eq. 14-15, we obtain the output voltage as

$$V_o = \frac{5 \times 2^0 + 0 \times 2^1 + 5 \times 2^2 + 0 \times 2^3 + 5 \times 2^4}{2^5}$$

$$= \frac{5 \times 1 + 0 + 5 \times 4 + 0 + 5 \times 16}{32}$$

$$= \frac{5 + 20 + 80}{32} = 3.281 \text{ V}$$

TABLE 14-1. / R-$2R$ Binary Ladder Output Voltages

Bit	Output Voltage	Binary Weight
MSB	$V/2$	1/2
2MSB	$V/4$	1/4
3MSB	$V/8$	1/8
4MSB	$V/16$	1/16
5MSB	$V/32$	1/32
6MSB	$V/64$	1/64
7MSB	$V/128$	1/128
8MSB	$V/256$	1/256
9MSB	$V/512$	1/512
10MSB	$V/1024$	1/1024

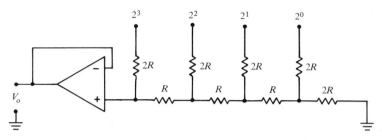

FIGURE 14-20   Terminating an *R-2R* ladder in a high resistance

It is generally desirable to terminate $R–2R$ ladders in a high resistance to reduce loading effects. One way of doing this is by using an op-amp connected as a unity-gain, noninverting amplifier, as shown in Fig. 14-20. The op-amp has a very high input impedance, which reduces loading to a negligible level. Since the gain of the op-amp is unity, its output is equal to the output of the ladder network; therefore, the op-amp simply serves as a buffer.

14-4
DIGITAL-TO-ANALOG CONVERTER

Either binary weighted or $R–2R$ ladders can be used as the heart of a D/A converter. However, additional circuitry, shown in Fig. 14-21, is required to make a practical, commercial product.

The logic circuits provide a means of setting data into the register at a precisely controlled time determined by the strobe pulse. The register, which may be made up of a number of kinds of flip-flops, stores the digital information that is to be converted to its analog equivalent. The level amplifiers provide a means of ensuring that all digital signals applied to the ladder network are of the same amplitude and are constant in amplitude. By expanding the block diagram of Fig. 14-21, we obtain a complete schematic diagram for a practical D/A converter, as shown in Fig. 14-22.

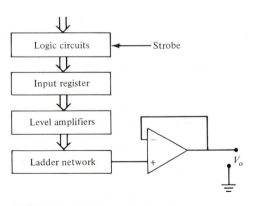

FIGURE 14-21   Block diagram of a D/A converter

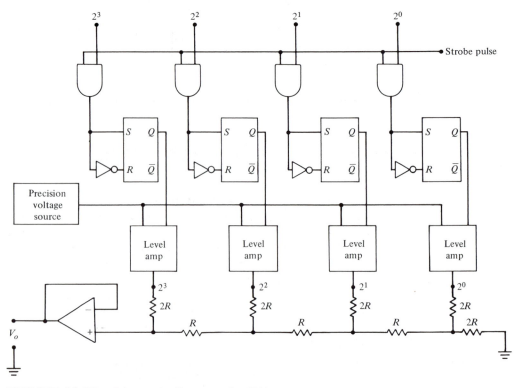

FIGURE 14-22   Schematic diagram of a D/A converter

14-5
ANALOG-TO-DIGITAL CONVERSION

The process of converting analog signals to an equivalent signal in digital form is called *analog-to-digital (A/D) conversion.* Most physical parameters are analog, or continuous, in nature. It will be necessary to convert from an analog to a digital form if the analog signal is to be measured, controlled, processed, or stored by using digital circuitry.

There are several methods by which conversion from analog to digital may be accomplished with varying degrees of accuracy, conversion rates, cost, and susceptibility to noise. In the following paragraphs we examine some of the methods in use today, including voltage-to-frequency, simultaneous, ramp, single-slope, dual-slope integration, and successive-approximation conversion.

14-5.1   Voltage-to-Frequency Conversion

Voltage-to-frequency conversion, as its name implies, is a conversion technique that converts an analog input voltage to a periodic waveform whose frequency is directly proportional to the input voltage. Voltage-to-frequency conversion can be achieved by using a very linear voltage-controlled oscillator (VCO), as shown in Fig. 14-23. The VCO must be designed so that the relationship between the output frequency and the input voltage is constant, which we can express as

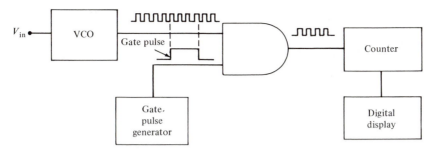

FIGURE 14-23 Block diagram of a basic voltage-to-frequency converter

$$k = \frac{f}{V_{in}}$$

or

$$V_{in} = \frac{f}{k} \qquad (14\text{-}17)$$

As can be seen in Fig. 14-23, the output signal from the VCO is applied to one input of a two-input AND gate. The second input to the AND gate is a "gating pulse." During the time that both signals are present at the inputs of the AND gate, the output of the AND gate is identical to the VCO output. Since there must be a linear relationship between the VCO input voltage and output frequency, the AND-gate output can be applied to a counting circuit to provide an indication of the analog voltage applied to the VCO input.

EXAMPLE 14-4 ▬▬▬▬▬▬▬▬▬▬▬▬▬▬▬▬▬▬▬▬▬▬▬▬▬▬▬▬▬▬

The constant of proportionality, k, which exists between the output frequency and input voltage for the VCO in Fig. 14-23, equals 50. If 530 pulses pass through the AND gate during a 0.1 s gating pulse, what is the amplitude of V_{in}?

SOLUTION

The output frequency of the VCO is

$$f = \frac{\text{pulses}}{\text{gate duration}} = \frac{530 \text{ pulses}}{0.1 \text{ s}} = 5300 \text{ pulses/s}$$

Therefore, the analog voltage is

$$V_{in} = \frac{f}{k}$$

$$\frac{f}{50} = \frac{5300 \text{ pulses/s}}{50} = 106 \text{ V}$$

❏

TABLE 14-2. / Input/ Output Relationship

Analog Input	Digital Output
0 to $V/4$	00
$V/4$ to $V/2$	01
$V/2$ to $3V/4$	10
$3V/4$ to V	11

14-5.2 Simultaneous Method of A/D Conversion

The simultaneous, or parallel, method of A/D conversion requires the use of a number of comparator circuits. As a general rule, the number of comparators required equals $2^n - 1$, where n is the number of digital bits in the output. The use of two digital bits allows us to define four ranges of an analog input voltage, in that we have the relationship shown in Table 14-2. A logic diagram for a basic two-bit simultaneous A/D converter is shown in Fig. 14-24. Table 14-3 tabulates analog and logic levels for the simultaneous A/D converter shown in Fig. 14-24.

Although simultaneous A/D converters are capable of very fast conversion

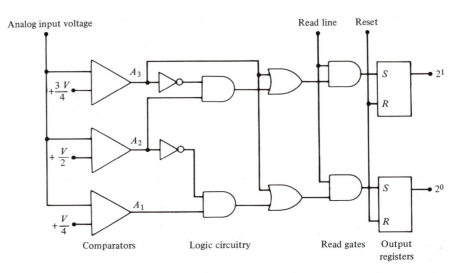

FIGURE 14-24 Logic diagram of a basic 2-bit simultaneous A/D converter

TABLE 14-3. / Analog and Logic Levels for the Simultaneous A/D Converter in Fig. 14-24

Analog Input Voltage	Comparator Outputs			Digital 2^1	Outputs 2^0
	A_1	A_2	A_3		
0 to $V/4$	Low	Low	Low	0	0
$V/4$ to $V/2$	High	Low	Low	0	1
$V/2$ to $3V/4$	High	High	Low	1	0
$3V/4$ to V	High	High	High	1	1

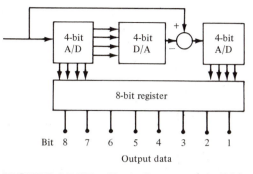

FIGURE 14-25 Block diagram of an 8-bit simultaneous A/D converter

rates, the number of comparators that are required quickly becomes prohibitive as the number of bits in the digital output increases. For example, a 4-bit converter requires 15 comparators, whereas an 8-bit converter requires 255. There are two ways to get around this problem. One way is to use one of the other methods of A/D conversion discussed in the following paragraphs. Another way is to implement an 8-bit simultaneous A/D converter with two 4-bit converters, as shown in Fig. 14-25.

14-5.3   Ramp-Type A/D Conversion

A slightly more sophisticated and accurate (better resolution) type of A/D converter that uses only one comparator is a *ramp converter*, also known as a stairstep-ramp or counter-type A/D converter. A block diagram for this type of converter is shown in Fig. 14-26. Using only one comparator is possible because we now have a stairstep reference voltage to which the analog input voltage is compared.

We can analyze the operation of the circuit by considering the counter to be reset and the output of the D/A converter to be zero. If the analog input voltage begins to increase, the comparator output will go HIGH when the analog voltage exceeds the reference voltage. The HIGH output state of the comparator enables the AND gate; therefore, the counter begins to store pulses from the clock. As the counter advances through its binary states, it produces the stairstep reference

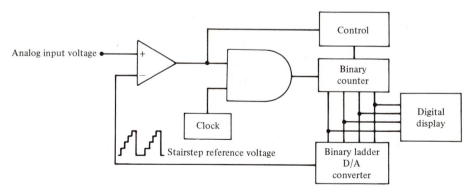

FIGURE 14-26 Ramp-type A/D converter

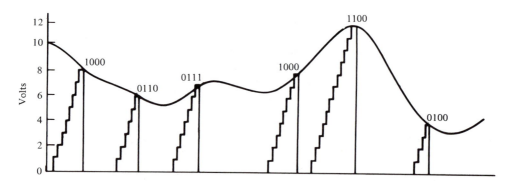

FIGURE 14-27   Graphical example of the conversion process of a ramp-type A/D converter

voltage at the output of the D/A converter. When the stairstep voltage exceeds the amplitude of the analog voltage, the comparator output is switched LOW, thus disabling the AND gate, which cuts off clock pulses to the counter. The binary number stored in the counter is displayed at the output. This display, of course, represents the amplitude of the analog voltage.

Figure 14-27 shows a conversion sequence for a 4-bit converter. As can be seen, the counter starts at zero and counts up to the point where the stairstep reference voltage is equal to the analog input voltage; therefore, the conversion time is directly related to the amplitude of the analog voltage.

EXAMPLE 14-5 ━━━

Determine the following for a 10-bit ramp-type A/D converter driven by a 1-MHz clock:

(a) The maximum conversion time
(b) The average conversion time
(c) The maximum rate of conversion

SOLUTION

(a) For a 10-bit converter the number of count states is given as

$$\text{count states} = 2^{10} = 1024$$

The period of clock signal is 1 μs. The counter advances one count state per clock pulse. The maximum conversion time is computed as

$$\text{maximum conversion time} = (1024 \text{ counts}) \frac{1 \ \mu s}{\text{count}} = 1024 \ \mu s$$

(b) The average conversion time is one-half the maximum conversion time, since the staircase increases linearly; therefore,

$$\text{average conversion time} = \frac{1024 \ \mu s}{2} = 512 \ \mu s$$

(c) The maximum rate of conversion is determined by the greatest time required for conversion and is computed as

$$\text{maximum conversion rate} = \frac{1}{\text{maximum conversion time}}$$

$$= \frac{1}{1024 \times 10^{-6} \text{ s}}$$

$$= 976 \text{ conversion/per s} \qquad \square$$

14-5.4   Single-Slope A/D Converter

Unlike ramp-type converters, which incorporate a D/A converter, single-slope A/D converters use a constant slope reference voltage generated with an integrating op-amp or similar technique. A schematic diagram and the associated waveforms for this type of circuit are shown in Fig. 14-28 on p. 392.

The operation of the circuit is as follows:

1. The analog input voltage causes the noninverting input of the comparator to be positive; therefore, the output of the comparator is at a positive potential as well. This positive potential causes the center terminal of the AND gate to be HIGH.

2. The main gate control circuit generates a positive pulse that activates the *S* drive and opens switch *S*. The positive pulse also sets the upper input of the AND gate HIGH.

3. When switch *S* opens, the integrator circuit capacitor, *C*, begins to charge linearly from zero in a positive direction. This ramp voltage is applied to the inverting input of the comparator.

4. The pulse train from the clock is applied to the lower input of the AND gate. Since the other two inputs of the AND gate are HIGH, the AND gate output is a series of clock pulses. These pulses are counted by the binary counter.

5. When capacitor *C* charges to a voltage level slightly higher (less than 1 mV higher) than V_{in}, the comparator switches abruptly and its output becomes zero. This causes one input of the AND gate to be LOW; therefore, it stops passing clock pulses into the counter.

6. The count stored in the binary counter at this time is directly proportional to the analog voltage; therefore, the output of the counter is the digital equivalent of the analog input voltage.

14-5.5   Dual-Slope A/D Converters

Dual-slope A/D converter use the linear-charge concept employed by single-slope converters, but several improvements are incorporated, primarily to improve long-term stability. Dual-slope converters not only charge the capacitor associated with the integrator, but discharge it as well during the conversion process. The charge/discharge cycle tends to significantly reduce long-term drift and stability problems usually associated with single-slope converters.

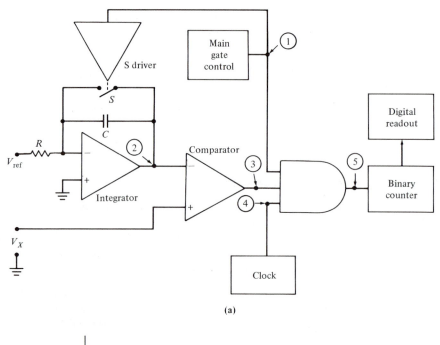

(a)

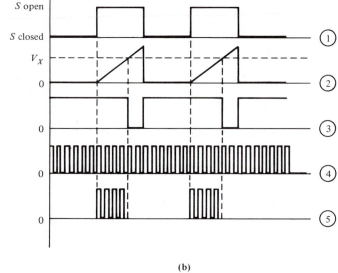

(b)

FIGURE 14-28   Circuit and timing diagram for a single-slope analog-to-digital converter

Dual-slope A/D converters are designed around an integrator consisting of op-amp A_1, resistor R, and capacitor C, shown in Fig. 14-29, plus a voltage comparator A_2 and high-speed electronic switches S_1 and S_2. The remaining circuitry is associated with the counter.

The output of the comparator is LOW if the output of the integrator is zero but will switch states to a HIGH output when the integrator output rises to approximately 1 mV. The initial steps in the operation of the converter is the momentary

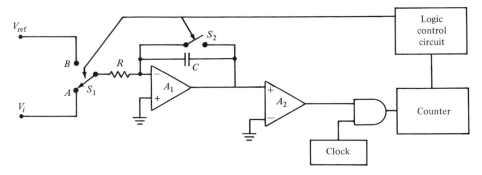

FIGURE 14-29   Basic dual-slope A/D converter

closure of switch S_2 and setting switch S_1 to position A. Both of these are done by the logic control circuit. When switch S_1 is in position A, the analog voltage is applied to the integrator. This causes the voltage at the output of the integrator to rise linearly with respect to time. As soon as the voltage at the output of the integrator rises very slightly, the comparator changes states, thereby setting its output HIGH. This enables the AND gate, which allows clock pulses to pass to the counter, which counts the pulses until it overflows at time t_2, shown in Fig. 14-30. When the counter overflows at time t_2, a pulse from the counter causes the control logic section to set switch S_1 to position B. With S_1 in position B, the reference voltage, V_{ref}, is connected to the input of the integrator, the reference voltage is connected to the input of the integrator. This negative voltage causes the capacitor to discharge at a constant rate. When the output of the integrator reaches zero at t_3, the comparator changes states, setting its output LOW, which disables the counter. The binary count stored by the counter at this time is directly proportional to the ratio of the input voltage to the reference voltage and is the digital equivalent of the analog input. During charge the voltage at the output of the integrator is given as

$$V_A = V_{in} \frac{t_2 - t_1}{RC} \qquad (14\text{-}18)$$

During discharge the amplitude of the voltage is expressed as

$$V_A = V_{ref} \frac{t_3 - t_2}{RC} \qquad (14\text{-}19)$$

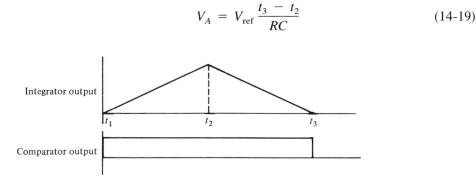

FIGURE 14-30   Integrator and comparator output waveforms for the circuit in Fig. 14-29

Since the left-hand side of both Eq. 14-18 and Eq. 14-19 are equal to V_A, we can set the right-hand sides equal to each other and write this as

$$V_{in} \frac{t_2 - t_1}{RC} = V_{ref} \frac{t_3 - t_2}{RC} \tag{14-20}$$

Solving this expression for $t_3 - t_2$ yields

$$t_3 - t_2 = (t_2 - t_1) \frac{V_{in}}{V_{ref}} \tag{14-21}$$

The product of the time interval $t_3 - t_2$ and the frequency of the clock is equal to the number of pulses stored in the counter. Since there is a linear relationship between the time interval $t_3 - t_2$ and the input voltage V_{in}, the binary count stored in the counter is a digital representation of the analog input voltage.

14-5.6 Successive-Approximation A/D Converters

The successive-approximation (SA) technique is one of the more widely used methods of A/D conversion due primarily to its short, constant conversion time. A basic block diagram for a 4-bit successive approximation A/D converter is shown in Fig. 14-31. As can be seen, it consists of a voltage comparator, D/A converter, successive-approximation register, and a clock.

In going through a conversion cycle, the system starts by enabling the bits of the D/A converter one bit at a time, starting with the MSB. As each bit is enabled, its amplitude is compared to the analog voltage, V_{in}, by the voltage comparator. The comparator then produces an output that indicates whether the analog voltage is greater or less in amplitude than the output of the D/A converter. If the output of the D/A converter is greater than the analog voltage, the MSB is reset to zero, since it will not be required in the digital representation of the analog input. If the D/A converter output is less than the analog input, the MSB is retained in the register.

The system makes this comparison with each bit, starting with the MSB, then the next MSB, and so on. As each bit of the D/A converter is compared, those that contribute toward the digital representation of the analog input are stored in the register, while those bits not required are reset.

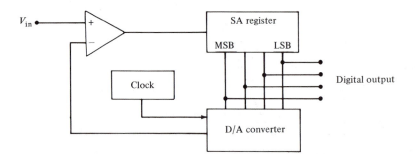

FIGURE 14-31 Block diagram of a basic 4-bit successive approximation A/D converter

EXAMPLE 14-6 ▬▬▬▬▬▬▬▬▬▬▬▬▬▬▬▬▬▬▬▬▬▬▬▬▬▬▬▬▬

A 12 V input signal is applied to the 4-bit successive approximation A/D converter shown in Fig. 14-20. There is a linear relationship between the digital input and the analog output of the D/A converter; that is, $2^0 = 1$ V, $2^1 = 2$ V, $2^2 = 4$ V, and $2^3 = 8$ V. What bits are stored in the SA register to represent the 12 V analog input voltage?

SOLUTION

The same relationship that exists between the digital input and analog output for the D/A converter exists between the analog input and digital output for the A/D converter. Therefore, 12 V analog equals 1100 digital so that the two most significant bits generated by the D/A converter are stored by the SA register. ❑

14-6
PERFORMANCE CHARACTERISTICS OF D/A CONVERTERS

The following paragraphs discuss several of the more important performance characteristics associated with digital-to-analog converters. These are

- Accuracy.
- Resolution.
- Linearity error.
- Monotonicity.
- Settling time.
- Quantum.
- Quantizing error.

Accuracy. *Accuracy* is a term used to indicate how close the *actual* output of a D/A converter is to the *expected* value and is usually expressed as a percentage of full-scale output voltage. As an example, for a converter that has a full-scale output voltage of 10 V and an accuracy expressed as ± 0.1 percent of full scale, any reading should be accurate within 10 mV, since 10 V $\times$ 0.001 = 10 mV. Since readings that are less than full scale may differ from the expected value by 10 mV, the greatest accuracy can be expected at full scale.

Ideally, the accuracy of D/A converters should be to within $\pm\frac{1}{2}$ of the least significant bit. For a 6-bit converter, the LSB = $\frac{1}{64}$ = 0.0156 (1.56 percent of full scale); therefore, the output should be accurate to within approximately ± 0.8 percent.

Resolution. The resolution of D/A converters is equal to the reciprocal of the number of discrete steps in the D/A output; however, this is directly related to the number of bits in the digital input. Therefore, resolution can be expressed as

$$resolution = \frac{1}{2^n}$$

where n is the number of bits in the digital input.

The most common number of input bits for commercial D/A converters are 8, 10, or 12. An 8-input-bit converter has a resolution of 1 part in $2^8(256)$, or

$$\text{resolution} = \frac{1}{2^8} = \frac{1}{256} = 0.0039$$

Expressed as a percentage,

$$\text{resolution} = 0.39 \text{ percent}$$

Linearity Error.   Linearity error is the deviation from the linear or straight-line output of a D/A converter. It is generally described in terms of differential linearity error and integral linearity error.

1. *Differential linearity error:* the maximum deviation of any LSB change in the output of a data converter from its ideal value.
2. *Integral linearity error:* the maximum deviation of the output of a data converter from the ideal straight line with offset and gain errors zeroed. It is generally expressed as a percent of full scale or in LSBs.

Monotonicity.   The term *monotonic,* as it applies to D/A converters, means that a converter does not skip a step or repeat a step when sequenced through its entire range of input bits.

Settling Time.   Settling time is defined as the time that is required for a D/A converter to settle to within $\pm\frac{1}{2}$ LSB of its final value when a change occurs at the input.

Quantum.   A quantum is the analog difference between two adjacent codes for an A/D or D/A converter and is equal in size to the LSB.

Quantizing Error.   The quantizing error is the inherent uncertainty in digitizing an analog value due to the finite resolution of the conversion process. The quantized value is uncertain by up to LSB/2. This error can be reduced only by increasing the resolution of the converter.

14-7
APPLICATIONS

There are many applications for both D/A and A/D converters. Applications for D/A converters include the conversion of computer outputs to an analog form to position the pen on recorders and plotters. Other applications deal with positioning the final element on numerically controlled machines, such as drill presses or milling machines, and in set-point adjustments in process control applications.

Analog-to-digital converters find applications in data acquisition systems, process control, automatic test systems, and measuring instruments. One very common application of A/D converters is in digital voltmeters such as the one shown in Fig. 14-32.

As long as the analog input voltage is greater than the reference voltage, the

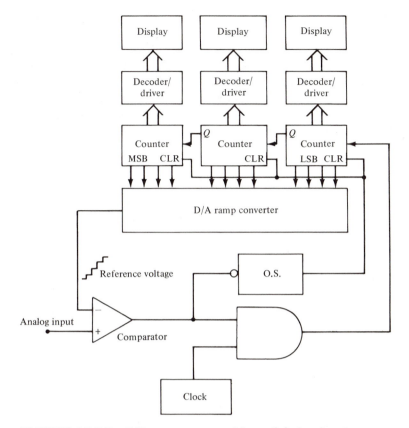

FIGURE 14-32 D/A converter used in a digital voltmeter

output of the comparator is HIGH; therefore, clock pulses will be stored in the counter. At the point where the reference voltage exceeds the amplitude of the analog input voltage, the comparator output switches LOW, which triggers the one-shot (OS) and clears the display after a period of time determined by the timing circuit of the one-shot.

14-8
SUMMARY

Digital-to-analog and analog-to-digital converters provide an interface network between digital systems and the analog world. In such circuits the versatile op-amp often plays an important role serving as a summing device, a buffer, or as the heart of an integrator, differentiator, or comparator.

D/A conversion is most easily accomplished by using resistance networks with R–$2R$ ladders offering several advantages over weighted-binary ladders. In addition to the resistive network, complete D/A converters usually have registers to momentarily store the digital data to be converted, level amplifiers, a precision voltage source, and logic circuitry for control.

There are many techniques used to convert an analog signal to an equivalent digital form, several of which incorporate a D/A converter. Simultaneous converters are very fast but require a prohibitive number of comparators when the digital output is to contain more than a few bits.

Voltage-to-frequency conversion is straightforward but less accurate than most other methods. Ramp-type converters are somewhat slower than simultaneous conversion techniques but are much more practical. Single-slope, dual-slope, and successive-approximation techniques each offer advantages for certain applications. However, the advantages offered by successive approximation are significant enough to account for its widespread use.

PROBLEMS

1. What is the output voltage for the following amplifier?

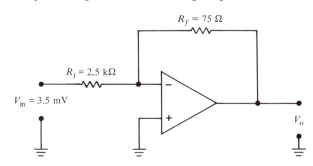

2. What is the output voltage for the following summing amplifier?

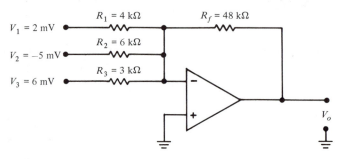

3. Compute the output voltage for the following circuit when $t = 15$ ms.

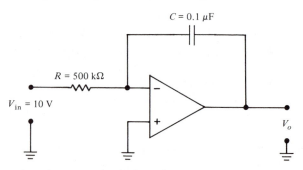

4. Determine the time required for the output voltage of the circuit to reach 7 V.

5. Determine the output voltage for the following circuit.

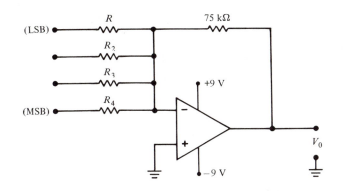

6. What minimum value can R have in the following circuit?

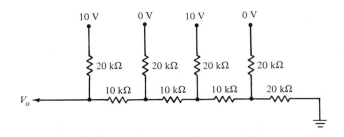

7. What is the value of R_2, R_3, and R_4 in the circuit of Problem 6?

8. Compute the analog voltage, V_{out}, for the following circuit.

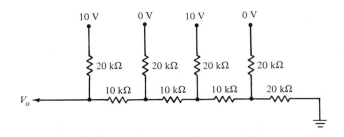

9. What is the resolution, expressed as a percentage, of a 12-bit R–$2R$ ladder?

10. If the maximum output voltage for the R–$2R$ ladder in Problem 9 is $+10$ V, what is the resolution in volts?

11. How many bits are required in a R–$2R$ ladder to achieve a resolution of 2 mV if the maximum output voltage is 10 V?

12. Compute the analog input voltage, V_{in}, for the following circuit if $k = 60$ and the counter stores 540 pulses during a 0.2-s gating pulse.

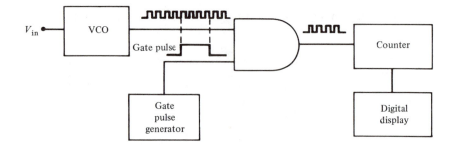

13. In the circuit shown, the counter starts counting at t_0 and stops counting at t_1. If the value of k for the VCO is 50 and the counter stores 1200 pulses, compute the frequency f.

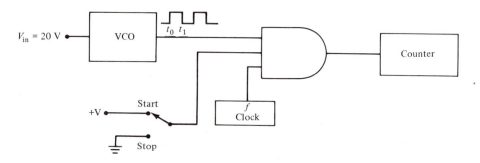

14. How many comparators are needed to construct a 60-bit simultaneous A/D converter?

15. Determine the following for an 8-bit ramp-type A/D converter that is driven by a 1-MHz clock:

(a) Maximum conversion time
(b) Average conversion time
(c) Maximum conversion rate

16. The maximum ladder output voltage for a 4-bit successive approximation A/D converter is 8 V. Determine the sequence of states for the SA register if an analog voltage of 5 V is applied to the input.

REFERENCES

Analog-to-Digital Conversion Techniques, Application Note AN-471. Phoenix, Ariz.: Motorola Semiconductor Products, Inc., 1974.

Data Conversion Components. Mansfield, Mass.: Datel-Intersil, Inc., 1983–1984.

Levine, Morris E., *Digital Theory and Practice Using Integrated Circuits*. Englewood Cliffs, N.J.: Prentice-Hall, Inc., 1978.

Malvino, Albert P., and Donald P. Leach, *Digital Principles and Applications*, 3rd ed. New York: McGraw-Hill Book Company, 1981.

A Single Ramp Analog-to-Digital Converter, Application Note AN-559. Phoenix, Ariz.: Motorola Semiconductor Products, Inc., 1972.

Successive Approximation A/D Conversion, Application Note AN-716. Phoenix, Ariz.: Motorola Semiconductor Products, Inc., 1974.

15

Memory and Memory Devices

INSTRUCTIONAL OBJECTIVES

In this chapter you are introduced to the concept of memory and to various memory devices. After completing the chapter, you should be able to:

1. Define terms related to memory.
2. Describe how magnetic cores are used to store binary information.
3. Explain the basic principles of coincident current as it relates to core memory.
4. Describe the primary difference between RAMs and ROMs.
5. Describe the purpose of each of the wires through a core in a core memory system.
6. Describe how memory systems are addressed.
7. Draw the family tree for ROM devices.
8. Describe how PROMs are programmed.
9. Describe the basic difference between static and dynamic RAMs.
10. Describe the difference between EPROMs and EEPROMs.
11. Describe the operation of charge-coupled devices.
12. Describe the operation of bubble memory.
13. Explain the basic principles of programmable logic arrays.
14. Make calculations related to the storage capacity of rotating magnetic storage devices.

SELF-EVALUATION QUESTIONS

The following questions relate to the material presented in this chapter. Read the questions prior to studying the chapter and, as you read through the material, watch for answers to the questions. After completing the chapter, return to this section and evaluate your comprehension of the material by answering the questions again.

1. Memory capacity is generally expressed as ''K words of memory.'' What is the value of K?
2. What does the term *volatile* mean?
3. How does a magnetic core store information?
4. What is a RAM, and how does it differ from a ROM?
5. What is the purpose of the inhibit wires in a core memory system?
6. How many bits must an address contain to be able to address a single core in a 128×128 core plane?

7. What is the basic difference between static RAMs and dynamic RAMs?
8. By what other acronym are EEPROMs known?
9. Where is charge stored in charge-coupled devices?
10. How are bubbles produced in magnetic bubble memory systems?
11. What is the logic structure of programmable logic arrays?
12. What three types of rotating magnetic systems are used for mass storage?

15-1
INTRODUCTION

One of the major differences between analog and digital systems is the ability of digital systems to store information. The information is stored by the memory unit, which is one of the major elements of a digital computer, as can be seen in Fig. 15-1. Modern digital computers have several types of memories, which are sometimes categorized as *working memories* and *auxiliary memories*. Working memories are contained within the computer's central processing unit, whereas auxiliary memories are peripheral devices such as magnetic tape or magnetic disk drives. Working memories often consist of one large memory and several smaller memories referred to as scratch pad memories, buffer memories, and special-function memories.

A more technical means of categorizing memory is as either *random-access memory* (RAM) or *read-only memory* (ROM). Memory designated as RAM might more appropriately be called ''read/write'' memory, since data may be written into or read from memory repeatedly. As opposed to this multiple read/write type of memory, data may be written into ROM only once; however, it can be read repeatedly. Some ROMs are ''programmed'' by the manufacturer, whereas others are ''programmed'' by the user.

The special-function memories mentioned above are generally ROM devices designed to store information permanently, even when power is turned off. Such storage can hold information that is immediately available when power is turned on. ROMs are used in computers and calculators for fixed instructions, such as subroutines, as well as for constants such as π or ϵ, and for look-up tables such as trigonometric and logarithmic tables.

Because of the tremendous strides made in semiconductor memory technology

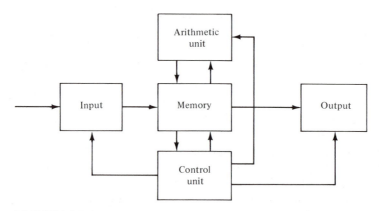

FIGURE 15-1   *Major elements of a digital computer*

in recent years, IC memory devices account for a substantial percentage of present computer memory. Therefore, a substantial portion of this chapter is devoted to semiconductor memory devices.

15-2
MEMORY CONCEPTS

The concept of memory was introduced in Chapter 8, where flip-flops were discussed. These devices function as 1-bit memories in that they remember whether their output was most recently SET or RESET. When flip-flops are cascaded to form registers, their memory capacity increases from one bit to 2^n bits, where n is the number of flip-flops in cascade. Even the smallest computer requires several thousand bits of memory; however, it is totally impractical to construct memories of this capacity by interconnecting discrete flip-flops. Memories of this capacity are obtained by using LSI semiconductor memory ICs or magnetic-type memories.

Whether memories use semiconductor ICs or magnetic devices, they are subdivided into groups of bits called *words*. When data are transferred into or out of memory, they are transferred a word at a time rather than a bit at a time; thus we can define a word as the number of bits involved in each data transfer. Word lengths vary between computer manufacturers and range from 8 to 32 bits. A word length of 8 bits is called a *byte*. Typical word sizes for mini- and micro-computers are 12 and 16 bits, with most registers in the computer designed with the same capacity as the length of the memory word.

The memory capacity of modern computers ranges from a few thousand to several million words. A typical memory size is 4K words, or an even multiple of 4K words. Most microcomputers have memory capacities of 16K, 32K, 48K, or 64K words. Since we are working in digital systems, K is defined in binary terms as

$$K = 2^{10} = 1024$$

Since 1K of memory has a word capacity of 1024 words, a 16K memory actually has a memory capacity of $16 \times 1024 = 16,384$ words rather than exactly 16K.

Memory can be organized as bit addressable or word addressable, with most being word addressable. A typical memory configuration is shown in Fig. 15-2.

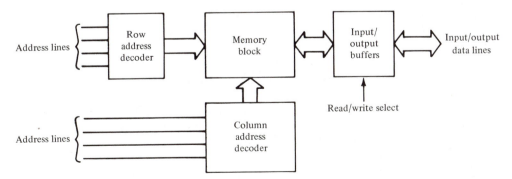

FIGURE 15-2 General block diagram of typical memory configuration

Address lines provide a means of identifying a specific memory location where data are to be written into or read from. Data enter or leave the memory block via the input/output buffers and the data lines. The memory block is made up of memory cells, which are defined in the following section.

15-3
MEMORY TERMINOLOGY

There are many new terms associated with memory. Rather than define the terms as they are introduced, we group the terms here in a single section for future reference.

Access Time. The access time is the time required to "read" one word out of memory. It is usually equal to the time required to "write" one word into memory.

Address. The address tells the exact location of a binary word that is to be entered into, or retrieved from, a large memory system.

Auxiliary Memory. Auxiliary memory is additional memory used in most large computers for the storage of large quantities of data in excess of those stored in the mainframe memory.

Bit-Organized Memory. Bit-organized memory is memory in which an identifiable address is assigned to each *bit*.

Byte. A byte is an even subdivision of a word, generally 8 bits.

Capacity. The capacity is the total number of bits that can be stored in a memory. Capacity can also be expressed as the number of words stored in a memory. For example, a memory with a capacity of 4096 words with 16 bits per word has a bit capacity of 65,536.

Content-Addressable Memory. Content-addressable memory is a type of memory in which a particular word, or a group of words, is selected according to a particular bit pattern.

Destructive Readout. Destructive readout occurs when the reading of a location in memory destroys the information stored at that location.

Direct Addressing. In direct addressing, each word has a separate address line. To select a particular word, its address line must be energized.

Mainframe Memory. Mainframe memory is the internal memory of a computer that stores data and instructions for the program being performed. Mainframe memory is also called working memory.

Memory Cell. A memory cell is a device that stores a single binary bit.

Random-Access Memory. Random-access memories (RAMs) are memories in which information can be written into or read from *any* storage location with the same access time.

Read. The term *read* refers to the act of retrieving a word from memory.

Read-Only Memory. Read-only memory is a logic circuit that functions as a memory whose stored information cannot be easily changed.

Sequential-Access Memory. Sequential-access memory is memory in which a particular address location is found by successively scanning all address locations. In this type of memory the access time depends on the memory address.

Static Storage. Static storage is a memory system in which the information does not physically change positions and does not need to be refreshed.

Volatile Memory. Volatile memories need electrical power in order to store information, and a loss of electrical power causes loss of stored data.

Word. A word is a group of binary bits, generally either 8, 16, or 32 bits.

Word-Organized Memory. Word-organized memory is memory in which an address is assigned to each *word* so that all bits of the word have the same address.

Write. The term *write* refers to entering a data word into memory.

15-4
MAGNETIC CORE MEMORY

Until recently, magnetic core memory was the dominant memory technology. Although LSI semiconductor ICs are rapidly replacing core memories in most applications, there are still sufficient numbers of core memories to warrant a brief discussion.

Magnetic cores are small toroidal pieces of ferromagnetic material called *ferrite cores*. Typical cores have outer diameters of approximately 0.030 inch. Ferrite cores have a high retentivity, which means that once magnetized they tend to remain magnetized for a long period of time. Information, in the form of a binary 1 or 0, is stored in a ferrite core by creating a magnetic flux within the core. This is accomplished by current flow through a wire that passes through the center of the core, as shown in Fig. 15-3. In Fig. 15-3a, current flow through the wire to the right produces counterclockwise magnetic flux within the core. In Fig. 15-3b, current flow through the wire has been reduced to zero, but the core is still magnetized. In Fig. 15-3c, current flow through the wire to the left produced clockwise magnetic flux within the core, whereas in Fig. 15-3d, the core remains magnetized after current flow through the wire has been reduced to zero. For purposes of illustration assume that clockwise flux within the core represents a binary 1, whereas counterclockwise flux represents a binary 0. In either case, the cores remained magnetized, thus retaining their memory content, after the magnetizing current was reduced to zero. With ferrite core memory, the memory

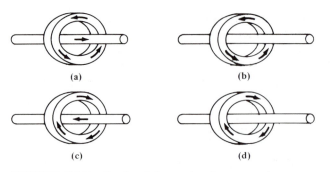

FIGURE 15-3   *Storing information in magnetic cores*

content is not lost even if system power is removed. This type of memory is said to be *nonvolatile* and is a major advantage of core memories. On the other hand, a memory system that loses its contents when power is removed is described as *volatile* memory.

The magnetic characteristics associated with ferrite cores are shown in Fig. 15-4. The figure shows a *hysteresis curve,* which ideally should be square, for a ferrite core. Sufficient current through the wire in Fig. 15-3 drives the ferrite core into saturation in one direction or the other, depending on the direction of current flow. The required level of current is called the *threshold magnetizing current value,* I_M. A current in one direction, which we will designate as $+I_M$, will drive the core into positive saturation, thus storing a binary 1 in memory. A current in the other direction, which we designate as $-I_M$, will drive the core into saturation in a negative direction, thus storing a binary 0 in memory.

15-4.1   Magnetic Core Planes

Since a single ferrite core can store a single binary 1 or 0, thousands of cores are required in a computer memory system. Any desired location in memory must be accessible in the shortest possible time; therefore, cores must be arranged so that any core can be addressed and accessed as quickly as possible. Figure 15-5 shows

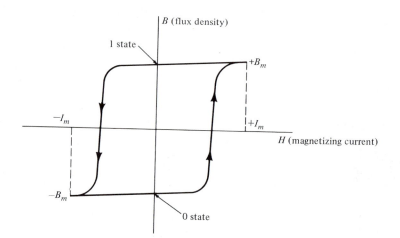

FIGURE 15-4   Hysteresis curve for a ferrite core

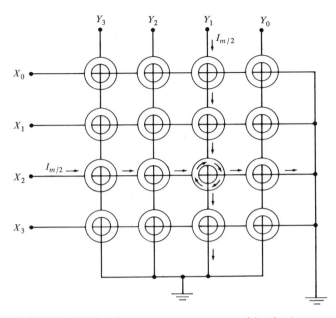

FIGURE 15-5   Core memory plane with single core
addressed

how core planes can be wired to permit a single core to be addressed. Since a
current equal to I_M is required to magnetize a core, if we pass a current equal to
$0.5I_M$ through one wire in both the X and Y planes, the sum of the two "half-
currents" at the point of intersection will be sufficient to magnetize that single
core.

The four horizontal rows in Fig. 15-5 are designated as X_0 through X_3 and the
four vertical columns are designated as Y_0 through Y_3. This permits us to *address*
any single core by specifying its X-Y coordinates. For example, the address of the
core in the upper right-hand corner of the memory plane is X_0, Y_0. These lines
are used to select a single core into which a binary 1 or 0 is to be set. Therefore,
the lines are called *select* lines or *write* lines. If we wish to write data at address
X_2 and Y_1, we momentarily place a half-current on lines X_2 and Y_1, as shown in
Fig. 15-5. The core where the currents are coincident is magnetized.

15-4.2 Sensing a Core

Once a binary bit is written into a ferrite core, how can we determine whether
the core contains a 1 or a 0? The ability to make this determination lies in the
characteristics of the hysteresis curve of Fig. 15-4. We can see that a substantial
transition of magnetic flux density occurs when a core is driven from $+B_M$ to
$-B_M$. If we pass a third wire in addition to the select lines through the center of
each core, current will be induced into this wire if the flux field on the core
changes. When this flux change occurs, it can be detected by measuring the current
induced into the third wire, or by measuring the voltage that is developed due to
the induced current.

When the contents of a core are to be determined, referred to as a *read* operation, a current pulse is applied to the appropriate X and Y lines to select a

particular core. The direction of the current is such as to cause the core to switch to the binary 0 state if it is not already in the "0" state. If the core is in its "1" state, it will switch states, which will induce a small current into the third wire. This wire senses the change of state of the core and is therefore called the *sense line*. The sense line and the associated sense amplifier are shown in Fig. 15-6.

Current induced into the sense line activates the sense amplifier. Sense amplifiers also have a *strobe* input that disabled them except when the *read* pulse is present, which prevents noise from producing an output signal from the *sense amplifier*. If a current is induced into the sense line, which indicates that the contents of the core being read was binary 1 and the read pulse is present, the *RS* flip-flop will be SET; thus it now contains the binary 1 that was in core memory.

After a core is read, its state will always be 0; therefore, the data at that address are destroyed. This is called *destructive readout;* however, it should be emphasized that the data are not lost but are simply transferred from the core to the flip-flop shown in Fig. 15-6. At the end of the clock cycle, data are written back into the core memory. To be able to write the correct data back into memory, a fourth wire through each of the cores in Fig. 15-7 is required. This wire is called the *inhibit* line. The inhibit line prevents a core from switching to the 1 state if a 0 is to be written into the core. This is accomplished by passing a current equal to $\frac{1}{2}I_m$ through the center of each core in the opposite direction of either the X or Y currents. In Fig. 15-7, the inhibit current is opposite in direction to the current in the X line, and therefore effectively cancels the magnetizing effect of the current in the X line. The Y-line current is not sufficient to switch the core to the 1 state;

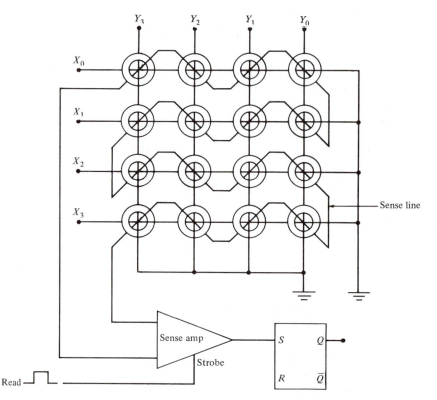

FIGURE 15-6   *Core memory plane with sense line added*

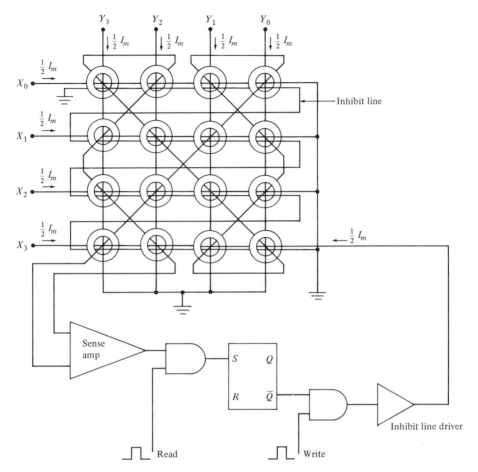

FIGURE 15-7 Core memory plane with inhibit line added

hence it remains in the 0 state. Suppose, for example, that a 0 is stored in the *RS* flip-flop; thus Q is LOW and $\overline{Q}$ is HIGH. The HIGH at $\overline{Q}$ provides the necessary current through the inhibit line to prevent a 1 from being stored by the core at X_2, Y_1; therefore, it remains in its 0 state. This is equivalent to writing, or restoring, the 0 that was in memory at X_2, Y_1 back into memory. Therefore, even though ferrite cores are destructive readout devices, magnetic core memory systems are nondestructive readout memory systems.

The waveforms associated with the process of reading a core and restoring the bit back into memory is shown in Fig. 15-8. Recall that during the read operation the direction of the current through the X and Y lines is always such as to store a 0 in the selected core, hence the $-\frac{1}{2}I_m$ currents. If a 1 is stored by the core at the start of the read operation, the core will switch states, thus inducing a current into the sense line. If a 0 was in the core at the start of the read operation, there would be no current induced into the sense line; therefore, we can see by observing the waveforms that the core was initially in its 1 state. The current in the sense line activates the sense amplifier and sets the Q output of the *RS* flip-flop HIGH while setting $\overline{Q}$ LOW. Since $\overline{Q}$ is LOW, there is no inhibit current, so the 1 is restored to memory location X_1, Y_1 during the write time.

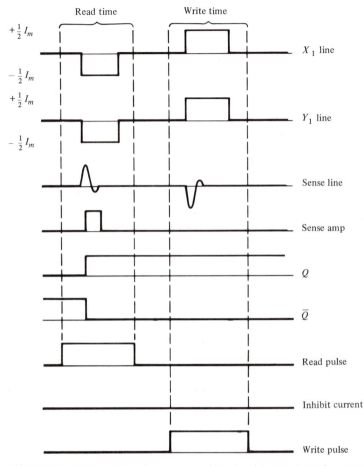

FIGURE 15-8 Waveforms associated with read and restore operation in a core memory

15-4.3   Multiple-Plane Core Memory System

Although the discussion regarding core memory has to this point dealt with a single binary bit written into or read from a single core, we are usually more concerned with reading and writing complete binary words. Core memory systems that permit us to accomplish this are multiple-plane systems such as the one shown in Fig. 15-9. A current of $\frac{1}{2}I_m$ through lines X_2 and Y_1 will magnetize the core at X_2, Y_1 in each of planes 1 through 4. As shown, the X and Y write lines are continuous through the four planes; however, there is a separate sense line and inhibit line for each plane.

15-4.4   Addressing Core Memory

In the preceding paragraphs we have given the address of specific cores by their X-Y coordinates. However, in digital computers, addresses are specified by binary codes, where a code group specifies one address location. This concept is illustrated in Fig. 15-10. There are eight rows and eight columns of cores in each plane, so there are 64 cores with 64 different addresses. To address 64 cores, a

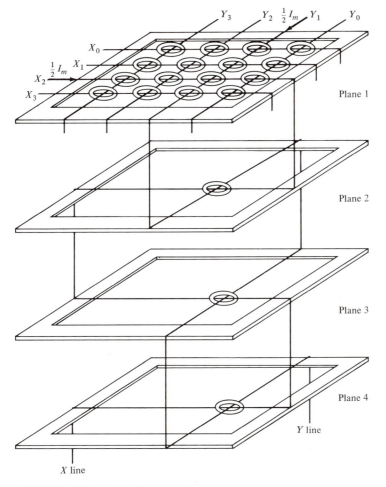

FIGURE 15-9   Multiple-plane core memory system

6-bit address code is required since $2^6 = 64$. This code is temporarily stored in the *memory address register* (MAR), shown in Fig. 15-10. The three lower-order bits of the address are used to select the correct Y line and the three higher-order bits are used to select the correct X line. Two 3-line-to-8-line decoders provide the necessary logic to make these selections.

EXAMPLE 15-1

A core memory plane contains 4096 ferrite cores. What size of memory address register is required to be able to address each core, and what are the dimensions of the plane?

SOLUTION

The square root of 4096 is 64; therefore, the memory plane is wired as a 64×64 array with 64 X lines and 64 Y lines. Since $2^{12} = 4096$, a 12-bit address code is required to address each core.   ❏

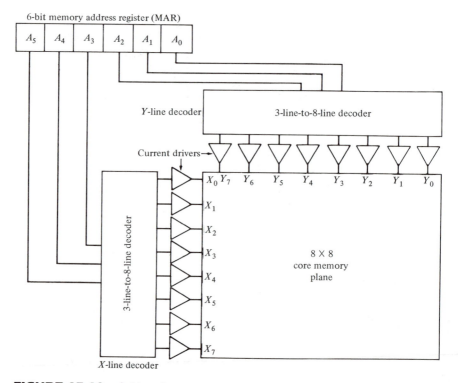

FIGURE 15-10   Addressing core memory

15-5
SEMICONDUCTOR MEMORY

Semiconductor memory devices have been available commercially for quite some time, but ferrite cores have been used extensively in memory systems because of their comparatively low cost. However, in recent years tremendous technical advancements have been made in the area of semiconductor memory devices, which have brought about greatly reduced prices.

Apart from competitive prices, semiconductor memories offer several advantages over core memories, including smaller size, lower power dissipation, greater speed, and more system design flexibility.

The technology associated with the manufacture of semiconductor memory devices is essentially the same as that associated with the manufacture of other integrated circuits. Semiconductor memories are constructed primarily from bipolar or MOS logic gates and store their information in flip-flops or on capacitors within the memory chip.

A basic static bipolar memory element is shown in Fig. 15-11. As can be seen, this TTL device is basically a bistable multivibrator. To select a particular memory cell, both the X and Y lines must be HIGH. If a binary 1 is stored in the cell, Q_1 is OFF and Q_2 is ON. This produces current flow in the sense 1 line, which is sensed by the sense amplifier. If a binary 0 is stored in the cell, Q_1 is ON and Q_2 is OFF. This produces current flow in the sense 0 line. To store a binary 1 in a cell, the X and Y lines are set HIGH. Since this reverse biases both transistors Q_1 and Q_2, some other means must be used to determine the state of the flip-flop.

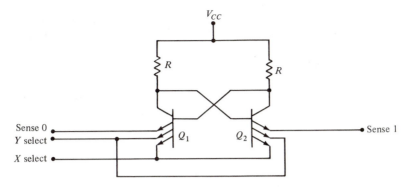

FIGURE 15-11 Basic static bipolar memory element

This is accomplished with the sense lines, which are used to both write (preset) and read (sense) the state of the flip-flop. If the sense 0 line is set HIGH, a 1 is written into the memory cell, whereas if the sense 1 line is set LOW, a 0 is written into the memory cell.

Figure 15-12 shows how 16 bipolar memory cells are connected to form a 4×4 memory plane, or array. A particular cell is addressed by placing a logic 1 on the proper X and Y lines. If a 1 is to be stored in the selected flip-flop, the write 1 line must be HIGH, whereas if a 0 is to be stored, the write 0 line must be HIGH.

To read the contents of any selected cell, the cell is addressed by placing a

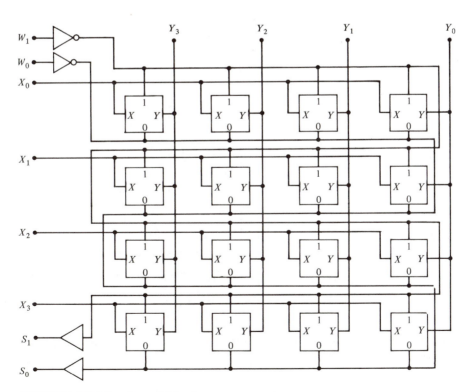

FIGURE 15-12 4×4 bipolar memory array

logic 1 on the proper *X* and *Y* lines. The contents of the addressed flip-flop is determined by the sense lines. If the sense 1 line is LOW, a 1 is stored. Reading the contents of a flip-flop does not affect its contents, so this is called *nondestructive readout* (NDRO).

The memory array shown in Fig. 15-13 is a random-access memory (RAM) system because any cell in the array can be randomly accessed to either read or write information.

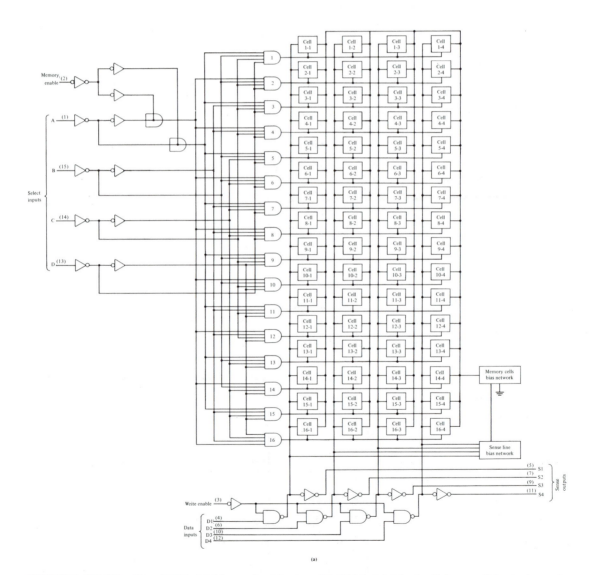

(a)

FIGURE 15-13 The 7489 64-bit static, bipolar, IC RAM: (a) block diagram; (b) pin layout; (c) function table

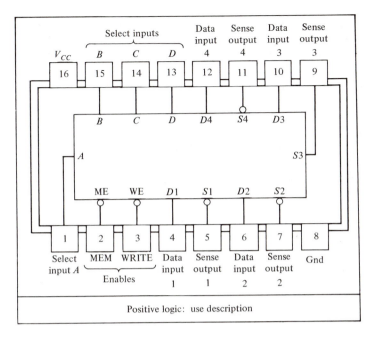

(b)

ME	WE	Operation	Condition of Outputs
L	L	Write	Complement of data inputs
L	H	Read	Complement of selected word
H	L	Inhibit storage	Complement of data inputs
H	H	Do nothing	High

(c)

FIGURE 15-13(b) and (c)

15-6
THE 7489 IC RAM

The 7489 is a 64-bit static bipolar IC RAM. Its functional block diagram, pin layout, and function table are shown in Fig. 15-13. The memory plane is arranged as 16 words of 4 bits each. The inputs to the 7489 are

- Four data input lines.
- Four data output lines.
- Four select lines to select 1 of 16 words.
- Two control lines, Write enable and Memory enable.

The 16 words are selected by placing binary numbers 0000_2 through 1111_2 on the select lines. The following steps are required to store a 4-bit binary word as memory word 16:

1. Set Write enable to logic 0.
2. Set Memory enable to logic 0.
3. Place the binary word to be stored on the data input pins.

4. Place the binary number corresponding to the desired memory location on the select lines.

5. The complement of the binary word applied to the data inputs is stored as memory 16.

The following steps are required to read a word out of memory:

1. Set Memory enable to logic 0.
2. Set Write enable to logic 1.
3. Place the binary number corresponding to the memory location to be read on the select lines.
4. The complement of the word stored in the memory location appears at the sense outputs.

Since the complement of the binary word that was initially applied to the data input pins was stored in memory, when the stored word is read from memory and complemented in the process, the original word appears at the sense outputs.

By observing the function table, we can see that by proper choice of input signal levels we can cause the 7489 to

- Read.

- Write.

- Invert data.

- Do nothing.

The *do nothing* option in effect *deselects* the IC; therefore, other 7489 chips can be placed in parallel with it and be *selected*. This increases the memory capacity by 16 words for each additional IC.

Figure 15-14 shows a 64-word by 4-bit memory fabricated with four 7489s. Since the four RAM chips are in parallel, three must be *deselected* and the fourth must be correctly *selected* and receive the proper Read and Write commands when data are to be stored or retrieved.

EXAMPLE 15-2 ━━━

A certain memory chip has a capacity of 64 words of 4 bits each.

(a) How many select lines are required to permit each word to be addressed?
(b) How many data input and data output pins are on the chip?
(c) How many memory cells are on the chip?

SOLUTION

(a) Since there are 64 different words, there must be 64 unique addresses that can be applied to the select lines. The number of select lines is 6 because $2^6 = 64$, which is the required number of addresses.

(b) There are four data input and four data output lines, since the words contain 4 bits.

(c) The number of memory cells is $4 \times 64 = 256$, since there are four cells per words and there are 64 words.   ❑

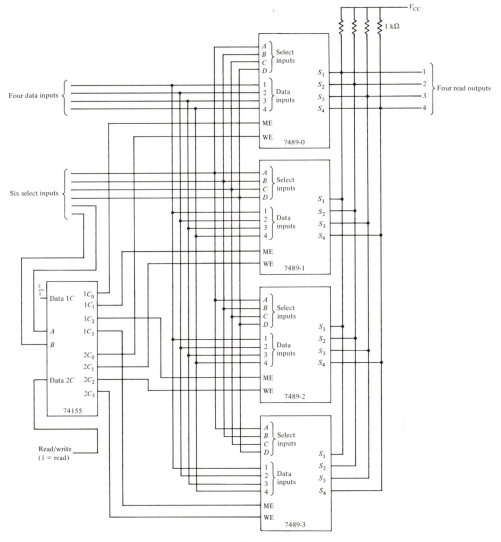

FIGURE 15-14 Parallel connected 7489 RAMs

15-7
READ-ONLY MEMORIES

Not all memory devices are designed so that their contents can be changed. Memory devices whose contents cannot be altered during the normal course of operation are called *read-only memories* (ROMs). ROMs are used to perform code conversions, as look-up tables such as logarithms, trigonometric functions, and exponential functions, and to control special-purpose programs for computers.

Small, simple ROMs can be constructed as a diode matrix or with logic gates and a decoder; however, most ROMs are in the form of integrated circuits. Figure 15-15 displays the family of ROM devices. As can be seen, they are divided into bipolar and MOS devices and then subdivided into mask and programmable ROMs.

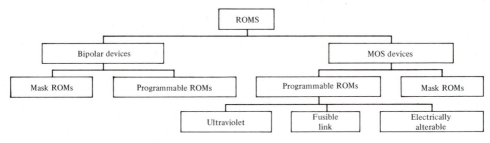

FIGURE 15-15   ROM family

Bipolar ROMs operate at higher speeds, but MOS ROMs have higher bit density and are less expensive. Consequently, MOS ROMs are much more widely used than are their bipolar counterparts.

15-8
BIPOLAR ROMs

Basic bipolar read-only memory cells are simply bipolar transistors, as shown in Fig. 15-16. The drawing shows six cells of a 256-bit ROM connected to form an array of 32 words by 8 bits. As can be seen, the collector of each transistor is

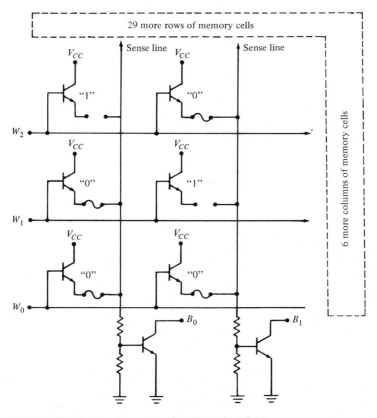

FIGURE 15-16   Six cells of a 256-bit ROM

connected to V_{CC} and a fuse link is placed in series with the emitter of each transistor. The base of each transistor is connected to an address decoder via the *word-select* lines.

ROM cells may be programmed by one of two different techniques. One technique, called *mask programming,* may be performed during the fabrication process. If a 1 is to be stored at a particular memory location, the emitter resistor for that cell is simply not connected.

The second method of ROM programming allows the user to program the device as desired. Such devices are called *field-programmable ROMs* (PROMs). In this type of device, the fuse links are made of some material such as titanium–tungsten or Nichrome and are connected to the emitter during fabrication. The user stores a 1 in a cell by addressing the cell and applying a high current pulse (approximately 30 mA) to it which "blows the fuse," as shown in Fig. 15-16. To read the contents of a particular cell, the cell is addressed by an address decoder that is usually an integral part of the PROM chip. The output of the decoder drives the word-select lines, which select the 8 bits of a single word. If a cell contains a 0, the sense line will be pulled HIGH by the addressed cell. This will forward bias the buffer transistor connected to the affected sense lines, thus turning it ON, which pulls its collector to logic 0. If a 1 is stored in the addressed cell, the sense line will not be affected because the fuse link is open.

15-9
MOS MEMORIES

MOS memory elements offer several advantages over the bipolar memories just discussed. Some of these advantages are

- Lower power dissipation.
- Less expensive per memory element.
- More economical in chip area.
- Fewer steps required in the manufacturing process.

Early MOS memory elements used PMOS technology, but they were difficult to interface with TTL. Recent technological developments have produced NMOS as well as complementary MOS ICs that can be interfaced directly with TTL.

MOS memory elements are classified as either static or dynamic. Static elements are more complex, dissipate more power, and operate at lower speeds. Dynamic elements are simpler in design, occupy less chip area, and dissipate less power at higher speeds than do their static counterparts.

15-10
THE MOS STATIC ELEMENT

The basic MOS static memory element is a flip-flop constructed with MOS field-effect transistors. Early MOS static memory cells consisted of eight MOSFETs; however, a six-transistor static cell, shown in Fig. 15-17, has been developed. As can be seen, transistors Q_1 and Q_2 form a bistable latch with Q_3 and Q_4 functioning as load resistances. Transistors Q_5 and Q_6 provide a means of reading

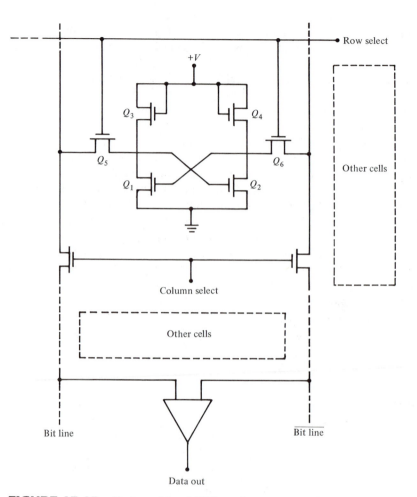

FIGURE 15-17   Six-transistor MOS static memory cell

or writing data from the BIT or $\overline{\text{BIT}}$ lines when the cell is selected by the row and column select lines.

15-11
INTEGRATED-CIRCUIT MOS STATIC RAMs

Very significant technical advancements have been made recently in the area of semiconductor memory devices. As a result, there is an impressive array of IC memory devices, including MOS static RAMs, commercially available. The following paragraphs briefly describe two such devices.

15-11.1   The 2114 Static RAM

The 2114 is a 4096-bit static RAM configured as a 1024-word by 4-bit device and organized in a 64-row by 64-column array. The device is fabricated using n-channel MOS technology. The 2114 is manufactured by several companies, including National Semiconductor. The pin diagram, logic symbol, and block diagram for the 2114 are shown in Fig. 15-18.

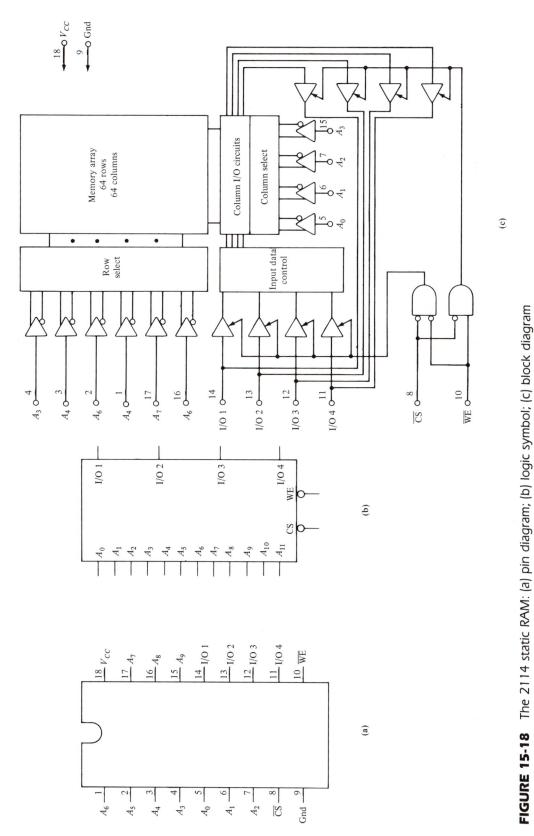

FIGURE 15-18 The 2114 static RAM: (a) pin diagram; (b) logic symbol; (c) block diagram

Some of the features of the 2114 are

- All internal circuits are fully static; therefore, no clock is required.
- All inputs and outputs are directly TTL compatible.
- The data are read out nondestructively.
- The device requires a single 5 V power supply.

15-11.2 The 2147 Static RAM

The 2147 is, as with the 2114, a 4096-bit static RAM. However, the 2147 is configured as a 4096-word by 1-bit memory device rather than as a 1024-word by 4-bit device. In all other respects the two devices are virtually identical with the exception of the number of address lines. The 2114 requires 10 lines to address 1024 4-bit words, whereas the 2147 requires 12 lines to address 4096 1-bit words.

15-12
THE MOS DYNAMIC ELEMENT

The number of static cells per IC RAM package is limited by the power dissipation per cell. Power dissipation per cell can be significantly reduced by using *dynamic* MOS RAM cells. This method requires that the device be clocked to *refresh* the memory.

Dynamic MOS RAM cells were originally designed with four transistors, but improved design techniques reduced the number of transistors per cell to three and finally to a single transistor. Most dynamic MOS RAMs of recent design use either the three-transistor or the one-transistor cell design, shown in Fig. 15-19. The three-transistor cell shown in Fig. 15-19a relies on the stored charge of capacitor C, which is actually the gate capacitance of transistor Q_2. The device has four control lines designated as *read data*, *write data*, *read select*, and *write*

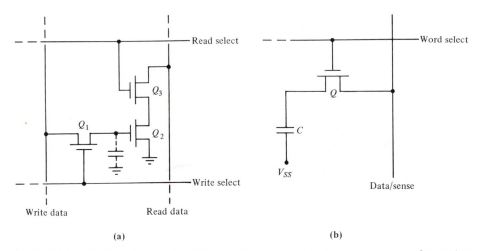

(a)   (b)

FIGURE 15-19 Dynamic MOS RAM cell: (a) 3-transistor configuration; (b) 1-transistor configuration

select. These lines are at 0 V potential except during *read* and *write* operations. The following steps are involved in the *read* and *write* operations.

1. To write, the *write-select* line must be at a negative potential to turn ON transistor Q_1.
   a. To write a 0 in the cell, the *write-data* line is set to a negative potential and the *read-select* line is held at 0 V. This charges capacitor C to a negative potential.
   b. To write a 1 in the cell, the *write-data* line is set to ground potential. This reduces any charge on the capacitor to zero and forces the voltage at the gate of Q_2 to ground potential.
2. To read, the *read-select* line must be at a negative potential to turn ON transistor Q_3. The inverted voltage at the gate of Q_2 is now present on the output *read-data* line.

Since transistor Q_2 inverts the signal that exists on its gate, another inverter is needed, external to the dynamic cell, to reinvert the signal. Also, since capacitor C is part of Q_2, there is some leakage current through the resistance associated with the gate of Q_2 that causes the voltage across C to decay; therefore, it is necessary to periodically refresh the charge on C.

The one-transistor dynamic MOS RAM cell shown in Fig. 15-19b probably represents the ultimate in the design of dynamic MOS memory cells. In this type of cell, the transistor Q simply acts as a switch. To write, the word-select line is activated and C is charge via transistor Q_1 to the level on the data/sense line. To read, the word-select line is again activated and charge is transferred from the capacitor to the data/sense line.

15-13
INTEGRATED-CIRCUIT DYNAMIC MOS RAMs

Because of the low-power dissipation associated with dynamic MOS memory cells, IC MOS RAMs using dynamic memory cells are available with at least 64K bits of memory capacity. A great deal of recent developmental work has produced a wide range of dynamic RAMs, and the product line is certain to continue to expand. The following paragraphs describe two such devices.

15-13.1   The 2118 Dynamic RAM

The 2118 is a 16,384-word by 1-bit dynamic MOS RAM. The device is fabricated using high-performance HMOS technology and is designed to operate from a single +5 V power supply.

The 2118 uses the one-transistor dynamic storage cell shown in Fig. 15-19b and requires refreshing every 2 ms. The device is packaged as a standard 16-pin DIP integrated circuit. The 2118, or an equivalent device, is manufactured by several companies. Figure 15-20 shows the pin configuration, logic symbol, and block diagram for the Intel 2118.

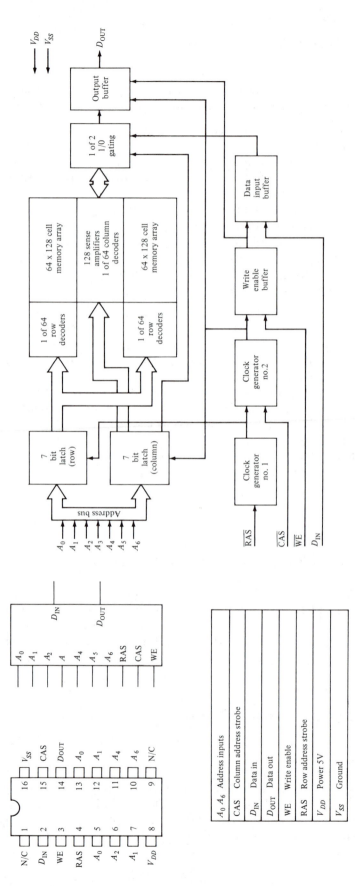

FIGURE 15-20 The 2118 dynamic MOS RAM: (a) pin layout; (b) logic symbol; (c) block diagram

15-14
EPROMs

Both mask and programmable ROMs have one very serious limitation in that a single incorrect bit renders the entire memory IC useless. This limitation can be overcome by using EPROMs, which are read-only memories that can be erased and reprogrammed many times.

The most popular EPROMs use MOSFET transistors. Any stored data can be erased by subjecting the IC chip to intense ultraviolet (UV) radiation through a transparent lid. The basic principle of operation of UV-EPROMs involves the storage of charge by trapping it at the gate of an enhancement-type MOSFET. Figure 15-21 shows the basic structure of a UV-EPROM cell. As can be seen, the cell consists of a MOSFET with an additional floating gate.

The cell is programmed by injecting high-energy electrons onto the floating gate. Once there, the charge remains indefinitely, since there are no electrical connections to the gate. After all the cells of a memory chip are programmed, the IC functions as a read-only memory device. To reprogram a UV-EPROM, it must be removed from the circuit and exposed to intense ultraviolet light of the correct wavelength and energy for a sufficient period of time to erase the contents. This typically requires 10 minutes or longer.

One popular UV-EPROM is the 2716, for which the pin configuration and block diagram are shown in Fig. 15-22. The memory capacity is 16K, arranged as 2K words by 8 bits.

The 2716 has six modes of operation, which are listed in Table 15-1. It should be noted that the inputs for all modes are TTL levels. The required power supplies are a $+5$ V V_{CC} and a V_{PP}. The V_{PP} supply output must be 25 V during the three programming modes and must be 5 V when in the other three modes.

Programmers for the 2716 and similar EPROMs are available from a number of manufacturers. A typical EPROM programmer, which also contains a source of ultraviolet light for erasure, is shown in Fig. 15-23. Using this type of programmer, we find that the total programming time for the 2716 is only 100 s.

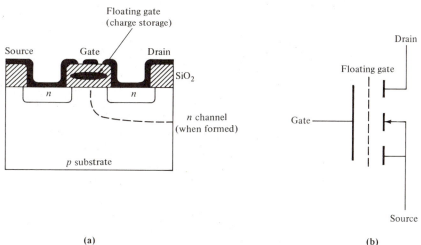

(a)   (b)

FIGURE 15-21   (a) Basic structure of a UV-EPROM cell; (b) logic symbol

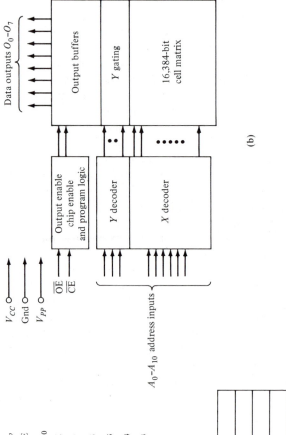

(b)

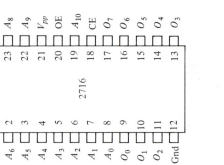

(a)

FIGURE 15-22 The 2716 UV-EPROM: (a) pin connection; (b) block diagram

TABLE 15-1 / Mode Selection for the 2716 UV-EPROM

Mode \ Pins	$\overline{CE}$	$\overline{QE}$	V_{PP}	V_{CC}	Outputs
Read	V_{IL}	V_{IL}	+5	+5	D_{Out}
Output disable	V_{IL}	V_{IH}	+5	+5	High Z
Standby	V_{IH}	X	+5	+5	High Z
Program	Pulsed V_{IL} or V_{IH}	V_{IH}	+25	+5	D_{In}
Verify	V_{IL}	V_{IL}	+25	+5	D_{Out}
Program inhibit	V_{IL}	V_{IH}	+25	+5	High Z

FIGURE 15-23 Typical EPROM programmer

15-15
EEPROMs

Electrically eraseable, programmable, read-only memories (EEPROMs) are the next technological steps in PROM technology beyond EPROMs. They use the same transistor structure as a storage cell that is used in EPROMs. The difference in the cell structures is the addition of a tunnel diode region above the drain of the floating-gate transistor, which allows the charge to move bidirectionally under the influence of an electric field.

EEPROM technology offers several significant benefits over EPROM technology. The charge transport mechanism has a very low current requirement, so only a modest power supply is required to program an EEPROM cell. Since only very low current is required, the necessary high-voltage programming pulses can be produced with the normal system power supplies. Therefore, programming, erasure, and reprogramming can be accomplished with the EEPROM in circuit and without the need for a PROM programmer unit.

Other advantages of EEPROMs include the ability to erase and reprogram individual bytes within an array and the ability to erase the memory contents very rapidly. The total contents of an EEPROM can be erased with one 10-ms pulse, compared to approximately 30 minutes required to erase the contents of an EPROM.

The 2815 is a typical EEPROM. Its pin configuration and functional block diagram are shown in Fig. 15-24. The 2815 has six modes of operation. All operational modes are designed to provide maximum microprocessor compatibility. All control input signals are TTL compatible with the exception of chip erase, which requires a 9- to 15-V signal at pin 20.

15-16
CHARGE-COUPLED DEVICES

Charge-coupled devices (CCDs) use a high-voltage *n*-channel silicon gate MOS fabrication process to store data as packets of electric charge that circulates in a shift register type of circuit. CCDs are serial storage devices made up of a linear array of closely spaced MOS capacitors and gates. Beneath the gates are *potential wells* formed by the closely spaced capacitors. The potential wells are the storage elements, or cells, of a CCD.

The basic structure of a single CCD cell is shown in Fig. 15-25 on p. 430. The substrate is *p*-doped silicon, the insulating layer is silicon oxide, and the gate is aluminum. When a positive voltage is applied to the gate, a potential well is formed beneath it, as delineated by the dashed line. Negative charge in the form of electrons is stored in the potential well. Typical commercial CCDs are the Fairchild 464, a 64K-bit device; and the Intel 2416, a 16K-bit device.

15-17
MAGNETIC BUBBLE MEMORY

Magnetic bubble memory (MBM) is a solid-state technology with high reliability, ruggedness, small size, light weight, and limited power dissipation. With this type of device, data are stored in an array of small magnetic *bubbles* that are embedded in a thin film of magnetic material. The presence of a bubble is interpreted as a binary 1, whereas the absence of a bubble is a 0. Bubbles are created from electrical signals and are reconverted to electrical signals by internal circuitry. Externally, magnetic bubble memory is TTL compatible.

During the manufacturing process of MBMs, a magnetic garnet film is deposited on a nonmagnetic garnet substrate. When the magnetic garnet film is viewed under a polarized light through a microscope, an irregular pattern of wavy magnetic strips can be seen. Part of these magnetic strips have their positive pole on the side of the garnet film nearest the microscope, while the remaining strips are of

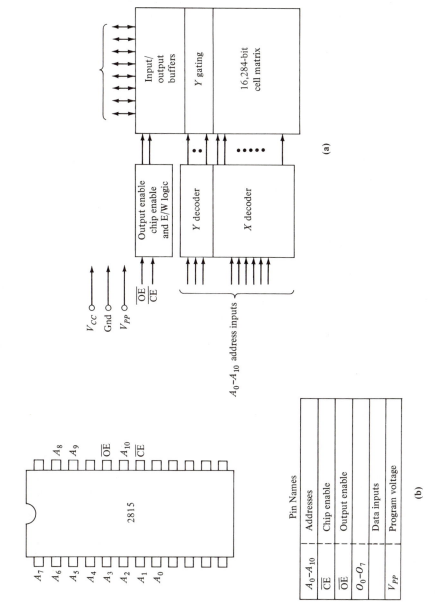

FIGURE 15-24 The 2815 EAROM: (a) block diagram; (b) pin configuration

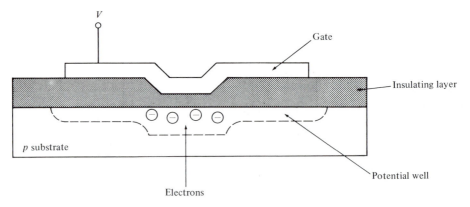

FIGURE 15-25   *Basic structure of a single CCD cell*

the opposite polarity. As a result, one set of magnetic strips appears bright and the other set dark when viewed under a polarized light. This light–dark pattern is shown in Fig. 15-26.

When an external magnetic field, called a *bias field,* is applied perpendicular to the garnet substrate and is slowly increased in strength, the wavy magnetic strips whose magnetization is at right angles to the external field begin to decrease in width as shown in Fig. 15-26b. If the external field is made sufficiently strong, the size of the magnetic strips decreases until their length and width are approximately equal. Each magnetic region is now circular, magnetized at right angles to the external field and immersed in a larger magnetic domain magnetized in the opposite direction. The small circular magnetic regions are the bubbles and are shown in Fig. 15-26c. If the external magnetic field is made sufficiently strong, all the bubbles will shrink and then disappear.

The bubbles will move if an additional *rotating* magnetic field parallel to the plane of the film is applied. The necessary rotating field can be produced by wrapping two coils that are perpendicular to each other around the magnetic film and applying sine waves that are 90° out of phase to each coil. The bubble transfer rate is approximately 100K bits per second.

The primary advantage of magnetic bubble memories is that they are nonvolatile. MBMs share this feature with semiconductor read-only memories, but unlike these devices, MBMs can have data written into them at any time, at speeds comparable to those at which the data are read. Unlike disk memories, bubble memories are quiet and very reliable, since there are no moving parts. Also they are compact, dissipate very little power, and have very high storage capacity. With a million or more bits per device, a bubble memory can store 16 to 64 times as much data as can be stored by semiconductor memories occupying the same space.

15-18
MAGNETIC-SURFACE STORAGE DEVICES

In addition to magnetic core memory, semiconductor memory, and magnetic bubble memory, there are several other types of magnetic memories in use, including disks, tapes, and drums. Each of these rely on a magnetic surface moving past a read/write head to store or retrieve data.

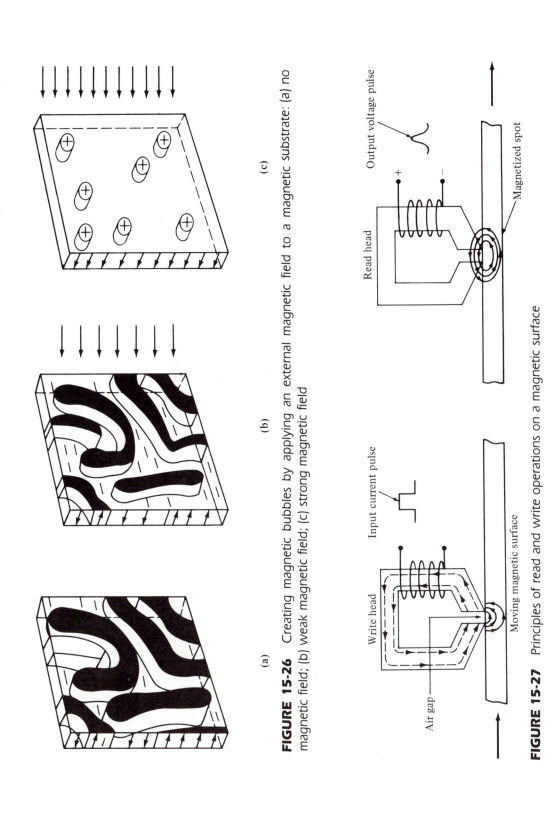

FIGURE 15-26 Creating magnetic bubbles by applying an external magnetic field to a magnetic substrate: (a) no magnetic field; (b) weak magnetic field; (c) strong magnetic field

FIGURE 15-27 Principles of read and write operations on a magnetic surface

A basic diagram illustrating the principles of the read and write operations is shown in Fig. 15-27. A binary bit is written on the magnetic surface by magnetizing a small segment of the surface as it moves past the *write head*. The direction of the magnetic flux lines across the air gap of the write head is determined by the direction of the current pulse in the windings, as shown in Fig. 15-27a. The magnetic flux magnetizes a small spot on the magnetic surface which is in the same direction as the flux field. The polarity of the magnetized spot determines whether it represents a binary 1 or a binary 0. Once a spot on the magnetic surface is magnetized, the data that the magnetic spot represents remain until it is written over with a magnetic field of opposite polarity.

When a magnetic surface passes a *read head*, each magnetized spot produces a magnetic field across the air gap of the read head, as shown in Fig. 15-27b. This in turn induces voltage pulses in the windings, where the polarity of the pulse is dependent on the direction of the magnetized spot.

There are several ways that digital data can be represented when doing magnetic-surface recording. Among these are techniques called *return-to-zero* (RZ), *nonreturn-to-zero* (NRZ), *biphase,* and *Manchester.*

15-18.1 Floppy Disks

Floppy disks, or diskettes, are small flexible Mylar disks with a magnetic surface. They are permanently encased in a square protective cover, as shown in Fig. 15-28. The two most widely used sizes are $5\frac{1}{4}$-inch disks for home systems and 8-inch disks for business applications and for use with minicomputers. Very recently, a $3\frac{1}{2}$-inch disk was introduced.

Floppy disks are an important mass storage device. Their low cost and direct access make them an ideal choice for use in microcomputer installations. They are very competitive with magnetic-tape storage systems in terms of cost per bit, ease of handling, reliability, and maintainability, and they are far superior to magnetic tape with regard to access time.

The surface of a floppy disk is coated with a thin magnetic film in which binary data are stored in the form of tiny magnetized regions. The protective cover is designed with several cutout areas for the read/write head, drive spindle, and

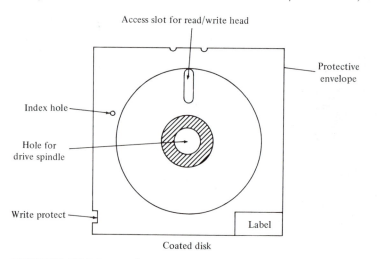

FIGURE 15-28   A $5\frac{1}{4}$-inch floppy disk in a protective cover

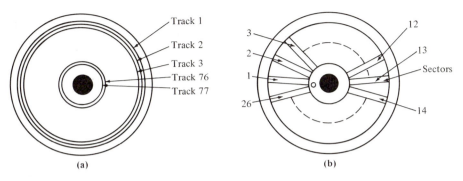

FIGURE 15-29 Track and sector organization of floppy disks

FIGURE 15-30 Typical sector format for one track of a floppy disk

index hole. The index hole establishes a reference point for all the tracks on the disk. As the disk rotates at 360 rpm within the stationary protective cover, the read/write head makes contact with the disk through the access window.

Regular 8-inch floppy disks are divided into 77 tracks numbered 0 through 76. The disk is divided into 26 sectors so that each of the 77 tracks is also divided into 26 equal-sized sectors, as shown in Fig. 15-29. Since the sectors are all equal in size, there is more unused space between sectors on the outside track than on the shorter tracks near the center. Each sector can store 128 bytes of data; therefore, the total storage capacity of an 8-inch disk is

$$\text{capacity} = \left(\frac{128 \text{ bytes}}{\text{sector}}\right)\left(\frac{26 \text{ sectors}}{\text{track}}\right)\left(\frac{77 \text{ tracks}}{1}\right)$$

$$= 256{,}256 \text{ bytes}$$

There are 35 tracks on $5\frac{1}{4}$-inch disks, sometimes referred to as minifloppies. The smaller disks use two formatting techniques, *hard sectoring* and *soft sectoring*. With hard-sectored disks, two formats are used: one with 16 sectors containing 128 bytes each and one with 10 sectors containing 256 bytes each. The total capacity of the 16-sector disk is

$$\text{capacity}_{(16 \text{ sect})} = \left(\frac{128 \text{ bytes}}{\text{sector}}\right)\left(\frac{16 \text{ sectors}}{\text{track}}\right)\left(\frac{35 \text{ tracks}}{1}\right) = 71{,}680 \text{ bytes}$$

The total capacity of the 10-sector disk is

$$\text{capacity}_{(10 \text{ sect})} = \left(\frac{256 \text{ bytes}}{\text{sector}}\right)\left(\frac{10 \text{ sectors}}{\text{track}}\right)\left(\frac{35 \text{ tracks}}{1}\right) = 89{,}600 \text{ bytes}$$

With soft-sectored minifloppies, the number of sectors is left to the manufacturer.

A typical sector format for a floppy disk is shown in Fig. 15-30. As can be seen, each sector is divided into fields. When in use, the address mark passes the read/write head first and identifies the next area of the sector as the ID field. The ID field identifies the data field by sector and track number, and the data mark indicates whether the upcoming data field contains a current, or active, record or a deleted record. The data field is the part of the sector that contains the stored data.

The average access time to a given sector of a floppy disk is about 300 ms. This is much slower than the access time for semiconductor memories but much faster than for magnetic tape.

15-18.2   Rigid Disks

Rigid disks, shown in Fig. 15-31a, are somewhat like phonograph records except that their surface is magnetically coated. They are generally larger in diameter and are much more rigid than floppy disks. Rigid disks are generally mounted in pack form, one above another on the same spindle. Stacks of these disks are used to provide mass storage for large computer systems.

Rigid disk, or magnetic disk, memory resembles a coin-operated automatic record player, or "jukebox," as can be seen in Fig. 15-31b. Data are recorded on concentric tracks in the same manner as with floppy disks. There is generally a read/write head for the upper and lower surface of each disk. The heads move in and out in the space between the disks as they rotate at speeds up to 6000 rpm.

A typical rigid disk memory may contain 32 disks with 256 tracks on the top surface and 256 more tracks on the bottom surface of each disk. The bytes per track may be on the order of 3000. The storage capacity of this memory system is

$$\text{capacity} = \left(\frac{3000 \text{ bytes}}{\text{track}} \right) \left(\frac{512 \text{ tracks}}{\text{disk}} \right) \left(\frac{32 \text{ disks}}{1} \right)$$

$$= 49,152,000 \text{ bytes}$$

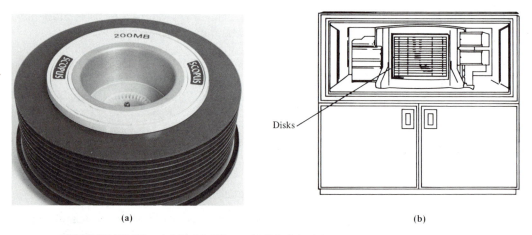

(a)   (b)

FIGURE 15-31   (a) Rigid disk pack; (b) disk drive system

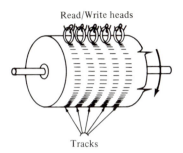

Read/Write heads

Tracks

FIGURE 15-32   Magnetic drum showing read/write heads and tracks

15-18.3 Magnetic Drums

A magnetic drum is a rotating cylinder whose surface is coated with a magnetic oxide. Magnetic drums were one of the first devices to provide a relatively inexpensive means of storing information while providing reasonably short access time. Data are stored on the surface of the drum in many tracks, as illustrated in Fig. 15-32. There is a read/write head for each track; therefore, as the drum rotates data can be read from each track simultaneously.

Memory capacity varies from less than 25,000 bits for drums with 15 to 25 tracks to approximately 10^8 bits for drums with 500 to 1000 tracks. Large drums may rotate as slowly as 120 rpm, and small drums rotate at speeds in excess of 10,000 rpm. The number of bits stored per inch of track, called the *packing density*, is inversely related to the rotating speed of the drum. Typical packing density for present-day drums is on the order of 200 to 300 bits per inch.

15-18.4 Magnetic Tape

Magnetic tape is a plastic tape coated with a magnetic oxide. The most common tape used in computer mass storage systems is about 2400 feet long and $\frac{1}{2}$ inch wide. Data are recorded on the tape in nine parallel tracks. Seven of the tracks are used to record 7 bits in ASCII code. One of the two remaining bits is used as a parity check and the other is generally used for timing purposes.

There is a separate read/write head for each track, as shown in Fig. 15-33a. Tracks are divided into records with gaps between records for starting and stopping of the tape. A typical format for a record is shown in Fig. 15-33b.

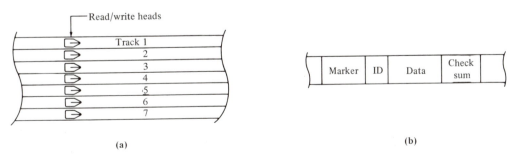

(a)

(b)

FIGURE 15-33   (a) Magnetic tape with read/write head for each track; (b) typical recording format for magnetic tape

Densities of 200, 556, 800, and 1600 bits per inch per track are used. Tape speeds vary from 50 to 200 inches per second. For storing smaller quantities of data, cassette tapes are sometimes used.

EXAMPLE 15-3 ━━━━━━━━━━━━━━━━━━━━━━━━━━━━━━━━━━━━━━

A magnetic tape with seven tracks of data has a recording density of 556 bits per inch, is 2400 feet long, and tape speed is 100 inches per second. Determine the following:

(a) Storage capacity of the tape
(b) Average access time
(c) Maximum access time
(d) Data rate (bits per second)

SOLUTION

(a) $\text{capacity} = \left(\dfrac{\text{bits}}{\text{in.}}\right)\left(\dfrac{\text{in.}}{\text{ft}}\right)\left(\dfrac{\text{total ft}}{1}\right)\left(\dfrac{\text{no. tracks}}{1}\right)$

$= \left(\dfrac{556 \text{ bits}}{\text{in.}}\right)\left(\dfrac{12 \text{ in.}}{\text{ft}}\right)\left(\dfrac{2400 \text{ ft}}{1}\right)\left(\dfrac{7 \text{ tracks}}{1}\right) = 1.12 \times 10^8 \text{ bits}$

(b) The average access time is the time to reach half the tape length:

$$\text{half tape length} = \left(\dfrac{2400 \text{ ft}}{2}\right)\left(\dfrac{12 \text{ in.}}{\text{ft}}\right) = 14{,}400 \text{ in.}$$

$$\text{average access time} = \dfrac{\text{distance}}{\text{in./s}} = \dfrac{14{,}400 \text{ in.}}{100 \text{ in./s}} = 144 \text{ s}$$

(c) The maximum access time is the time to reach the opposite end of the tape:

$$\text{maximum access time} = \left(\dfrac{\text{length}}{1}\right)\left(\dfrac{\text{s}}{100 \text{ in.}}\right)$$

$$= \left(\dfrac{28{,}800 \text{ in.}}{1}\right)\left(\dfrac{\text{s}}{100 \text{ in.}}\right) = 288 \text{ s}$$

(d) The data rate is

$$\text{data rate} = \dfrac{\text{total bits}}{\text{maximum access time}}$$

$$= \dfrac{1.12 \times 18^8 \text{ bits}}{288 \text{ s}} = 389{,}200 \text{ bits per second}$$ ❑

15-19
PROGRAMMABLE ARRAY LOGIC

Programmable logic devices consist of an array of logic gates that can be programmed to implement an almost unlimited number of logic designs. Programmable logic devices are manufactured in three basic architectures:

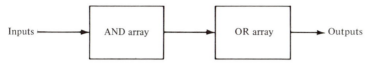

FIGURE 15-34   Basic architecture of programmable logic devices

- Programmable read-only memory (PROM).
- Programmable logic arrays (PLAs).
- Programmable array logic (PALs).

All three architectures share the same basic internal AND–OR structure, which is shown in Fig. 15-34, but vary in allocation of logic features and amount of programmability.

Our interest in this section focuses on PALs. Programmable array logic devices are two-level combinational logic networks. The first logic level consists of inverters and programmable AND gates, and the second level consists of nonprogrammable OR gates. The AND gates are programmable to the extent that their AND function can be varied by burning fuses on the appropriate AND-gate inputs. By constructing PALs with AND gates followed by OR gates, we generate product terms with the AND gates and sum terms with the OR gates. Therefore, PAL output equations are in the sum-of-products form.

Programmable array logic offers some very attractive advantages compared with fixed-function logic devices. One of the major advantages is the reduction in the number of IC chips required in a digital circuit because chip count is the most significant factor in reducing costs associated with manufacturing a product. By reducing chip count, inventory is reduced, incoming inspections and test are minimized, and printed-circuit-board failures are less likely because the boards are simpler and have fewer connections.

One measure of the efficiency of a programmable logic device is an expression of how many fixed-function devices it can replace. In existing circuits, a single programmable device can often replace four to six 7400 series IC chips. However, when designing new circuits incorporating programmable logic devices, a single programmable logic chip can eliminate the need for 10 or more fixed-function IC chips.

The circuit shown in Fig. 15-35 depicts a basic PAL array. As can be seen, each of the inputs to the AND gates is fused. This permits one to produce the desired product terms by burning the fuse on the unwanted inputs.

Producing a completely functional IC from an unprogrammed logic array chip incorporates three phases:

- *Logic design:* defining the logic functions that the circuit must perform.
- *Programming:* writing those logic functions into the unprogrammed device.
- *Testing:* ensuring that the programmed logic array functions exactly as specified.

The logic design phase consists of defining the required logic functions, expressing them in a convenient form, and converting them to a *binary fuse map*. A binary fuse map is a representation of the states of fuses in a programmable IC.

Figure 15-36 depicts the rules associated with the notation commonly used

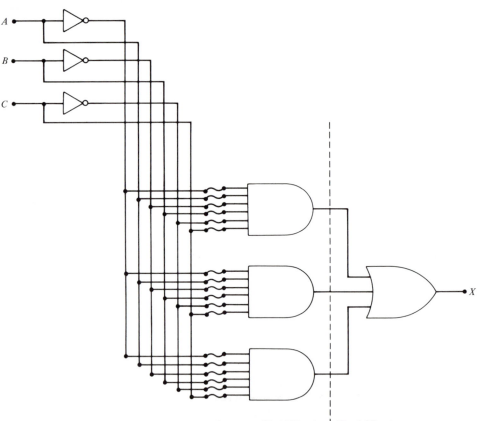

FIGURE 15-35 Programmable array logic device

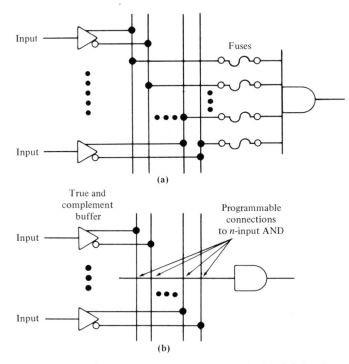

FIGURE 15-36 Programmable array logic: (a) Logic
diagram notation; (b) logic equivalent

438

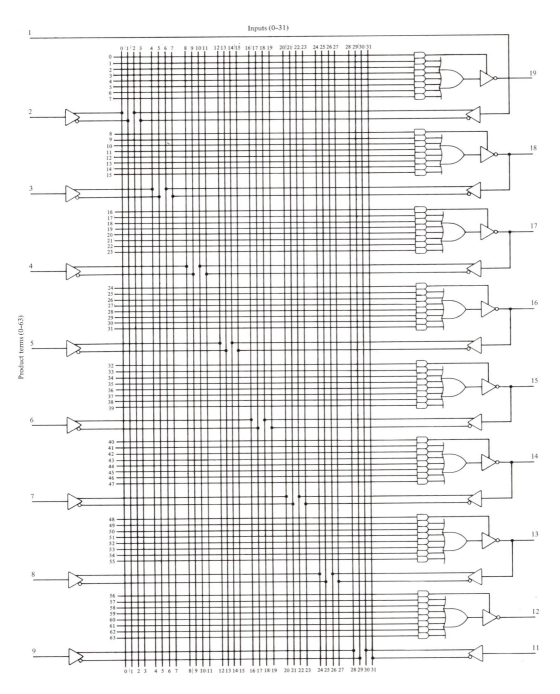

FIGURE 15-37 PAL 16L8 logic diagram

in logic diagrams to describe programmable array logic. As can be seen in Fig. 15-36a, all array inputs are shown connecting to a single-input AND gate. In reality, each array input is an input to an AND gate, as shown in Fig. 15-36b. Since each variable and its complement are applied to each AND gate, a device with n inputs will have AND gates with $2n$ inputs.

Figure 15-37 on p. 439 shows the logic diagram for a 16L8 PAL, which is an industry standard. The grid pattern between the input terminals on the left and the AND gates is the same programmable array notation mentioned in association with Fig. 15-35. The 20-pin device has 10 inputs, 2 outputs, and 6 bidirectional terminals, which may be used as either inputs or outputs. If all 6 bidirectional terminals are used as inputs, there will be 16 inputs; therefore, each of the 64 AND gates shown in Fig. 15-37 must have 32 inputs.

15-20
SUMMARY

Memory is that portion of many digital systems where information is stored. The need for, and use of, memory is most easily associated with computer systems where memories are often categorized as working memories or as auxiliary memories. Working memories are an integral part of a computer's central processing unit, whereas auxiliary memories are peripheral devices.

Working memories are directly accessible to achieve fast access time. Both core memory and semiconductor memory are used, but core memory is rapidly giving way to semiconductor memories fabricated using either bipolar transistors or MOSFETs. Bipolar memories are static devices, whereas MOS memories can be either static or dynamic. MOS memory ICs are available in the form of RAMs, ROMs, PROMs, EPROMs, EEPROMs, and CCDs.

An additional solid-state memory device is the magnetic bubble memory IC. MBMs are very rugged, small in size, and lightweight, and dissipate limited amounts of power. Of greater importance is the fact that they are nonvolatile and have very high storage capacity.

Auxiliary memories consist primarily of magnetic-surface devices, including floppy disks, rigid disks, magnetic drums, and magnetic tape. Although these systems have much slower access times than working memories, they provide very large storage capacities at a lower cost per bit than is presently possible with working memories.

PROBLEMS

1. Briefly define the following terms:
   (a) Volatile memory
   (b) Byte
   (c) Memory cell
   (d) Destructive readout
   (e) Access time

2. How many words storage capacity does a memory system described as a 64K memory actually have?

3. If a core memory system has a storage capacity of 32,768 bits, what is the word capacity expressed in K if each word contains 8 bits?

4. What do the following acronyms stand for?
  (a) RAM
  (b) ROM
  (c) PROM
  (d) CCD
  (e) EAPROM

5. How many memory cells are there in a 16K-word memory system with 16-bit words?

6. In a 64×64 core memory plane, how many input lines and how many output lines must the X-line and Y-line decoders have? Describe the decoder as an n-line-to-m-line decoder, where m and n are the number of input and output lines.

7. The core memory system of a certain computer is capable of storing 1024 8-bit words.
  (a) How many planes does the memory system contain?
  (b) How many cores does the memory system contain?
  (c) How many X lines and Y lines are required?
  (d) How many sense lines are required?
  (e) How many inhibit lines are required?

8. What size address register is required for the memory system in Problem 7?

9. If a threshold magnetizing current of 60 mA is required to write a binary 1 into a core, what magnitude of current through the inhibit line is necessary to prevent the 1 from being stored?

10. If an address register used with core memory has a 16-bit capacity, how many cores are in a plane?

11. In a core memory system with a 12-bit address register, 16 sense lines, and 16 inhibit lines, how many bytes can be stored?

12. What is the primary disadvantage of semiconductor RAMS compared to ROMs?

13. A certain RAM has a memory capacity of 256 8-bit words. How many select lines are required for addressing? How many data input and data output lines are required.

14. The binary number 1001 is to be stored as the third word in memory in a 7489 RAM. List the necessary signals and their logic levels to accomplish this.

15. If the application of binary number 1100 to the select lines of a 7489 RAM in the process of reading a word from memory caused the number 1010 to appear at the sense outputs, what number was stored in the memory location addressed?

16. What is the least number of transistors that can be used to fabricate an MOS dynamic RAM cell?

17. Compare the erase time for EPROMs and EEPROMs.

18. Compare the storage capabilities of magnetic bubble memories and semiconductor memories.

19. What are the three major magnetic memory systems used for mass storage?

20. If a seven-track magnetic tape has a storage capacity of 1 million bits and a recording density of 200 bits per inch, how many feet long is the tape?

21. Determine which of the following magnetic tapes has the shorter maximum access time: tape A, which is 2000 feet long and has a tape speed of 80 inches per second; or tape B, which is 2400 feet long and has a tape speed of 100 inches per second.

22. What type of magnetic material is normally used in bubble memories?

23. Describe the storage cell of a charge-coupled device.

24. Determine the storage capacity of a magnetic disk storage system.

25. If the symbol + represents programmable connections and the symbol + represents fixed connections, redraw the PAL array architecture in its logic equivalent and show what fuse links must be blown to program the array to provide the Boolean expressions shown.

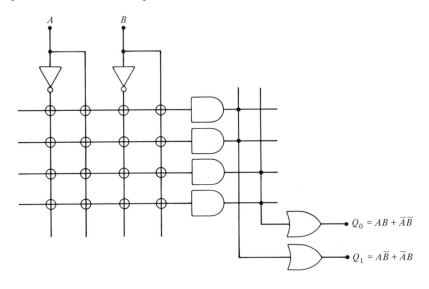

$$Q_0 = AB + \overline{A}\,\overline{B}$$

$$Q_1 = A\overline{B} + \overline{A}B$$

REFERENCES

Bipolar Memory Data Book, Mountain View, Calif.: Fairchild Camera and Instrument Corp., 1983.

Field-Programmable Logic Data Book. Dallas, Tex.: Texas Instruments, Inc., 1983.

Memory Components Handbook. Santa Clara, Calif.: Intel Corp., 1983.

Memory Databook. Santa Clara, Calif.: National Semiconductor Corp., 1980.

PAL Databook. Santa Clara, Calif.: National Semiconductor Corp., 1982.

Programmable Array Logic. Sunnyvale, Calif.: Advanced Micro Devices, 1983.

''Special Report on Semiconductor Memories,'' *Computer Design,* June 1983, pp. 153–196.

Triebel, W. A., and A. E. Chu, *Handbook of Semiconductor and Bubble Memories*. Englewood Cliffs, N.J.: Prentice-Hall, Inc., 1982.

16

Microprocessors and Microcomputers

INSTRUCTIONAL OBJECTIVES

This chapter concludes your study of digital logic and brings together many of the topics discussed to this point. In this chapter you are introduced to microprocessors and microcomputers in preparation for a follow-on course dealing with these systems. After completing the chapter, you should be able to:

1. List the three categories of digital computers.
2. List five factors that are used to determine in which category a particular computer belongs.
3. Define or describe a microprocessor.
4. List five functions that microprocessors perform as part of their normal operation.
5. List and describe the subsystems of a microcomputer.
6. List and describe the subsystems of a microprocessor.
7. List and describe the buses in a microcomputer.
8. List and describe the three most common registers in a microprocessor.
9. Describe why the program counter is often divided into a PCH and a PCL.
10. List and describe the register that uses LIFO-type memory.
11. List and describe the three most common flags used to indicate the status of microprocessor conditions.
12. List the four basic arithmetic operations performed by the ALU.
13. List the three logical operations typically performed by ALUs.
14. Describe the purpose of interrupt signals.

SELF-EVALUATION QUESTIONS

The following questions relate to the material presented in this chapter. Read the questions prior to studying the chapter and, as you read through the material, watch for answers to the questions. After completing the chapter, return to this section and evaluate your comprehension of the material by answering the questions again.

1. What are five factors used to determine in which category of computer a particular computer belongs?
2. Draw a block diagram of a microprocessor and label the major subsystems.

3. Draw a block diagram of a microcomputer and label the major subsystems and buses.
4. What development brought about the microprocessor?
5. What are the three most frequently used registers in a microprocessor?
6. What function does the microprocessor serve in a microcomputer?
7. What are the three most common flags used in a microprocessor, and what is their purpose?
8. Describe the sequence of events that occurs when an interrupt signal is received by a microprocessor.
9. List five widely used microprocessor chips by manufacturer and part number.
10. Describe the purpose of nonmaskable interrupt signals.

16-1
INTRODUCTION

Although there is a tremendous range of applications for digital circuitry, all of them are overshadowed by digital computers, which represents the principal application. Today's digital computers fall into the following three categories:

- Large mainframe computers.
- Minicomputers.
- Microcomputers.

Some of the factors used to determine in which of these categories a computer belongs are physical size, cost, speed, memory capacity, computing power, and application.

Mainframe computers are the largest, fastest, and most powerful systems and can therefore handle large numbers of instructions and process tremendous amounts of data at very high speeds. Until the mid-1960s, most computers fell into this category. In 1965, smaller, slower computers designed for specific rather than general applications were introduced. These computers, which became known as *minicomputers*, were used primarily in laboratories and for control in processing industries.

By the late 1960s the use of computers had grown to the point that there was a tremendous market for small computer systems in small businesss. To meet this need, Intel introduced the computerlike 8080 chip in 1969, and the *microcomputer* era was born. This type of IC chip is now referred to as a *microprocessor* (μP) and is the heart of a microcomputer (μC). Our primary interest in this chapter relates to microprocessors and microcomputers.

The Microprocessor. The microprocessor is a single IC chip containing the arithmetic and control circuitry for a computer. Early units were designed to handle 4 bits of data, but the technology soon developed to the point that 8 bits were commonplace. State-of-the-art microprocessor chips now handle 16, and in some cases 32, bits of data.

The capabilities of a computer are determined primarily by the capabilities of the microprocessor around which it is built. Hence the speed of the microprocessor determines the speed of the computer; its word length determines the word length of the computer; the number of address lines determines the maximum size of

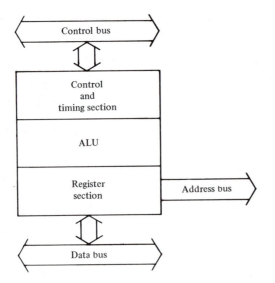

FIGURE 16-1 Block diagram of a microprocessor and bus structure

directly addressable memory for the computer; and its control configuration determines the type of I/O interfacing that must be used.

Microprocessors perform many functions, including

- Synchronizing all microcomputer operations by providing timing and control
  signals.
- Decoding instructions.
- Performing arithmetic operations.
- Fetching instructions and data from memory.
- Transferring data to or from I/O devices.
- Responding to control signals from I/O devices.

To perform these functions, signals in the form of instructions are transmitted
via the buses shown in Fig. 16-1 to and from other subsystems of the microcomputer. Understanding the significance of these instructions is essential for understanding the microprocessor.

16-2
CONTROL AND TIMING SECTION

Although microprocessors contain an incredible amount of digital circuitry in the
form of logic gates, registers, clocks, and so on, one must keep in mind that the
circuitry is accessible only to the extent that the 40 or so external pins on the IC
chip provide accessibility. Most of the circuitry within the microprocessor is only
software accessible—by means of a set of instructions called a *program*.

A basic operation of all computers is a two-step cycle of *fetching* an instruction
and then *executing* the instruction. The control and timing signals that are required
by the ALU and register section of the microprocessor, as well as by other sub-

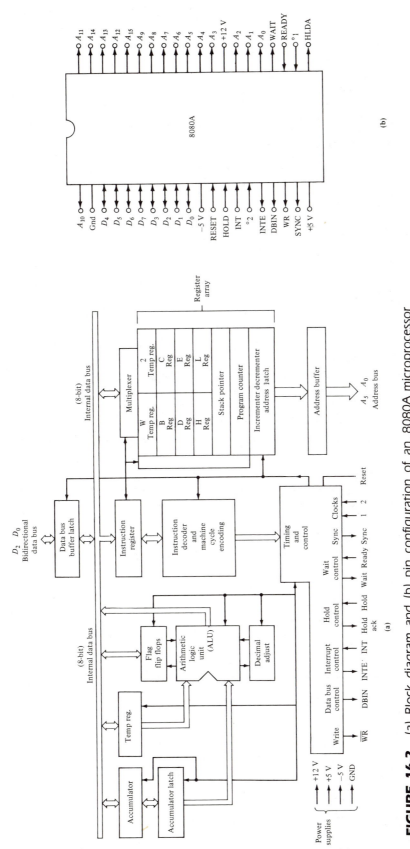

FIGURE 16-2 (a) Block diagram and (b) pin configuration of an 8080A microprocessor

systems of the computer, to complete the *instruction cycle* are generated by the control and timing section shown in Fig. 16-1.

Control signals that are required by other subsystems of a microcomputer are transmitted via the control bus. Approximately 10 to 12 pins of a 40-pin microprocessor chip are used for control purposes. Figure 16-2 shows the block diagram and pin configuration for the Intel 8080A microprocessor with 16 address lines. The 12 lines leading out of the timing and control block provide the control and timing signals and therefore connect to the control bus.

16-3
MICROPROCESSOR REGISTERS

In addition to the timing and control section, microprocessor architecture includes a *register section*. Registers are used to store instruction codes, addresses, data, and information on the status of various microprocessor operations. Others are used as counters for such things as providing sequential stepping through a set of instructions or for sequentially reading memory locations.

Although each microprocessor manufacturer has its own unique architecture for the register section, some of the basic functions performed by the various registers are essentially identical. The three most common and widely used registers are the *instruction register, program counter,* and the *accumulator*. These, together with several other common registers, are discussed in the following paragraphs.

Instruction Register.   When the microprocessor, or CPU, fetches an instruction from memory, it transfers the instruction to the *instruction register* (IR). The IR is automatically used by the microprocessor during each instruction cycle; therefore, this is not a programming step. The size of the IR will be the same as the word size for the microprocessor.

Program Counter.   The *program counter* (PC) always contains the address in memory of the next instruction that the CPU is to fetch. The microcomputer automatically increments the program counter after each instruction, so the stored program is executed sequentially unless it contains an instruction that alters the sequence.

The size of the program counter is determined by the number of address lines.

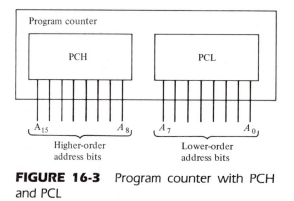

FIGURE 16-3   Program counter with PCH and PCL

Most of the popular microprocessor chips use 16-bit addresses; therefore, the PC must be a 16-bit register. However, many microprocessors use two 8-bit registers, one to hold the eight higher-order address bits and one to hold the eight lower-order address bits. These are called the program counter high (PCH) and the program counter low (PCL) and are illustrated in Fig. 16-3. The reason that two 8-bit registers are used is to permit the contents of the program counter to be stored in memory as two 8-bit words.

Accumulator. The *accumulator* is the most frequently used register. It is the register that is used in most of the operations performed by the ALU. In most microprocessors, when addition or subtraction is performed, one of the numbers comes from the accumulator. Generally, the result of an arithmetic operation is automatically stored in the accumulator. As with the program counter, the size of the accumulator, which is sometimes referred to as the A register, is determined by the word size of the microprocessor.

In addition to its use in arithmetic operations, the accumulator is used to store data from memory or from I/O devices and to send data to memory or to I/O devices.

Memory Address Register. The *memory address register* (MAR) is sometimes called the *address latch register* and is identified as such in Fig. 16-2. The MAR is used to hold the address of *data* that the CPU is writing into or reading from memory.

When executing a program, the microprocessor goes to memory to obtain instructions and data. There are two sources of addresses for the address bus, the MAR and the PC. The MAR is used for *data* addresses and the PC is used for *instruction* addresses. The output signals from the two registers are multiplexed onto the address bus, as shown in Fig. 16-4.

Stack Pointer Register. The *stack* is an area in random-access memory used to store data temporarily. The *stack pointer* (SP) is a register in the microprocessor

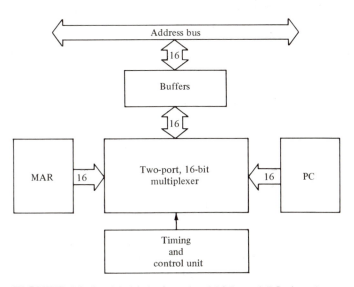

FIGURE 16-4 Multiplexing the MAR and PC signals

that holds the address where data can be stored on, or retrieved from, the stack. Whenever a word is to be stored on the stack, it is stored at the address contained in the stack pointer and whenever a word is to be read from the stack, it is read from the address specified by the stack pointer. The location of the pointer is initiated by the user at the start of a program; thereafter, it is decremented automatically *after* a word is stored on the stack and incremented *before* a word is read from the stack. This type of memory is referred to as *last in/first out* (LIFO) *memory* or as a *push-down stack*. Thus the stack pointer counts down to fill the stack and counts up as the stack empties.

Status Register. The *status register* is referred to by several names, including condition code register, processor status register, P register, and flag register. In the 8080A block diagram of Fig. 16-2, it is designated as the flag flip-flop.

The status register consists of individual bits called *flags*. Each flag is used to indicate the status of some particular microprocessor condition. Three of the most common flags are the *carry* flag, the *zero* flag, and the *sign* flag. The carry flag is used in the arithmetic operations of addition and subtraction to indicate whether there is a carry or a borrow, respectively. The zero flag is used to indicate whether the results of an instruction is zero, and the sign flag is used to indicate whether a number is positive or negative.

General-Purpose Registers. General-purpose registers are used to provide temporary storage for functions performed within the CPU. This speeds up program execution, since the CPU does not have to access memory as frequently as would otherwise be required.

The number of general-purpose registers varies depending on the particular microprocessor. Some microprocessors have none, whereas others have 12 or more. The 8080A shown in Fig. 16-2 has 6 such registers.

16-4
THE ARITHMETIC-LOGIC UNIT

The arithmetic-logic unit (ALU) was discussed in some detail in Chapter 13, so here we will only briefly summarize its functions. The primary function of the ALU is to perform arithmetic and logical operations. Arithmetic instructions allow microprocessors to compute and manipulate data. The basic arithmetic instructions are the add, subtract, increment, and decrement instructions. The increment and decrement instructions add 1 to or subtract 1 from the contents of a microprocessor register or a memory location. The contents of a memory location are incremented or decremented as follows:

1. The data in a memory location are moved from memory to the ALU.
2. The data are incremented, or decremented.
3. The incremented, or decremented, data are transferred back to the original memory location.

This type of instruction, called a read/modify/write instruction, produces a "rollover" result. This means that if the contents of a register or memory location are all 1's and the register is incremented, its contents will be all 0's and the carry flag is not affected. Similarly, if the contents of a register or memory location are

TABLE 16-1. / Microprocessor Data

Manufacturer	Part Number	Clock Rate (MHz)	Word Size (bits)	Direct Addressing Capability (words)	Number of Pins
Intel	8080	2	8	65K	40
Commodore	6502		8	65K	40
Intel	8085	5	8	65K	40
Motorola	6800	2	8	65K	40
Motorola	6809	5	8	65K	40
Texas Instruments	TMS7000	8	8	65K	40
Zilog	Z80	4	8	65K	40
Intel	8086	10	16	1M	40
Motorola	68000	12	16	16M	64
Texas Instruments	TMS99000	12	16	32K	40
Zilog	Z8001	10	16	48M	48

all 0's and the register is incremented, its contents will be all 1's and, again, the carry flag is not affected.

Arithmetic-logic units also perform logical operations. The logical operations of AND, OR, and Exclusive-OR can be performed by microprocessors on 8 bits at a time. The contents of the accumulator and data from memory can be combined by any of these logical operations. As with the arithmetic instructions for adding and subtracting, logical operations are performed in the ALU and the results are automatically placed in the accumulator.

16-5
COMMERCIALLY AVAILABLE MICROPROCESSOR CHIPS

Some of the more popular microprocessor ICs are listed in Table 16-1.

16-6
BASIC MICROCOMPUTER SYSTEM

Microcomputers are small, relatively slow digital information processing systems made possible by LSI technology. Microcomputers contain several subsystems, one of which is the microprocessor, as shown in Fig. 16-5. As shown, the microprocessor is the central processing unit (CPU) for the microcomputer. The microprocessor contains all the control and arithmetic circuits for the microcomputer.

The memory unit shown in Fig. 16-5 contains both RAM and ROM devices, which is typical of most microcomputers, although one or the other might be omitted for certain applications. The RAM section of the memory unit is used to temporarily store programs and data that change often during the course of operation. It is also used to store intermediate and final results of operations performed in the process of executing a program.

The ROM section is used to store instructions and data that do not change. For example, it might be used to store look-up tables, such as trigonometric tables, log tables, and tables of ASCII codes or other codes, or to store programs related to inputting and outputting information.

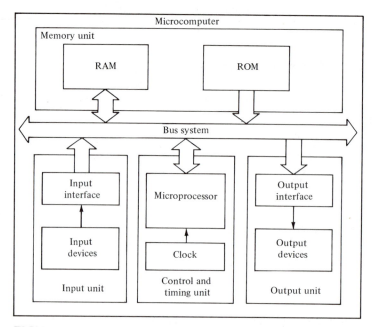

FIGURE 16-5 Microcomputer block diagram showing sub-systems

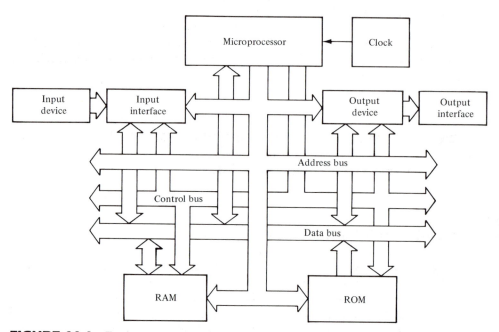

FIGURE 16-6 Typical microcomputer system with innerconnecting buses

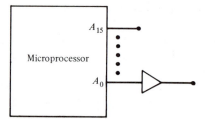

FIGURE 16-7 Buffered address line

The input and output sections contain the input or output device, as well as the necessary interface circuits, to permit the input or output device to communicate with the rest of the computer. The interface circuits range in complexity from simple buffer registers to a system of LSI chips designed to permit a particular microcomputer to interface with a wide range of input/output devices.

16-7
MICROCOMPUTER ARCHITECTURE

Architecture is the term used to describe how a microcomputer is configured internally to permit the microprocessor to communicate with the other subsystems that make up the microcomputer. Figure 16-6 shows a typical microcomputer system. Its architecture includes three buses that tie the system together. The following paragraphs describe the operation of the bus system as information is transmitted via the buses to the various subsystems that make up the microcomputer system.

16-7.1 The Bus System

Microcomputers have three buses, as shown in Fig. 16-6, which connect the microprocessor or CPU with the other subsystems and which carry all information involved in the operation of the system. In many, but not all microcomputers, the CPU is involved in all information transfers; therefore, if subsystem A wishes to transfer information to subsystem B, it must make the transfer via the CPU. That is, subsystem A transfers the information to the CPU, which in turn transfers it to subsystem B. [Some microcomputers use direct memory access (DMA), where peripheral devices gain access to memory without the CPU being involved.] These information transfers are called *read* and *write operations* with reference to the operation of the CPU. If the CPU is receiving information from another subsystem, the operation is called a read operation. If the CPU is transferring information to another subsystem, the operation is called a write operation. The three buses that carry all information during read and write operations are the *address bus, data bus,* and *control bus*. Each of these is discussed in the following paragraphs.

Control Bus. As its name implies, signals transmitted via the control bus are used to synchronize and control the activities and operations of the subsystems of the microcomputer. Although the total set of control signals varies widely among microcomputer manufacturers, certain control signals or instructions are common to virtually all microcomputers. Some of the common control signals are shown on the microprocessor block diagram in Fig. 16-2.

Address Bus. The address lines, of which there are typically 16, are *undirectional* in that an address originates inside the microprocessor chip, is placed on the address line, and is sent to memory or to an I/O port.

Since address lines are often required to drive two or more loads, they usually require some form of buffer circuit, especially if a microprocessor that uses MOS technology is required to drive TTL loads. Figure 16-7 shows a single microprocessor address line buffered with a noninverting buffer such as a 7407.

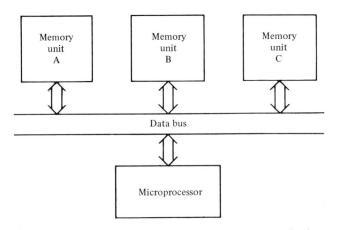

FIGURE 16-8 Microprocessor and memory units interconnected via the data bus

Data Bus.   The same considerations regarding buffering apply to the *data bus* as to the *address bus,* but in addition, the data bus is *bidirectional.* Figure 16-8 shows three memory units connected to the eight-line data bus of a microprocessor. Although the three memory units share a common data bus, they cannot operate simultaneously on the bus. Therefore, each memory unit must be enabled separately so that only one operates at a time. The selected unit is enabled by use of the address bits and the necessary control bits from the microprocessor.

Since the data bus is bidirectional, buffering must be done using tri-state devices, as shown in Fig. 16-9. The buffering shown is for data line D_0. The same type of buffering is required on each of the remaining data lines. This allows data transfer from the microprocessor to a selected memory unit or from a selected memory unit to a microprocessor without interfering with other units connected to the bus.

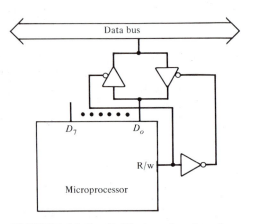

FIGURE 16-9 Bufffering the data bus with tri-state devices

16-8
INTERRUPTS

Most microprocessors and microcomputers include a feature that allows them to recognize a signal from an external device, stop the program being executed, and defer temporarily to a secondary program in memory. The signal is called an *interrupt signal* or an *interrupt request* because the main program is interrupted. The secondary program is a subroutine program and is called an *interrupt service routine*. The steps that occur when an interrupt request is received are shown in Fig. 16-10. The microprocessor saves the address on the stack and then defers to the interrupt service routine. After completing the service routine, the microprocessor executes the return instruction. The return address is then removed from the stack and loaded into the program counter. The microprocessor then continues with the main program.

The interrupt feature is very important in microprocessors because without it, the only way that an external device could transmit a signal to the microprocessor would be for the program to contain a procedure that would check periodically to see whether an external device needed to be serviced. This is not a desirable approach because the external event is usually not synchronized with the microprocessor's internal instructions. This would mean that the external signal would not be accepted until the microprocessor queried the external device. In the meantime the external signal may disappear, in which case the information is lost. In addition to the possibility of losing information, such systems would operate more slowly and waste memory space because of the need for additional instructions to tell the microprocessor what to do and when to do it.

Although most microprocessors incorporate the interrupt feature, the operation of the microprocessor is not dictated by an external device connected to its input/output terminals. Manufacturers have made provisions to override the interrupt signal, thus providing the microprocessor with the option of accepting or not accepting the signal.

Some microcomputers incorporate two interrupt mechanisms: an *interrupt request* and a *nonmaskable interrupt*. Signals on the interrupt request line can be

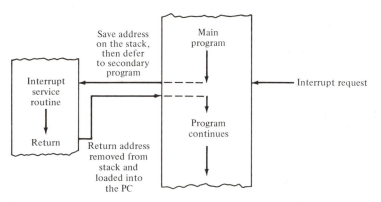

FIGURE 16-10 Steps that occur when an interrupt request is received

accepted or ignored according to the programmer's discretion, but a signal on the nonmaskable interrupt line must always be acknowledged by the microprocessor. A signal on this line indicates that an emergency exists.

16-9
SUMMARY

Although there are many applications for digital techniques, the primary application is in computers. Since the introduction of the microprocessor chip in 1969, the microcomputer segment of the computer business has expanded at an incredible rate. The microprocessor chip is used widely as both the heart of a minicomputer and as a "stand-alone" device for control purposes.

The major subsystems of a microprocessor are the control and timing section, the arithmetic logic unit, and the register section. When used in a microcomputer, the microprocessor serves as the central processing unit. Signals from it to the other subsystems within the microcomputer or to input/output devices are transmitted via three buses: the control bus, the address bus, and the data bus. Interaction between microprocessors or microcomputers and external devices is enhanced by the interrupt signal feature, which allows an external device to be recognized and serviced.

PROBLEMS

1. Draw a block diagram of a microprocessor. Include at least five registers, · including the three most widely used. Label all blocks.

2. Draw a block diagram of a microcomputer. Label all subsystems, including the buses.

3. What function does a microprocessor serve in a microcomputer?

4. List five functions that microprocessors perform.

5. How much memory capacity is possible when using an 8080 microprocessor?

6. Which register is used to store the results of an ALU operation?

7. How many registers does the 8080A have?

8. Look up and list the pin numbers of the data bus lines of a 6800 microprocessor chip.

9. Look up and list the pin numbers of the data bus lines of a Z80 microprocessor chip.

10. Look up and list the interrupt pin or pins of a 6809 microprocessor chip.

11. Which buses in a microcomputer are bidirectional?

12. Describe the primary use of a nonmaskable interrupt pin on a microprocessor chip.

13. Explain why it is necessary to buffer the address and data lines of a microprocessor.

14. List and describe the function of the three most common flags in a microprocessor.

15. Describe the sequence of events that occur due to an interrupt signal.

REFERENCES

Daley, Henry O., *Fundamentals of Microprocessors*. New York: Holt, Rinehart and Winston, 1983.

Driscoll, Frederick F., *Microprocessor–Microcomputer Technology*. North Scituate, Mass.: Breton Publishers, 1983.

Microelectronic Data Book. Carrolton, Tex.: Mostek Corp., 1982.

Microprocessor and Peripheral Handbook. Santa Clara, Calif.: Intel Corp., 1983.

Microprocessor Applications Reference Book. Cupertino, Calif.: Zilog, Inc., 1981.

TMS7000 Family Data Manual. Dallas, Tex.: Texas Instruments, Inc., 1983.

Tocci, Ronald J., and Lester P. Laskowski, *Microprocessors and Microcomputers: Hardware and Software,* 2nd ed. Englewood Cliffs, N.J.: Prentice-Hall, Inc., 1982.

Appendix A
Table of Powers of 2

2^n	n	2^{-n}
1	0	1.0
2	1	0.5
4	2	0.25
8	3	0.125
16	4	0.0625
32	5	0.03125
64	6	0.01562 5
128	7	0.00781 25
256	8	0.00390 625
512	9	0.00195 3125
1024	10	0.00097 65625
2048	11	0.00048 82812 5
4096	12	0.00024 41406 25
8192	13	0.00012 20703 125
16,384	14	0.00006 10351 5625
32,768	15	0.00003 05175 78125
65,536	16	0.00001 52587 89062 5
131,072	17	0.00000 76293 94531 25
262,144	18	0.00000 38146 97265 625
524,288	19	0.00000 19073 48632 8125
1,048,576	20	0.00000 09536 74316 40625
2,097,152	21	0.00000 04768 37158 20312 5
4,194,304	22	0.00000 02384 18579 10156 25
8,388,608	23	0.00000 01192 09289 55078 125
16,777,216	24	0.00000 00596 04644 77539 0625
33,554,432	25	0.00000 00298 02322 38769 53125
67,108,864	26	0.00000 00149 01161 19384 76562 5
134,217,728	27	0.00000 00074 50580 59692 38281 25
268,435,456	28	0.00000 00037 25290 29846 19140 625
536,870,912	29	0.00000 00018 62645 14923 09570 3125
1,073,741,824	30	0.00000 00009 31322 57461 54785 15625
2,147,483,648	31	0.00000 00004 65661 28730 77392 57812 5
4,294,967,296	32	0.00000 00002 32830 64365 38696 28906 25

Appendix B
Number Conversion Table

Decimal	Binary	Octal	Hex	Decimal	Binary	Octal	Hex
0	000000	0	0	26	011010	32	1A
1	000001	1	1	27	011011	33	1B
2	000010	2	2	28	011100	34	1C
3	000011	3	3	29	011101	35	1D
4	000100	4	4	30	011110	36	1E
5	000101	5	5	31	011111	37	1F
6	000110	6	6	32	100000	38	20
7	000111	7	7	33	100001	39	21
8	001000	10	8	34	100010	40	22
9	001001	11	9	35	100011	41	23
10	001010	12	A	36	100100	42	24
11	001011	13	B	37	100101	43	25
12	001100	14	C	38	100110	44	26
13	001101	15	D	39	100111	45	27
14	001110	16	E	40	101000	46	28
15	001111	17	F	41	101001	47	29
16	010000	20	10	42	101010	48	2A
17	010001	21	11	43	101011	49	2B
18	010010	22	12	44	101100	50	2C
19	010011	23	13	45	101101	51	2D
20	010100	24	14	46	101110	52	2E
21	010101	25	15	47	101111	53	2F
22	010110	26	16	48	110000	54	30
23	010111	27	17	49	110001	55	31
24	011000	30	18	50	110010	56	32
25	011001	31	19	51	110011	57	33

Appendix C
Manufacturers of
Digital ICs

The following companies are some of the primary manufacturers of digital integrated circuits. These companies offer a wide variety of technical literature in the form of application notes, data books, technical briefs, design manuals, and other material. This literature can usually be obtained at a nominal cost by writing to the manufacturer or by contacting a local distributor.

Advanced Micro Devices, Inc.
901 Thompson Place
Sunnyvale, California 94086

American Microsystems, Inc.
3800 Homestead Road
Santa Clara, California 95051

Fairchild Camera and Instrument Corp.
464 Ellis St.
Mountain View, California 94042

Fujitsu Microelectronics
2985 Kifer Road
Santa Clara, California 95051

Harris Semiconductors, Inc.
P.O. Box 883
Melbourne, Florida 32901

Intel Corporation
3065 Bowers Ave.
Santa Clara, California 95051

Mostek Corporation
1215 W. Crosby Road
Carrolton, Texas 95006

Motorola Semiconductor Products, Inc.
3501 Ed Bluestein Blvd.
Austin, Texas 78721

National Semiconductor Corp.
2900 Semiconductor Drive
Santa Clara, California 95051

NEC Electronics, U.S.A. Inc.
One Natick Executive Park
Natick, Massachusetts 01760

Precision Monolithics, Inc.
1500 Space Park Drive
Santa Clara, California 95050

RCA Solid State Division
P.O. Box 3200
Somerville, New Jersey 08876

Rockwell Semiconductor Products Div.
4311 Jamboree Road
Newport Beach, California 92660

Siemens Corp.
19000 Homestead Road
Cupertino, California 95014

Signetics Corp.
811 East Arques Ave.
Sunnyvale, California 94086

Standard Microsystems Corp.
35 Marcus Blvd.
Hauppague, New York 11788

Teledyne Semiconductor
1300 Terra Bella Ave.
Mountain View, California 94043

Texas Instruments, Inc.
P.O. Box 225012
Dallas, Texas 75265

Western Digital Corp.
2445 McCabe Way
Irvine, California 92714

Zilog Corp.
10460 Bulb Road
Cupertino, California 95014

Appendix D
Glossary

Access Time. The time to gain access to stored information in an electronic memory.

Accumulator. A register in a microprocessor in which the result of a given operation is stored temporarily.

Active Edge. The triggering edge that causes an edge-sensitive device to respond.

Active Level. The logic state in which a device is enabled.

A/D Conversion. The process of converting an analog signal into digital form.

Addend. In addition, the number added to a second number called the augend.

Adder. A digital circuit that performs the addition of numbers.

Address. The location of a given storage cell in a memory.

Adjacent Cells. Karnaugh map cells that have a common border or cells whose addresses differ by only 1 bit.

Algorithm. A step-by-step procedure that outlines the sequence of actions necessary to solve a problem.

Alphanumeric. A system of symbols consisting of both numerals and alphabetic characters.

ALU. Arithmetic logic unit, generally a part of the central processing unit in computers and microprocessors.

Amplitude. In terms of pulse waveforms, the height of maximum value of the pulse.

Analog. Being continuous or having a continuous range of values, as opposed to a discrete set of values.

AND Gate A digital logic circuit in which a HIGH output occurs if and only if all the inputs are HIGH.

ANSI. American National Standards Institute.

Arithmetic. Related to the four operations of add, subtract, multiply, and divide.

Astable. Having no stable state. A type of multivibrator that oscillates between two quasistable states.

Asynchronous. Having no fixed time relationship.

Augend. In addition, the number to which the addend is added.

Base. The number of symbols in a number system. The decimal system has a base of 10 because there are 10 digits.

BCD. Binary Coded Decimal, a digital code.

BCD Counter. A counter that counts in a BCD code.

Bidirectional Bus. A set of conductors capable of transmitting information in two directions via tri-state buffers.

Binary. Having two values or states.

Binary Counter. An interconnection of flip-flops whose outputs progress through a natural binary sequence when a periodic signal is applied to its clock input.

Binary Digital Signal. A digital signal having exactly two discrete levels. (See "Digital Signal.")

Binary Number. A numerical quantity expressed as a weighted combination of two symbols, 0 and 1.

Binary Point. The point that separates the integer and fractional parts of a binary number.

Bipolar. Referring to a junction type of semiconductor device. A *pnp* or *npn* transistor.

Bipolar RAM. A random-access memory in which the memory cells are made up of bipolar transistors.

Biquinary Counter. A counter that counts in the biquinary code.

Bistable. Having two stable states. A type of multivibrator commonly known as a flip-flop.

Bit. Binary digit. A 1 or a 0.

Boolean Algebra. A mathematics of logic.

Borrow Digit. A digit produced from the MSB stage in subtraction when the subtrahend is larger than the minuend.

Bubble Memory. A type of memory that uses tiny magnetic bubbles to store 1's and 0's.

Buffer Gate. A gate whose output can drive substantially more inputs than a standard gate, thereby providing increased fanout.

Bulk Storage. The storage of large quantities of binary information on an inexpensive medium, usually magnetic tape or disk.

Byte. A group of 8 binary bits.

Carry Digit. A digit produced from the MSB state in arithmetic addition when the sum cannot be expressed with the allotted number of digits.

Carry Input. In cascadable modules, an input accepted from a previous stage.

Carry Output. In cascadable modules, an output intended for a successive stage, which, when asserted, indicates overflow in the current stage.

Cascade. A configuration in which one device drives another.

CCD. Charge-coupled device. A type of semiconductor technology.

Cell. A single storage element in a memory.

Character. A symbol, letter, or numeral.

Circuit. A combination of electrical and/or electronic components connected together to perform a specified function.

Clear. To reset, as in the case of a flip-flop, counter, or register.

Clock. The basic timing signal in a digital system.

Clocked S-R Latch. A type of latch that contains "set," "reset," and "enable" (or clock) inputs, and requires an enable signal to load data.

CMOS. Complementary metal oxide semiconductor.

Code. A combination of binary digits that represents information such as numbers, letters, and other symbols.

Code Converter. An electronic digital circuit that converts one type of coded information into another coded form.

Combinational Logic. A combination of gate networks, having no storage capability, used to generate a specified function.

Comparator. A digital device that compares the magnitudes of two digital quantities and produces an output indicating the relationship of the quantities.

Complement. In Boolean algebra, the inverse function. The complement of a 1 is a 0, and vice versa.

Computer. A digital electronic system that can be programmed to perform various tasks, such as mathematical computations, at extremely high speed, and that can store large amounts of data.

Core. A magnetic memory element.

Counter. A digital circuit capable of counting electronic events, such as pulses, by progressing through a sequence of binary states.

CPU. Central processing unit, a main component in all computers.

Cycle Time. The minimum time between successive read or write cycles in an electronic memory.

D/A Conversion. A process whereby information in digital form is converted into analog form.

Data. Information in numeric, alphabetic, or other form.

Data Bus. A set of conductors capable of transmitting and receiving data between the CPU of a computer and other systems within the computer or I/O devices.

D Flip-Flop. A type of bistable multivibrator in which the output follows the state of the D input.

Decade Counter. A digital counter having 10 states.

Decode. To determine the meaning of coded information.

Decoder. A digital circuit that converts coded information into a familiar form.

Decrement. To decrease the contents of a register or counter by one.

Delay. The time interval between the occurrence of an event at one point in a circuit and the corresponding occurrence of a related event at another point.

De Morgan's Theorems. (1) The complement of a product of terms is equal to the sum of the complements of each term. (2) The complement of a sum of terms is equal to the product of the complements of each term.

Demultiplexer. A logic circuit in which a single input is switched onto one of many output lines.

Difference. The result of a subtraction.

Digit. A symbol representing a given quantity in a number system.

Digital. Related to digits or discrete quantities.

DIP. Dual in-line package. A type of integrated circuit package.

Dividend. In a division operation, the quantity that is being divided.

Divisor. In a division operation, the quantity that is divided into the dividend.

DMA. Direct memory access.

Don't Care. A condition in a logic network in which the output is independent of the state of a given input.

Down Count. A counter sequence in which each successive state is less than the previous state in binary value.

DTL. Diode-transistor logic.

Duplex. Bidirectional transmission of data along a transmission line.

Dynamic Memory. A memory having cells that tend to lose stored information over a period of time and therefore must be "refreshed." Typically, the storage elements are capacitors.

EAPROM. Electrically alterable programmable read-only memory.

ECL. Emitter-coupled logic.

Edge-Triggered Flip-Flop. A type of flip-flop in which input data are entered and appear on the output on the same clock edge.

EEPROM. Electrically erasable programmable read-only memory.

Enable. To activate or put into an operational mode.

Enclosure. An encirclement on a Karnaugh map that denotes an implicant.

Encode. To convert information into coded form.

Encoder. A digital circuit that converts information into coded form.

End-Around Carry. The final carry that is added to the result in a 1's or 9's complement addition.

EPROM. Erasable programmable read-only memory.

Equivalent States. Two separate states in a sequential circuit that lead to the same new state under identical input conditions and with identical outputs.

Error Correction. The process of correcting bit errors occurring in a digital code.

Error Detection. The process of detecting bit errors occurring in a digital code.

Essential Prime Implicant. A prime implicant that encloses one or more "1"-cells, and cannot be enclosed by any other implicants.

Even Parity. A characteristic of a group of bits having an even number of 1's.

Excess-3. A digital code where each of the decimal digits is represented by a 4-bit code derived by adding 3 to each of the digits.

Exclusive-NOR. A logic circuit that has a logic 1 output when the two inputs have the same binary value.

Exclusive-OR. A logic circuit that has a logic 1 output when one but not both inputs are a logic 1.

Fall Time. The time interval between the 10 percent point to the 90 percent point on the negative-going edge of a pulse.

Fan Out. The number of equivalent gate inputs that a logic gate can drive.

FET. Field-effect transistor.

Fetch. The cycle of a CPU in which an instruction or data byte is retrieved from the memory.

FIFO. First-in-first-out memory.

Fixed Point. A binary point having a fixed location in a binary number.

Flash Converter. A type of A/D converter utilizing a resistive divider (for the analog input), a comparator for each quantization level, and a binary encoder.

Flat Pack. A type of integrated circuit package.

Flip-Flop. A bistable device used for storing a bit of information.

Floating Point. A binary point having a variable location in a binary number.

Floppy Disk. A magnetic storage device. Typically, a $5\frac{1}{4}$-inch or an 8-inch flexible Mylar disk.

FPLA. Field programmable logic array.

Frequency. The number of pulses in 1 second for a periodic waveform. Expressed in hertz (Hz) or pulses per second (p/s).

Frequency Divider. A counter whose clock input is derived from a periodic signal, and whose Q output frequencies are submultiples of the clock frequency.

Full Adder. A digital circuit that adds two binary digits and an input carry to produce a sum and an output carry.

Full Duplex. A digital circuit that adds two binary digits and an input carry to produce a sum and an output carry.

Full-Duplex. Simultaneous bidirectional transmission of data on a transmission line.

Function Table. An abbreviated state table.

Fusible Link. A fine wire that, when intact, represents logic 0, and when burned open, represents logic 1. Fusible-link matrices are used in some ROM circuits.

Gate. A logic circuit that performs a specified logical operation such as AND, OR, NAND, NOR, and Exclusive-OR.

Generator. An energy source for producing electrical or magnetic signals.

Glitch. A voltage or current spike of short duration and usually unwanted.

Gray Code. A type of digital code characterized by a single-bit change from one code word to the next.

Ground Return. A connection that completes the electrical circuit in logic gate circuits.

Half-Adder. A digital circuit that adds 2 bits and produces a sum and an output carry. It cannot handle input carries.

Hamming Code. A type of error detection and correction code.

Hexadecimal. A number system consisting of 16 characters. A number system with a base of 16.

Hexadecimal Code. A simplified means of representing binary numbers in a radix-16 number systems.

Hexadecimal Point. The point that separates the integer and fractional parts of a hexadecimal number.

Hold Time. The time interval required for the control levels to remain on the inputs to a flip-flop after the triggering edge of the clock in order to reliably activate the device.

Hysteresis. A condition in which the threshold between two discrete states varies, and is determined by the state that is currently occupied. Hysteresis operates so as to allow the widest possible analog variation within a discrete state prior to switching. Having switched to the next state, the threshold of that new state widens to include points that were previously stable in the last state. A characteristic of magnetic core or a threshold triggered circuit.

IC. Integrated circuit. A type of circuit where all the components are integrated on a single silicon chip of very small size.

II'L. Integrated injection logic.

Implicant. A product term in a sum-of-products expression. (See ''Enclosure,'' ''Prime Implicant,'' and ''Essential Prime Implicant.'')

Increment. To increase the contents of a register or counter by 1.

Indexing. The modification of the address of the operand contained in the instruction of a microprocessor.

Index Register. A register used for indexed addressing in a microprocessor.

Indirect Address. The address of the address of an operand.

Information. In a digital system, the data as represented in binary form.

Inherent Address. An address that is implied or inherent in the instruction itself.

Initialize. To put a logic circuit in a beginning state, such as to clear a register.

Input. The signal or line going into a circuit. A signal that controls the operation of a circuit.

Input/Output Register (IOR). A parallel array of flip-flops in the CPU of a computer used to hold data that is being received from, or transmitted to, the input/output bus.

Instruction.   In a microprocessor or computer system, the information that tells the machine what to do. One step in a computer program.

Instruction Register (IR).   A parallel array of flip-flops in the CPU of a computer used to hold the instruction code after the fetch process is complete.

Interrupt.   The process of stopping the normal execution of a program in a computer in order to handle a higher priority task.

Inversion.   Conversion of a HIGH level to a LOW level, or vice versa.

Inverter.   The digital circuit that performs inversion.

JFET.   Function field-effect transistor.

JK **Flip-Flop.**   A flip-flop with two conditioning inputs (*J* and *K*) and one clock input. The output is determined by the level of the signals applied to the *J* and *K* terminals prior to when a clock pulse occurs.

Johnson Counter.   A divide-by-*n* counter where *n* equals twice the number of flip-flops. It is built by connecting the complementary output of the last flip-flop in a shift register to the opposite input (*D* or *J*) of the first flip-flop.

Karnaugh Map.   A display of a truth table in a way that facilitates reduction (simplification or minimization) of a Boolean expression. It consists of a rectangular or square array (depending on the number of variables) of cells, where the "address" of the cell corresponds to truth table inputs.

Kilo.   A prefix representing 1000, or 10^3. Abbreviated K.

Kilobit.   1000 bits.

Ladder Network.   A resistor network in which the resistors have values of *R* and 2*R*.

Latch.   The simplest type of memory element, possessing asynchronous "set" and "reset" inputs.

LCD.   Liquid crystal display.

Leading Edge.   The first edge to occur on a pulse.

LED.   Light-emitting diode.

Level-Sensitive Device.   A device whose state may change only when its Enable input is asserted.

LIFO.   Last-in-first-out memory.

Line Printer.   An output device used in computer systems to print program and data information on paper at a high rate (one line in parallel at one time).

Linearity.   The degree to which the output of an A/D converter is exactly proportional to changes in its digital input.

Logic.   In digital electronics, the decision-making capability of gate circuits in terms of yes/no or on/off type of operation.

Logic Analyzer.   A digital diagnostic instrument that may include word triggering, digital delays, and storage registers.

Logic Circuit.   An electronic circuit that operates on digital signals in accordance with a logic function.

Logic Clip.   A digital diagnostic tool that displays the logic states at all pins of an IC package.

Logic Family.   A group of logic gates, modules, and components built around a standardized integrated circuit transistor configuration. Example families: TTL (Transistor-Transistor Logic), CMOS (Complementary Metal-Oxide Semiconductor Logic), ECL (Emitter Coupled Logic), and IIL (Integrated Injection Logic).

Logic Function.   A well-defined relationship between inputs and outputs of a logic circuit. A logic function predicts the output of a logic circuit given a specific set in input conditions.

Logic Gate. An electronic circuit that implements a specific logic function and can be treated as a simple, cascadable block element.

Logic Inverter. The simplest logic gate. The output voltage of a logic inverter is HIGH when the input voltage is LOW; conversely, it is LOW when the input voltage is HIGH.

Logic Operator. A word required to form a complex logic sentence. For example, AND and OR are logic operators.

Logic Probe. A digital diagnostic tool that displays the logic state applied at the tip of the probe.

Logic Proposition. A sentence that can be answered by TRUE or FALSE.

Logic Pulser. A pulse generator that can override the low-impedance output of a digital circuit.

Logic State. One of two possible conditions that a logical quantity may have, represented numerically as ''0'' or ''1'' and electrically as ''low'' or ''high'' voltage levels.

Look-Ahead Carry Adder. A three-stage adder that is a compromise in complexity and propagation delay between a ripple adder and a two-stage adder.

Low Output Voltage. The more negative of the two output voltage levels.

Low-Power TTL. TTL with propagation delays of $\approx$ 35 ns and power dissipations of $\approx$ 1 mW/gate.

Low-Power Schottky TTL. Schottky-diode clamped TTL with propagation delays of $\approx$ 10 ns and power dissipations of $\approx$ 2 mW/gate.

LSB. Least significant bit.

LSD. Least significant digit.

LSI. Large-scale integration.

Magnetic Bubble Memory (MBM). An array of small magnetic domains (bubbles) embedded in a thin magnetizable material. The bubbles are moved by application of two signals that are 90° out of phase to each other.

Magnetic Core. A memory element made of magnetic materials and capable of existing in either of two states.

Magnitude. The size or value of a quantity.

Magnitude Relation. One of the five relations of greater than, less than, equal, greater than or equal, less than or equal.

Majority Voter. A circuit in which the output always agrees with the majority of the inputs.

Manchester. A format for recording digital data on a magnetic surface.

Map Address. In a Karnaugh map, the row and column numbers that identify a particular cell.

Master-Slave Flip-Flop. A clocked flip-flop consisting of two serial *S-R* latches. Input data are transferred to the master when the clock is asserted, and from the master to the slave when the clock is deasserted.

Maximum Counting Speed. The maximum frequency of the input waveform at which the counter counts correctly.

Memory Address. The location of a storage cell in a memory array.

Memory Array. An arrangement of memory cells.

Memory Address Register (MAR). A parallel array of flip-flops in the CPU of a computer, used to hold the address of the RAM cell currently being accessed. The MAR outputs drive the address bus.

Memory Cell. An individual storage element in a memory.

Memory Data Register (MDR). A parallel array of flip-flops in the CPU of a computer used to hold data that are being received from, or transmitted to, RAM.

Memory Matrix. An array of memory cells.

Metal-Oxide Semiconductor (MOS). A *p*- or *n*-doped semiconductor channel, the transconductance of which is modulated by a voltage applied to an insulated metal gate.

Microcomputer. A computer system utilizing a microprocessor CPU.

Microprocessor. The central processing unit (CPU) of a computer, implemented as a single integrated circuit containing 5000 or more transistors.

Minimal Circuit. A logic circuit that has been implemented with the minimum number of logic gates.

Minimal Equation. A Boolean equation containing the fewest number of sum and product terms.

Minimization. Generally thought of as the act of reducing the number of terms in a Boolean equation or the number of gates in a logic circuit.

Minimum Sum of Products (MSP). A sum-of-products expression that contains the fewest possible terms.

Minuend. In subtraction, the quantity from which the subtrahend is subtracted.

Modified Modulus Counter. A counter that does not sequence through all of its ''natural'' states.

Mod-2 Addition. Exclusive-OR addition. A sum of 2 bits with the carry dropped.

Modulus. The maximum number of states in a counter sequence.

Monostable. Having only one stable state. A multivibrator characterized by one stable state and commonly called a one-shot.

MOS. See metal-oxide semiconductor.

MOSFET. Metal-oxide semiconductor field-effect transistor.

MOS Logic. Logic circuits using *n*-channel or *p*-channel MOS transistors.

MSB. An acronym for Most Significant Bit, which is the digit of a binary number weighted most heavily.

MSI. Medium-scale integration.

Multiple Output Circuit. A single-logic circuit that has two or more primary outputs.

Multiplex. To put information from several sources onto a single line or transmission path.

Multiplexer. A digital circuit capable of multiplexing digital data.

Multiplicand. The number being multiplied.

Multiplier. The number used to multiply the multiplicand.

NAND. A logic operator that is equivalent to an AND operator followed by an inversion.

NAND Gate. A logic gate performing the same function as an AND gate followed by a logic inverter.

Natural Binary Sequence. A progression of binary numbers in which successive entries correspond to successive outputs.

Natural Count. The maximum possible modulus of a counter.

Negative Logic. The system of logic where a LOW represents a 1 and a HIGH represents a 0.

Nine's Complement. The value obtained by subtracting a decimal digit from nine, or a decimal number from a succession of nine's.

NMOS.   An *n*-channel metal oxide semiconductor.

Noise Immunity.   The ability of a circuit to reject unwanted signals.

Noise Margin.   The difference between the maximum low output of a gate and the maximum acceptable low-level input of an equivalent gate. Also, the difference between the minimum high output of a gate and the minimum acceptable high-level input of an equivalent gate.

Nonvolatile Memory.   A type of electronic memory that does not require power to retain stored data. Example: Read-only memory. (See "Volatile Memory.")

NOR.   A logic operator that is equivalent to an OR operator followed by an inversion.

NOR Gate.   A logic gate performing the same function as an OR gate followed by a logic inverter.

Normally Closed (n.c.) Contact.   A relay contact that is closed when the relay is not energized.

Normally Open (n.o.) Contact.   A relay contact that is open when the relay is not energized.

NOT Circuit.   An inverter.

npn.   Referring to a junction structure of a bipolar transistor.

Number System.   A scheme whereby counted or measured quantities are symbolically expressed. The decimal number system uses 10 symbols: 0, 1, 2, 3, 4, 5, 6, 7, 8, 9, while the binary number system uses only two: 0 and 1.

Numeric.   Related to numbers.

Octal.   A number system having a base of 8 consisting of eight digits.

Octal Code.   A simplified means of representing binary numbers in a radix-eight number system.

Octal Point.   The point that separates the integer and fractional parts of an octal number.

Odd Parity.   Referring to a group of binary digits having an odd number of 1's.

One's Complement.   The value obtained by reversing the logic of each bit in a given binary number. The one's complement of "0110111" is "1001000." (See "Two's Complement.")

One-Shot.   A monostable multivibrator.

Open-Collector Output.   The collector of a transistor that is available as output and that has no other internal connections.

Open-Loop Gain.   The gain of an amplifier without feedback.

Operational Amplifier.   A dc-coupled, high-gain, differential amplifier.

Optical Isolator.   A device that transmits and receives digital information via light, permitting complete isolation between the transmitter and the receiver.

OR Gate.   An electronic circuit whose output is at logic 1 if at least one input is at logic 1.

Oscillator.   An electronic circuit that switches back and forth between two states. The astable multivibrator is an example.

Output Device.   A device through which data are retrieved from a computer.

Output Function.   The binary value of the output that depends on the states of the input variables.

PAL.   Programmable array logic.

Parallel Input.   Input through which external data can be jammed into an individual FF in a counter or shift register.

Parallel Loading.   Jamming data into the individual FFs in a counter or shift register.

Parallel Output.   The output of a FF in a shift register.

Parallel-to-Serial Converter.   A register whose memory elements can be simultaneously loaded with parallel data and then caused to shift in one direction until all bits have been transmitted from the last stage. (See ''Serial-to-Parallel Converter.'')

Parity.   Referring to the oddness or evenness of the number of 1's in a specified group of bits.

Parity Bit.   A bit deliberately added to a binary word to make the total number of bits either odd or even.

Parity Detector.   A circuit that inspects received information bits for even or odd parity.

Parity Generator.   A circuit that inspects transmitted information bits and generates an even or odd parity bit.

Period.   The time required for a periodic waveform to repeat itself.

Periodic.   Repeating at fixed intervals.

PLA.   Programmable array logic.

PMOS.   A p-channel metal oxide semiconductor.

pnp.   Referring to a junction structure of a bipolar transistor.

Positive Logic.   The system of logic where a HIGH represents a 1 and a LOW represents a 0.

Positional Notation.   The ordering of digits (of a number in any number base) whereby the leftmost digit carries the highest weight and the rightmost digit carries the lowest weight.

p/s.   Pulses per second. A measure of frequency of a pulse waveform.

Preset.   To initialize a digital circuit to a predetermined state.

PRF.   Pulse repetition frequency.

Primary Input.   An input to a sequential circuit derived from external sources, such as switches and sensors.

Primary Output.   An output of a sequential circuit that is transmitted to external devices, such as indicator lamps and actuators.

Prime Implicant.   An implicant that cannot be fully enclosed by another implicant on a Karnaugh map.

Priority Encoder.   A digital logic circuit that produces a coded output corresponding to the highest-valued input.

Product.   The result of a multiplication.

Product-of-Sums.   A form of Boolean expression that is the ANDing of ORed terms.

Program.   A sequence of instructions in a computer that exactly define a specific activity, and stored in memory as a sequence of binary numbers.

Program Counter (PC).   A natural binary up counter in the CPU of a computer whose output indicates the memory address of the next instruction.

Program Loop.   A sequence of computer instructions that are executed many times. A branch instruction is required at the end of the sequence to close the loop.

Programmable Read-Only Memory (PROM).   A read-only memory that in the unprogrammed state has all cells either at logic 0 or at logic 1. To program the memory, links are selectively broken to obtain the desired binary patterns.

Programmed Counter.   A counter that utilizes special combinational logic to cycle through a nonnatural binary sequence.

Proof.   A demonstration that a property is true by showing that no contradictions exist for all possible conditions of the component variables.

Propagation Delay. The intrinsic delay existing in a logic gate or device, beginning with the application of an input and ending with the device's response. Propagation delay is due to unavoidable parasitic capacitance or inductance in an electronic circuit. NAND gate propagation delays in TTL are typically 10 ns.

Pulse. A sudden change from one level to another followed by a sudden change back to the original level.

Pulse Duration. The time interval that a pulse remains at its high level (positive-going pulse) or at its low level (negative-going pulse). Typically measured between the 50 percent points on the leading and trailing edges of the pulse.

Pulse-Triggered Device. Another term for a master-slave device.

Pulse Width. Pulse duration.

Quantization. (1) Time: the subdivision of the time axis of an analog signal into a sequence of discrete equally spaced sample points. (2) Amplitude: the subdivision of a smooth range of voltage or current values into a finite number of encodable levels.

Quiescent. At rest. The condition of a circuit when no input signal is being applied to it.

Quiescent Power Dissipation. Power dissipation when the circuit is not switched.

Quotient. The result of a division.

R-2R Ladder. A resistor network constructed using only two values of resistance, one of which is twice the value of the other and applied in digital-to-analog converter circuits where currents must be exponentially weighted and summed.

Race. A condition that exists in sequential circuits when simultaneous changes in flip-flop states cause temporary false excitations, resulting from output skew. A noncritical race is one where the correct final state is reached regardless of intermediate false excitations. A critical race leads to an erroneous final state or produces false outputs.

Radix. A quantity that expresses the number of unique symbols used in a number system. Same as "Base."

Random-Access Memory (RAM). A memory in which the access time to each cell is the same as to any other cell.

RC Oscillator. An astable circuit whose frequency-determining elements are a resistor and capacitor.

Read. The process of retrieving information from a memory.

Read Cycle. The sequence of signal changes needed to constitute a read operation in an electronic memory. The "read-cycle time" is the minimum duration between successive read cycles. (See "Write Cycle.")

Read-Only Memory (ROM). A type of RAM whose data is permanently recorded.

Read/Write Input. The input found on an electronic memory that determines whether data will be read from or written into the addressed cell.

Recirculate. The process of retaining information in a register as it is shifted out.

Redundant States. Binary combinations that are not utilized.

Redundant Terms. In a Boolean expression these are additional terms the presence of which does not change the function.

Refresh. The process of renewing the contents of a dynamic memory.

Regenerative. Having feedback so that an initiated change is automatically continued, such as when a multivibrator switches from one state to the other.

Register.   A digital circuit capable of storing and moving (shifting) binary information. Typically used as a temporary storage device.

Reset.   The state of a flip-flop, register, or counter when 0's are stored. Equivalent to the clear function.

Resolution.   As applied to an ADC or DAC, it is the value of the least significant bit relative to the full scale.

Ring-Counter.   A divide-by-n counter where n equals the number of FFs. It is built by connecting the output of the last FF in a shift register to the D input of the first FF.

Ripple Adder.   A multibit adder built from 1-bit adders.

Ripple Comparator.   A multibit comparator built from 1-bit comparators.

Ripple Counter.   A counter in which the clock input of each FF is connected to an output of the preceding FF, in some cases via logic gates; the clock input of the first FF is the input of the counter.

Rise Time.   Transition time of the positive-going transient between stated levels.

Sample-and-Hold.   A circuit that samples an analog voltage for a short duration and holds the sample during conversion by an ADC.

Schmitt Trigger.   A circuit with hysteresis that is suitable for reshaping noisy signals.

Schottky-TTL.   Schottky-diode clamped TTL with propagation delays of ≈ 3 ns and power dissipations of ≈ 20 mW/gate.

Secondary Input.   An input to a sequential circuit derived from internal feedback paths. (See ''Primary Input.'')

Secondary Output.   An output of a sequential circuit that is used for internal feedback.

Self-Complementing Code.   A binary code in which the one's complement of a BCD quantity equals the nine's complement of the corresponding decimal digit.

Semiconductor.   A material used to construct electronic devices, such as integrated circuits, transistors, and diodes. Silicon is the most common semiconductor material.

Sequential Logic.   A broad category of digital circuits whose logic states depend on a specified time sequence.

Serial.   An in-line arrangement where one element follows another, such as in a serial shift register. Also, the occurrence of events, such as pulses, in a time sequence rather than simultaneously.

Serial Access Memory (SAM).   A type of electronic memory in which the data are stored serially in a medium, and where the access time depends upon where the data is located in the device or on the medium. (See ''Random Access Memory.'')

Serial Addition.   Addition that is performed bit by bit on two serial-format binary numbers, from LSB to MSB, using a single flip-flop to store carries from one iteration to the next.

Serial Input.   Input to a shift register through which data can be shifted in serially.

Serial-to-Parallel Converter.   A register that can accept data in the first stage, shift it to later stages upon the receipt of clock pulses, and present the accumulated data on parallel outputs.

Set.   The state of a flip-flop when it is in the binary 1 state.

Setting Time.   The time elapsed from the application of a step function to the time when an operational amplifier has entered and remained within a specified error band symmetrical about the final value.

Set-Up Time. The time interval required for the control levers to be on the inputs to a digital circuit, such as a flip-flop, prior to the triggering edge of the clock pulse.

Seven-Segment Display. A display of the decimal numerals and selected characters of the alphabet. The display is obtained through illuminating selected segments of a basic seven-segment pattern.

Shift. To move binary data within a shift register or other storage device.

Shift-Register. A chain of FFs through which data can be shifted serially.

Sign Bit. A bit appended to a binary number for the purpose of indicating the sign of that number. Typically, the sign bit is the MSB.

Sign-Magnitude Representation. Identifying positive and negative numbers by a magnitude and a sign.

Simplification. The act of reducing a Boolean equation to (a) a single value when all variables are exactly specified, or (b) to an expression with fewer terms, using algebraic variables only.

Single-Shot. See monostable multivibrator.

Sink Current. Current flowing into the output of a circuit.

Slope Converter. A type of analog-to-digital converter in which a comparator measures a constant input voltage against increasing reference voltage. The time at which the comparator's output switches is a measure of the analog input and is used to gate a binary counter.

Speed-Power Product. The product of the propagation delay and the power dissipation.

Square Wave. A pulse train with equal "0" and "1" periods.

SR Flip-Flop. A set-reset flip-flop.

SSI. Small-scale integration.

Stable State. A condition in a sequential circuit where the input states of the delay elements (or flip-flops) are equal to their corresponding output states.

Stack. A LIFO memory consisting of registers or memory locations.

Stage. One storage element in a register or counter.

Standard Product Term. A Boolean product of all the variables, complemented or uncomplemented, of a function.

Standard Sum Term. A Boolean sum of all the variables complemented or uncomplemented, or a function.

Standard TTL. TTL with propagation delays of $\approx$ 10 ns and power dissipations of $\approx$ 10 mW/gate.

State. (1) The condition of a digital signal, being either logic "1" or logic "0". (2) A particular pattern of "1"s and "0"s observed on the outputs of a multibit device.

State Diagram. A diagram showing the states, transitions, and outputs of a digital circuit or system.

State Table. Describes all possible states of a circuit.

Steering Logic. The added gates that are present in a natural binary counter to cause the component flip-flop to reenter a previous state, responsive only when a "detect" signal is asserted from a detector gate. (See "Detect and Steer.")

Static Memory. A memory composed of storage elements such as flip-flops or magnetic cores that are capable of retaining information indefinitely.

Storage.   The memory capability of a digital device. The process of retaining digital data for later use.

Strobe.   The memory capability of a digital device. The process of retaining digital data for later use.

Strobe.   A pulse used to sample the occurrence of an event at a specified point in time in relation to the event.

Subroutine.   A program that is normally used to perform specialized or repetitive operations during the course of a main program. A subprogram.

Subtractor.   One of the operands in a subtraction.

Subtrahend.   The other operand in a subtraction.

Sum.   The result of an addition.

Sum-of-Products.   A form of Boolean expression that is the Oring and ANDed terms.

Symbolic Logic.   A mathematical structure used to express reasoning.

Synchronous.   Having a fixed time relationship.

Synchronous Counter.   A counter in which the states of all FFs change simultaneously.

Synthesis.   For a switching circuit, it results in a logic diagram realizing a given function.

T **flip-flop.**   A type of flip-flop that toggles or changes state on each clock pulse.

Ten's Complement.   The value obtained by adding one to the nine's complement of a number.

Three-State Logic.   A type of logic circuit having the normal two-state (HIGH, LOW) output and, in addition, an open state in which it is disconnected from its load.

Threshold.   The borderline between a valid logic-state voltage and a transition zone.

Time-Division Multiplexing.   The sharing of a single transmission channel (wire) by two or more signals at specific and periodic times. No two signals are ever on line at the same time.

Timing Diagram.   A waveform drawing showing the time relationship between two or more digital signals.

Toggle.   The action of a flip-flop when it changes back and forth between its two states on each clock pulse.

Toggle FF ($T = $ 1FF).   An FF with control input T permanently connected to logic 1.

Totem-Pole Output.   An output stage consisting of two stacked transistors.

Trailing Edge.   The second transition of a pulse.

Transducer.   A device that converts one type of signal (such as an electrical one) into another type of signal (such as a mechanical one).

Transfer Characteristic.   The output voltage of a gate as a function of the input voltage.

Transistor.   A semiconductor device exhibiting current gain or voltage gain. When used as a switching device, it can approximate an open or a closed switch.

Transistor-Transistor Logic (TTL).   Logic circuits using *npn* transistors that are either cut off or saturated.

Transition.   A change from one level to another.

Transition Table.   A table particularly useful in asynchronous sequential circuit analysis to show the permissible patterns of change between states and to reveal race conditions. It is derived from the excitation table during analysis, and from the state table during design.

Trigger.   A pulse used to initiate a change in the state of a logic circuit.

Truth Table.   A systematic listing of all possible combinations of the states of the variables, including inputs and outputs.

Twisted-Ring Counter.   (See ''Johnson Counter.'')

Two's (2's) Complement.   Obtained by adding 1 to the 1's complement.

Two's-Complement Subtraction.   A method of subtraction in which the ''difference'' is obtained by adding the minuend and the two's complement of the subtrahend.

Undirectional Bus.   A parallel array of conductors transmitting in one direction only. (See ''Bidirectional Bus.'')

Unique Function.   A Boolean function that can be described by only one MSP equation. A ''nonunique'' function is represented by two or more MSP equations.

Unit Load.   One gate input represents a unit load to a gate output within the same logic family.

Universal Counter Stage.   A multibit synchronous up/down counter module, usually 4 to 8 bits long and parallel-loadable that can be cascaded to any desired length.

Universal Gate.   A logic gate that can be connected with copies of itself to obtain any desired logic function. NAND and NOR gates are examples of universal gates.

Unstable State.   A condition in a sequential circuit where the input state of one or more delay elements (or flip-flops) is unequal to the corresponding output states. (See ''Stable State.'')

Upcount.   A counter sequence in which each binary state has a successively higher value.

UV EPROM.   Ultraviolet erasable programmable read-only memory.

Variable Modulus Counter.   A counter in which the maximum number of states can be changed.

VLSI.   Very large-scale integration.

Volatile.   The characteristic of a memory whereby it loses stored information if power is removed.

Weight.   The value of a digit in a number based on its position in the number.

Weighted Code.   A digital code that utilizes weighted numbers as the individual code words.

Wired-AND.   An AND function attained by connecting outputs of logic gates in parallel.

Wired-OR.   An OR function attained by connecting outputs of logic gates in parallel.

Word.   A group of bits representing a complete piece of digital information.

Write.   The process of storing information in a memory.

Write Cycle.   The sequence of signal changes needed to constitute a write operation in an electronic memory. The ''write cycle time'' is the minimum duration between successive write cycles. (See ''Read Cycle.'')

***X*-NOR.**   An exclusive-NOR gate.

***X*-OR.**   An exclusive-OR gate.

*Data Sheets for TTL Logic Devices

*Reprinted by permission of the Fairchild Camera and Instrument Corp., Digital Logic Division. (Fairchild reserves the right to make changes in the circuitry or specifications at any time without notice.)

54/7400
54H/74H00
54S/74S00
54LS/74LS00
QUAD 2-INPUT NAND GATE

CONNECTION DIAGRAMS
PINOUT A

ORDERING CODE: See Section 9

PKGS	PIN OUT	COMMERCIAL GRADE V_{CC} = +5.0 V ±5%, T_A = 0°C to +70°C	MILITARY GRADE V_{CC} = +5.0 V ±10%, T_A = -55°C to +125°C	PKG TYPE
Plastic DIP (P)	A	7400PC, 74H00PC 74LS00PC, 74S00PC		9A
Ceramic DIP (D)	A	7400DC, 74H00DC 74LS00DC, 74S00DC	5400DM, 54H00DM 54LS00DM, 54S00DM	6A
Flatpak (F)	A	74LS00FC, 74S00FC	54LS00FM, 54S00FM	3I
	B	7400FC, 74H00FC	5400FM, 54H00FM	

PINOUT B

INPUT LOADING/FAN-OUT: See Section 3 for U.L. definitions

PINS	54/74 (U.L.) HIGH/LOW	54/74H (U.L.) HIGH/LOW	54/74S (U.L.) HIGH/LOW	54/74LS (U.L.) HIGH/LOW
Inputs	1.0/1.0	1.25/1.25	1.25/1.25	0.5/0.25
Outputs	20/10	12.5/12.5	25/12.5	10/5.0 (2.5)

DC AND AC CHARACTERISTICS: See Section 3*

SYMBOL	PARAMETER	54/74 Min	54/74 Max	54/74H Min	54/74H Max	54/74S Min	54/74S Max	54/74LS Min	54/74LS Max	UNITS	CONDITIONS
I_{CCH}	Power Supply		8.0		16.8		16		1.6	mA	V_{IN} = Gnd V_{CC} = Max
I_{CCL}	Current		22		40		36		4.4		V_{IN} = Open
t_{PLH}	Propagation Delay		22		10	2.0	4.5		10	ns	Figs. 3-1, 3-4
t_{PHL}			15		10	2.0	5.0		10		

*DC limits apply over operating temperature range; AC limits apply at T_A = +25°C and V_{CC} = +5.0 V.

54/7402
54S/74S02
54LS/74LS02

QUAD 2-INPUT NOR GATE

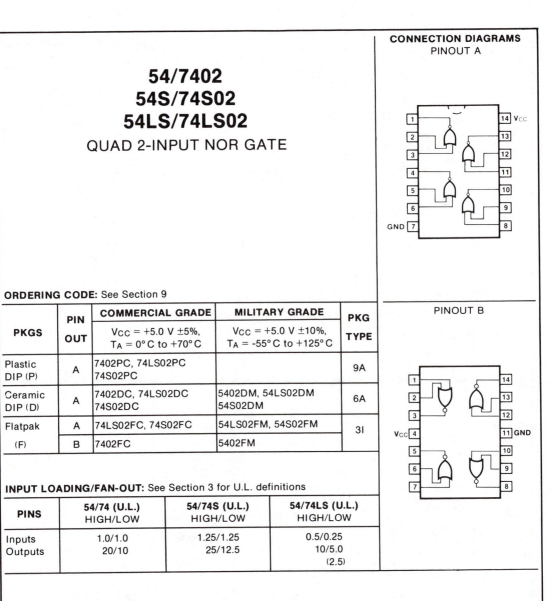

CONNECTION DIAGRAMS
PINOUT A

PINOUT B

ORDERING CODE: See Section 9

PKGS	PIN OUT	COMMERCIAL GRADE $V_{CC} = +5.0$ V ±5%, $T_A = 0°$C to +70°C	MILITARY GRADE $V_{CC} = +5.0$ V ±10%, $T_A = -55°$C to +125°C	PKG TYPE
Plastic DIP (P)	A	7402PC, 74LS02PC 74S02PC		9A
Ceramic DIP (D)	A	7402DC, 74LS02DC 74S02DC	5402DM, 54LS02DM 54S02DM	6A
Flatpak (F)	A	74LS02FC, 74S02FC	54LS02FM, 54S02FM	3I
	B	7402FC	5402FM	

INPUT LOADING/FAN-OUT: See Section 3 for U.L. definitions

PINS	54/74 (U.L.) HIGH/LOW	54/74S (U.L.) HIGH/LOW	54/74LS (U.L.) HIGH/LOW
Inputs	1.0/1.0	1.25/1.25	0.5/0.25
Outputs	20/10	25/12.5	10/5.0 (2.5)

DC AND AC CHARACTERISTICS: See Section 3*

SYMBOL	PARAMETER	54/74 Min	54/74 Max	54/74S Min	54/74S Max	54/74LS Min	54/74LS Max	UNITS	CONDITIONS	
I_{CCH}	Power Supply Current		16		29		3.2	mA	V_{IN} = Gnd	V_{CC} = Max
I_{CCL}			27		45		5.4		V_{IN} = Open	
t_{PLH}	Propagation Delay		15	2.0	5.5		10	ns	Figs. 3-1, 3-4	
t_{PHL}			15	2.0	5.5		10			

*DC limits apply over operating temperature range; AC limits apply at T_A = +25°C and V_{CC} = +5.0 V.

54/7404
54H/74H04
54S/74S04
54S/74S04A
54LS/74LS04
HEX INVERTER

CONNECTION DIAGRAMS
PINOUT A

PINOUT B

ORDERING CODE: See Section 9

PKGS	PIN OUT	COMMERCIAL GRADE Vcc = +5.0 V ±5%, TA = 0°C to +70°C	MILITARY GRADE Vcc = +5.0 V ±10%, TA = -55°C to +125°C	PKG TYPE
Plastic DIP (P)	A	7404PC, 74H04PC 74S04PC, 74S04APC 74LS04PC		9A
Ceramic DIP (D)	A	7404DC, 74H04DC 74S04DC, 74S04ADC 74LS04DC	5404DM, 54H04DM 54S04DM, 54S04ADM 54LS04DM	6A
Flatpak (F)	A	74S04FC, 74S04AFC 74LS04FC	54S04FM, 54S04AFM 54LS04FM	3I
	B	7404FC, 74H04FC	5404FM, 54H04FM	

INPUT LOADING/FAN-OUT: See Section 3 for U.L. definitions

PINS	54/74 (U.L.) HIGH/LOW	54/74H (U.L.) HIGH/LOW	54/74S (U.L.) HIGH/LOW	54/74LS (U.L.) HIGH/LOW
Inputs	1.0/1.0	1.25/1.25	1.25/1.25	0.5/0.25
Outputs	20/10	12.5/12.5	25/12.5	10/5.0 (2.5)

DC AND AC CHARACTERISTICS: See Section 3*

SYMBOL	PARAMETER	54/74 Min	54/74 Max	54/74H Min	54/74H Max	54/74S Min	54/74S Max	54/74LS Min	54/74LS Max	UNITS	CONDITIONS
ICCH	Power Supply		12		26		24		2.4	mA	VIN = Gnd, Vcc = Max
ICCL	Current		33		58		54		6.6		VIN = Open
tPLH / tPHL	Propagation Delay		22 / 15		10 / 10	2.0 / 2.0	4.5 / 5.0		10 / 10	ns	Fig. 3-1, 3-4
tPLH / tPHL	Propagation Delay (54/74S04A only)					1.0 / 1.0	3.5 / 4.0			ns	Fig. 3-1, 3-4

*DC limits apply over operating temperature range; AC limits apply at TA = +25°C and Vcc = +5.0 V.

54/7408
54H/74H08
54S/74S08
54LS/74LS08

QUAD 2-INPUT AND GATE

ORDERING CODE: See Section 9

PKGS	PIN OUT	COMMERCIAL GRADE $V_{CC} = +5.0$ V $\pm5\%$, $T_A = 0°$ C to $+70°$ C	MILITARY GRADE $V_{CC} = +5.0$ V $\pm10\%$, $T_A = -55°$ C to $+125°$ C	PKG TYPE
Plastic DIP (P)	A	7408PC, 74H08PC 74S08PC, 74LS08PC		9A
Ceramic DIP (D)	A	7408DC, 74H08DC 74S08DC, 74LS08DC	5408DM, 54H08DM 54S08DM, 54LS08DM	6A
Flatpak (F)	A	7408FC, 74S08FC 74LS08FC	5408FM, 54S08FM 54LS08FM	3I
	B	74H08FC	54H08FM	

INPUT LOADING/FAN-OUT: See Section 3 for U.L. definitions

PINS	54/74 (U.L.) HIGH/LOW	54/74H (U.L.) HIGH/LOW	54/74S (U.L.) HIGH/LOW	54/74LS (U.L.) HIGH/LOW
Inputs	1.0/1.0	1.25/1.25	1.25/1.25	0.5/0.25
Outputs	20/10	12.5/12.5	25/12.5	10/5.0 (2.5)

DC AND AC CHARACTERISTICS: See Section 3*

SYMBOL	PARAMETER	54/74 Min	54/74 Max	54/74H Min	54/74H Max	54/74S Min	54/74S Max	54/74LS Min	54/74LS Max	UNITS	CONDITIONS
I_{CCH}	Power Supply		21		40		32		4.8	mA	V_{IN} = Open V_{CC} = Max
I_{CCL}	Current		33		64		57		8.8		V_{IN} = Gnd
t_{PLH}	Propagation Delay		27		12	2.5	7.0		13	ns	Fig. 3-1, 3-5
t_{PHL}			19		12	2.5	7.5		11		

*DC limits apply over operating temperature range; AC limits apply at $T_A = +25°$ C and $V_{CC} = +5.0$ V.

54/7420
54H/74H20
54S/74S20
54LS/74LS20
DUAL 4-INPUT NAND GATE

ORDERING CODE: See Section 9

PKGS	PIN OUT	COMMERCIAL GRADE $V_{CC} = +5.0$ V $\pm5\%$, $T_A = 0°$C to $+70°$C	MILITARY GRADE $V_{CC} = +5.0$ V $\pm10\%$, $T_A = -55°$C to $+125°$C	PKG TYPE
Plastic DIP (P)	A	7420PC, 74H20PC 74S20PC, 74LS20PC		9A
Ceramic DIP (D)	A	7420DC, 74H20DC 74S20DC, 74LS20DC	5420DM, 54H20DM 54S20DM, 54LS20DM	6A
Flatpak (F)	A	74S20FC, 74LS20FC	54S20FM, 54LS20FM	3I
	B	7420FC, 74H20FC	5420FM, 54H20FM	

PINOUT B

INPUT LOADING/FAN-OUT: See Section 3 for U.L. definitions

PINS	54/74 (U.L.) HIGH/LOW	54/74H (U.L.) HIGH/LOW	54/74S (U.L.) HIGH/LOW	54/74LS (U.L.) HIGH/LOW
Inputs	1.0/1.0	1.25/1.25	1.25/1.25	0.5/0.25
Outputs	20/10	12.5/12.5	25/12.5	10/5.0 (2.5)

DC AND AC CHARACTERISTICS: See Section 3*

SYMBOL	PARAMETER	54/74 Min Max	54/74H Min Max	54/74S Min Max	54/74LS Min Max	UNITS	CONDITIONS	
I_{CCH}	Power Supply	4.0	8.4	8.0	0.8	mA	V_{IN} = Gnd	V_{CC} = Max
I_{CCL}	Current	11	20	18	2.2		V_{IN} = Open	
t_{PLH}	Propagation Delay	22	10	2.0 4.5	15	ns	Figs. 3-1, 3-4	
t_{PHL}		15	10	2.0 5.0	15			

*DC limits apply over operating temperature range; AC limits apply at T_A = +25°C and V_{CC} = +5.0 V.

54/7432
54S/74S32
54LS/74LS32

QUAD 2-INPUT OR GATE

ORDERING CODE: See Section 9

PKGS	PIN OUT	COMMERCIAL GRADE V_{CC} = +5.0 V ±5%, T_A = 0°C to +70°C	MILITARY GRADE V_{CC} = +5.0 V ±10%, T_A = -55°C to +125°C	PKG TYPE
Plastic DIP (P)	A	7432PC, 74S32PC 74LS32PC		9A
Ceramic DIP (D)	A	7432DC, 74S32DC 74LS32DC	5432DM, 54S32DM 54LS32DM	6A
Flatpak (F)	A	7432FC, 74S32FC 74LS32FC	5432FM, 54S32FM 54LS32FM	3I

INPUT LOADING/FAN-OUT: See Section 3 for U.L. definitions

PINS	54/74 (U.L.) HIGH/LOW	54/74S (U.L.) HIGH/LOW	54/74LS (U.L.) HIGH/LOW
Inputs	1.0/1.0	1.25/1.25	0.5/0.25
Outputs	20/10	25/12.5	10/5.0 (2.5)

DC AND AC CHARACTERISTICS: See Section 3 for U.L. definitions

SYMBOL	PARAMETER	54/74 Min	54/74 Max	54/74S Min	54/74S Max	54/74LS Min	54/74LS Max	UNITS	CONDITIONS
I_{CCH}	Power Supply Current		22		32		6.2	mA	V_{IN} = Open V_{CC} = Max
I_{CCL}			38		68		9.8		V_{IN} = Gnd
t_{PLH}	Propagation Delay		15	2.0	7.0		15	ns	Figs. 3-1, 3-5
t_{PHL}			22	2.0	7.0		15		

*DC limits apply over operating temperature range; AC limits apply at T_A = +25°C and V_{CC} = +5.0 V.

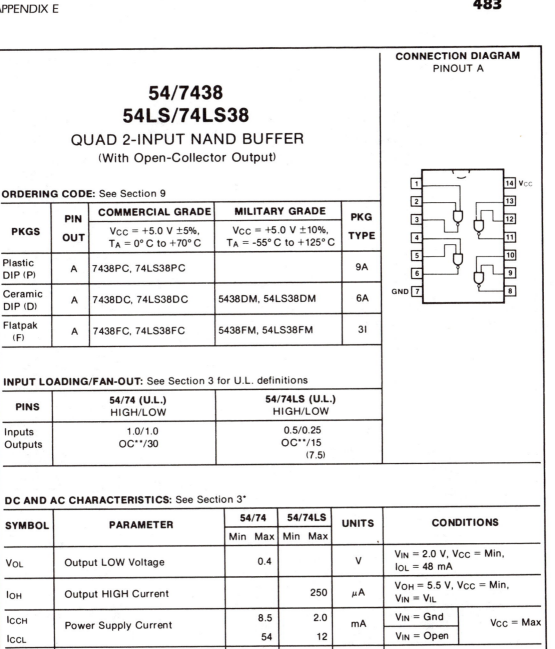

54/7438
54LS/74LS38

QUAD 2-INPUT NAND BUFFER
(With Open-Collector Output)

CONNECTION DIAGRAM
PINOUT A

ORDERING CODE: See Section 9

PKGS	PIN OUT	COMMERCIAL GRADE V_{CC} = +5.0 V ±5%, T_A = 0° C to +70° C	MILITARY GRADE V_{CC} = +5.0 V ±10%, T_A = -55° C to +125° C	PKG TYPE
Plastic DIP (P)	A	7438PC, 74LS38PC		9A
Ceramic DIP (D)	A	7438DC, 74LS38DC	5438DM, 54LS38DM	6A
Flatpak (F)	A	7438FC, 74LS38FC	5438FM, 54LS38FM	3I

INPUT LOADING/FAN-OUT: See Section 3 for U.L. definitions

PINS	54/74 (U.L.) HIGH/LOW	54/74LS (U.L.) HIGH/LOW
Inputs	1.0/1.0	0.5/0.25
Outputs	OC**/30	OC**/15 (7.5)

DC AND AC CHARACTERISTICS: See Section 3*

SYMBOL	PARAMETER	54/74 Min Max	54/74LS Min Max	UNITS	CONDITIONS	
V_{OL}	Output LOW Voltage	0.4		V	V_{IN} = 2.0 V, V_{CC} = Min, I_{OL} = 48 mA	
I_{OH}	Output HIGH Current		250	μA	V_{OH} = 5.5 V, V_{CC} = Min, V_{IN} = V_{IL}	
I_{CCH}	Power Supply Current	8.5	2.0	mA	V_{IN} = Gnd	V_{CC} = Max
I_{CCL}		54	12		V_{IN} = Open	
t_{PLH}	Propagation Delay	22	22	ns	Figs. 3-2, 3-4	
t_{PHL}		18	22			

*DC limits apply over operating temperature range; AC limits apply at T_A = +25° C and V_{CC} = +5.0 V.
**OC — Open Collector

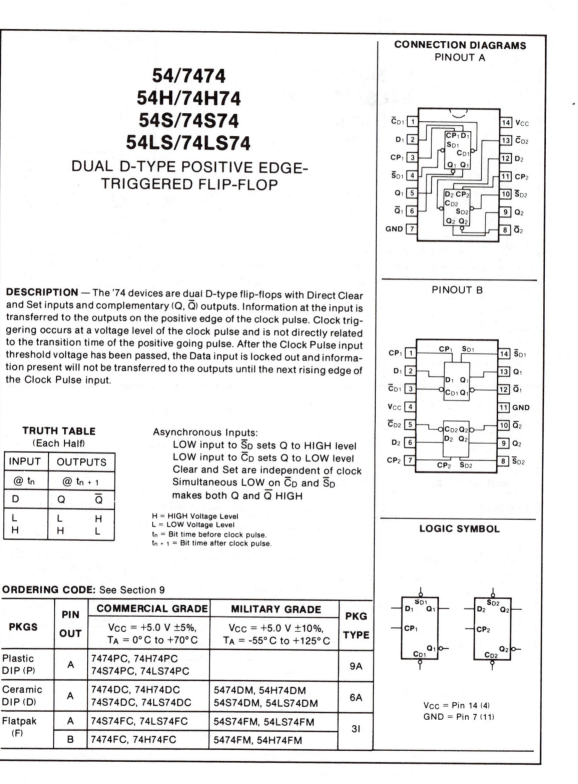

54/7474
54H/74H74
54S/74S74
54LS/74LS74

DUAL D-TYPE POSITIVE EDGE-TRIGGERED FLIP-FLOP

DESCRIPTION — The '74 devices are dual D-type flip-flops with Direct Clear and Set inputs and complementary (Q, $\overline{Q}$) outputs. Information at the input is transferred to the outputs on the positive edge of the clock pulse. Clock triggering occurs at a voltage level of the clock pulse and is not directly related to the transition time of the positive going pulse. After the Clock Pulse input threshold voltage has been passed, the Data input is locked out and information present will not be transferred to the outputs until the next rising edge of the Clock Pulse input.

TRUTH TABLE
(Each Half)

INPUT	OUTPUTS	
@ t_n	@ t_{n+1}	
D	Q	$\overline{Q}$
L	L	H
H	H	L

Asynchronous Inputs:
LOW input to $\overline{S}_D$ sets Q to HIGH level
LOW input to $\overline{C}_D$ sets Q to LOW level
Clear and Set are independent of clock
Simultaneous LOW on $\overline{C}_D$ and $\overline{S}_D$
makes both Q and $\overline{Q}$ HIGH

H = HIGH Voltage Level
L = LOW Voltage Level
t_n = Bit time before clock pulse.
t_{n+1} = Bit time after clock pulse.

CONNECTION DIAGRAMS
PINOUT A

PINOUT B

LOGIC SYMBOL

V_{CC} = Pin 14 (4)
GND = Pin 7 (11)

ORDERING CODE: See Section 9

PKGS	PIN OUT	COMMERCIAL GRADE V_{CC} = +5.0 V ±5%, T_A = 0°C to +70°C	MILITARY GRADE V_{CC} = +5.0 V ±10%, T_A = -55°C to +125°C	PKG TYPE
Plastic DIP (P)	A	7474PC, 74H74PC 74S74PC, 74LS74PC		9A
Ceramic DIP (D)	A	7474DC, 74H74DC 74S74DC, 74LS74DC	5474DM, 54H74DM 54S74DM, 54LS74DM	6A
Flatpak (F)	A	74S74FC, 74LS74FC	54S74FM, 54LS74FM	3I
	B	7474FC, 74H74FC	5474FM, 54H74FM	

INPUT LOADING/FAN-OUT: See Section 3 for U.L. definitions

PIN NAMES	DESCRIPTION	54/74 (U.L.) HIGH/LOW	54/74H (U.L.) HIGH/LOW	54/74S (U.L.) HIGH/LOW	54/74LS (U.L.) HIGH/LOW
D_1, D_2	Data Inputs	1.0/1.0	1.25/1.25	1.25/1.25	0.5/0.25
CP_1, CP_2	Clock Pulse Inputs (Active Rising Edge)	2.0/2.0	2.5/2.5	2.5/2.5	1.0/0.5
$\overline{C}_{D1}, \overline{C}_{D2}$	Direct Clear Inputs (Active LOW)	3.0/2.0	3.75/2.5	3.75/3.75	1.5/0.75
$\overline{S}_{D1}, \overline{S}_{D2}$	Direct Set Inputs (Active LOW)	2.0/1.0	2.5/1.25	2.5/2.5	1.0/0.5
$Q_1, \overline{Q}_1, Q_2, \overline{Q}_2$	Outputs	20/10	12.5/12.5	25/12.5	10/5.0 (2.5)

LOGIC DIAGRAM (one half shown)

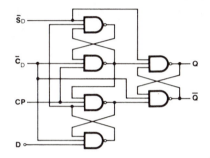

DC CHARACTERISTICS OVER OPERATING TEMPERATURE RANGE (unless otherwise specified)

SYMBOL	PARAMETER		54/74 Min	54/74 Max	54/74H Min	54/74H Max	54/74S Min	54/74S Max	54/74LS Min	54/74LS Max	UNITS	CONDITIONS
I_{CC}	Power Supply Current	XM		30		42		50		8.0	mA	V_{CC} = Max, V_{CP} = 0 V
		XC		30		50		50		8.0		

AC CHARACTERISTICS: V_{CC} = +5.0 V, T_A = +25° C (See Section 3 for waveforms and load configurations)

SYMBOL	PARAMETER	54/74 C_L = 15 pF R_L = 400 Ω Min	54/74 Max	54/74H C_L = 25 pF R_L = 280 Ω Min	54/74H Max	54/74S C_L = 15 pF R_L = 280 Ω Min	54/74S Max	54/74LS C_L = 15 pF Min	54/74LS Max	UNITS	CONDITIONS
f_{max}	Maximum Clock Frequency	15		35		75		30		MHz	Figs. 3-1, 3-8
t_{PLH} t_{PHL}	Propagation Delay CP_n to Q_n or $\overline{Q}_n$		25 40		15 20	9.0 11			25 35	ns	Figs. 3-1, 3-8
t_{PLH} t_{PHL}	Propagation Delay $\overline{C}_{Dn}$ or $\overline{S}_{Dn}$ to Q_n or $\overline{Q}_n$		25 40		20 30	6.0 13.5			15 35	ns	$V_{CP} \geq$ 2.0 V Figs. 3-1, 3-10
t_{PLH} t_{PHL}	Propagation Delay $\overline{C}_{Dn}$ or $\overline{S}_{Dn}$ to Q_n or $\overline{Q}_n$		25 40		20 30	6.0 8.0			15 24	ns	$V_{CP} \leq$ 0.8 V Figs. 3-1, 3-10

AC OPERATING REQUIREMENTS: V_{CC} = +5.0 V, T_A = +25° C

SYMBOL	PARAMETER	54/74		54/74H		54/74S		54/74LS		UNITS	CONDITIONS
		Min	Max	Min	Max	Min	Max	Min	Max		
t_s (H)	Setup Time HIGH D_n to CP_n	20		10		3.0		10		ns	Fig. 3-6
t_h (H)	Hold Time HIGH D_n to CP_n	5.0		0		0		5.0		ns	
t_s (L)	Setup Time LOW D_n to CP_n	20		15		3.0		20		ns	Fig. 3-6
t_h (L)	Hold Time LOW D_n to CP_n	5.0		0		0		5.0		ns	
t_w (H) t_w (L)	CP_n Pulse Width	30 37		15 13.5		6.0 7.3		18 15.5		ns	Fig. 3-8
t_w (L)	$\overline{C}_{Dn}$ or $\overline{S}_{Dn}$ Pulse Width LOW	30		25		7.0		15		ns	Fig. 3-10

54/7486
54S/74S86
54LS/74LS86
QUAD 2-INPUT EXCLUSIVE-OR GATE

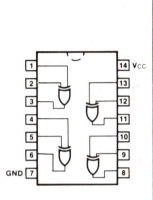

ORDERING CODE: See Section 9

PKGS	PIN OUT	COMMERCIAL GRADE $V_{CC} = +5.0$ V ±5%, $T_A = 0°$C to $+70°$C	MILITARY GRADE $V_{CC} = +5.0$ V ±10%, $T_A = -55°$C to $+125°$C	PKG TYPE
Plastic DIP (P)	A	7486PC, 74S86PC 74LS86PC		9A
Ceramic DIP (D)	A	7486DC, 74S86DC 74LS86DC	5486DM, 54S86DM 54LS86DM	6A
Flatpak (F)	A	7486FC, 74S86FC 74LS86FC	5486FM, 54S86FM 54LS86FM	3I

INPUT LOADING/FAN-OUT: See Section 3 for U.L. definitions

PINS	54/74 (U.L.) HIGH/LOW	54/74S (U.L.) HIGH/LOW	54/74LS (U.L.) HIGH/LOW
Inputs	1.0/1.0	1.25/1.25	1.0/0.375
Outputs	20/10	25/12.5	10/5.0 (2.5)

DC AND AC CHARACTERISTICS: See Section 3*

SYMBOL	PARAMETER		54/74 Min	54/74 Max	54/74S Min	54/74S Max	54/74LS Min	54/74LS Max	UNITS	CONDITIONS
I_{CC}	Power Supply Current	XM		43		75		10	mA	V_{CC} = Max, V_{IN} = Gnd
		XC		50		75		10		
t_{PLH}	Propagation Delay			23	3.5	10.5		12	ns	Other Input LOW Figs. 3-1, 3-5
t_{PHL}				17	3.0	10		17		
t_{PLH}	Propagation Delay			30	3.5	10.5		13	ns	Other Input HIGH Figs. 3-1, 3-4
t_{PHL}				22	3.0	10		12		

*DC limits apply over operating temperature range; AC limits apply at $T_A = +25°$C and $V_{CC} = +5.0$ V.

54/7490A
54LS/74LS90
DECADE COUNTER

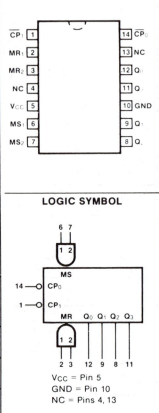

CONNECTION DIAGRAM
PINOUT A

DESCRIPTION — The '90 is a 4-stage ripple counter containing a high speed flip-flop acting as a divide-by-two and three flip-flops connected as a divide-by-five counter. It can be connected to operate with a conventional BCD output pattern or it can be connected to provide a 50% duty cycle output. In the BCD mode, HIGH signals on the Master Set (MS) inputs set the outputs to BCD nine. HIGH signals on the Master Reset (MR) inputs force all outputs LOW. For a similar counter with corner power pins, see the 'LS290; for dual versions, see the 'LS390 and 'LS490.

LOGIC SYMBOL

ORDERING CODE: See Section 9

PKGS	PIN OUT	COMMERCIAL GRADE V_{CC} = +5.0 V ±5%, T_A = 0°C to +70°C	MILITARY GRADE V_{CC} = +5.0 V ±10%, T_A = -55°C to +125°C	PKG TYPE
Plastic DIP (P)	A	7490APC, 74LS90PC		9A
Ceramic DIP (D)	A	7490ADC, 74LS90DC	5490ADM, 54LS90DM	6A
Flatpak (F)	A	7490AFC, 74LS90FC	5490AFM, 54LS90FM	3I

INPUT LOADING/FAN-OUT: See Section 3 for U.L. defintions

PIN NAMES	DESCRIPTION	54/74 (U.L.) HIGH/LOW	54/74LS (U.L.) HIGH/LOW
$\overline{CP}_0$	÷2 Section Clock Input (Active Falling Edge)	2.0/2.0	0.125/1.5
$\overline{CP}_1$	÷5 Section Clock Input (Active Falling Edge)	3.0/3.0	0.250/2.0
MR_1, MR_2	Asynchronous Master Reset Inputs (Active HIGH)	1.0/1.0	0.5/0.25
MS_1, MS_2	Asynchronous Master Set (Preset 9) Inputs (Active HIGH)	1.0/1.0	0.5/0.25
Q_0	÷2 Section Output*	20/10	10/5.0 (2.5)
$Q_1 - Q_3$	÷5 Section Outputs	20/10	10/5.0 (2.5)

*The Q_0 output is guaranteed to drive the full rated fan-out plus the $\overline{CP}_1$ input.

FUNCTIONAL DESCRIPTION — The '90 is a 4-bit ripple type decade counter. It consists of four master/slave flip-flops which are internally connected to provide a divide-by-two section and a divide-by-five section. Each section has a separate clock input which initiates state changes of the counter on the HIGH-to-LOW clock transition. State changes of the Q outputs do not occur simultaneously because of internal ripple delays. Therefore, decoded output signals are subject to decoding spikes and should not be used for clocks or strobes. The Q_0 output of each device is designed and specified to drive the rated fan-out plus the $\overline{CP}_1$ input. A gated AND asynchronous Master Reset (MR_1, MR_2) is provided which overrides the clocks and resets (clears) all the flip-flops. A gated AND asynchronous Master Set (MS_1, MS_2) is provided which overrides the clocks and the MR inputs and sets the outputs to nine (HLLH). Since the output from the divide-by-two section is not internally connected to the succeeding stages, the devices may be operated in various counting modes.:

A. BCD Decade (8421) Counter — The $\overline{CP}_1$ input must be externally connected to the Q_0 output. The $\overline{CP}_0$ input receives the incoming count and a BCD count sequence is produced.

B. Symmetrical Bi-quinary Divide-By-Ten Counter — The Q_3 output must be externally connected to the $\overline{CP}_0$ input. The input count is then applied to the $\overline{CP}_1$ input and a divide-by-ten square wave is obtained at output Q_0.

C. Divide-By-Two and Divide-By-Five Counter — No external interconnections are required. The first flip-flop is used as a binary element for the divide-by-two function ($\overline{CP}_0$ as the input and Q_0 as the output). The $\overline{CP}_1$ input is used to obtain binary divide-by-five operation at the Q_3 output.

MODE SELECTION

MR₁	MR₂	MS₁	MS₂	Q₀	Q₁	Q₃	Q₃
H	H	L	X	L	L	L	L
H	H	X	L	L	L	L	L
X	X	H	H	H	L	L	H
L	X	L	X	Count			
X	L	X	L	Count			
L	X	X	L	Count			
X	L	L	X	Count			

H = HIGH Voltage Level
L = LOW Voltage Level
X = Immaterial

BCD COUNT SEQUENCE

COUNT	Q₀	Q₁	Q₂	Q₃
0	L	L	L	L
1	H	L	L	L
2	L	H	L	L
3	H	H	L	L
4	L	L	H	L
5	H	L	H	L
6	L	H	H	L
7	H	H	H	L
8	L	L	L	H
9	H	L	L	H

NOTE: Output Q_0 is connected to Input $\overline{CP}_1$ for BCD count.

LOGIC DIAGRAM

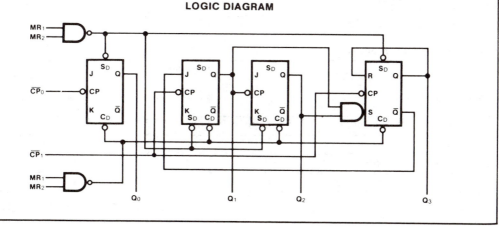

DC CHARACTERISTICS OVER OPERATING TEMPERATURE RANGE (unless otherwise specified)

SYMBOL	PARAMETER	54/74 Min Max	54/74LS	UNITS	CONDITIONS
I_{IH}	Input HIGH Current, $\overline{CP}_0$	1.0	0.2	mA	V_{CC} = Max, V_{IN} = 5.5 V
I_{IH}	Input HIGH Current $\overline{CP}_1$	1.0	0.4	mA	V_{CC} = Max, V_{IN} = 5.5 V
I_{CC}	Power Supply Current	42	15	mA	V_{CC} = Max

AC CHARACTERISTICS: V_{CC} = +5.0 V, T_A = +25° C (See Section 3 for waveforms and load configurations)

SYMBOL	PARAMETER	54/74 C_L = 15 pF R_L = 400 Ω Min Max	54/74LS C_L = 15 pF Min Max	UNITS	CONDITIONS
f_{max}	Maximum Count Frequency, $\overline{CP}_0$	32	32	MHz	Figs. 3-1, 3-9
f_{max}	Maximum Count Frequency, $\overline{CP}_1$	16	16	MHz	Figs. 3-1, 3-9
t_{PLH} t_{PHL}	Propagation Delay $\overline{CP}_0$ to Q_0	16 18	16 18	ns	Figs. 3-1, 3-9
t_{PLH} t_{PHL}	Propagation Delay $\overline{CP}_0$ to Q_3	48 50	48 50	ns	Figs. 3-1, 3-9
t_{PLH} t_{PHL}	Propagation Delay $\overline{CP}_1$ to Q_1	16 21	16 21	ns	Figs. 3-1, 3-9
t_{PLH} t_{PHL}	Propagation Delay $\overline{CP}_1$ to Q_2	32 35	32 35	ns	Figs. 3-1, 3-9
t_{PLH} t_{PHL}	Propagation Delay $\overline{CP}_1$ to Q_3	32 35	32 35	ns	Figs. 3-1, 3-9
t_{PLH}	Propagation Delay MS to Q_0 and Q_3	30	30	ns	Figs. 3-1, 3-17
t_{PHL}	Propagation Delay MS to Q_1 and Q_3	40	40	ns	Figs. 3-1, 3-17
t_{PHL}	Propagation Delay MR to Q_n	40	40	ns	Figs. 3-1, 3-17

AC OPERATING REQUIREMENTS: V_{CC} = +5.0 V, T_A = +25° C

SYMBOL	PARAMETER	54/74 Min Max	54/74LS Min Max	UNITS	CONDITIONS
t_w (H)	$\overline{CP}_0$ Pulse Width HIGH	15	15	ns	Fig. 3-9
t_w (H)	$\overline{CP}_1$ Pulse Width HIGH	30	30	ns	Fig. 3-9
t_w (H)	MS Pulse Width HIGH	15	15	ns	Fig. 3-17
t_w (H)	MR Pulse Width HIGH	15	15	ns	Fig. 3-17
t_{rec}	Recovery Time, MS to $\overline{CP}$	25	25	ns	Fig. 3-17
t_{rec}	Recovery Time, MR to $\overline{CP}$	25	25	ns	Fig. 3-17

54/7494
4-BIT SHIFT REGISTER

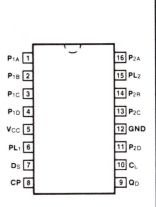

DESCRIPTION — The '94 contains four dc coupled RS master/slave flip-flops with serial data entry into the first stage for synchronous Serial-in/Serial-out operation, and with a common asynchronous Clear and two sets of individual asynchronous Preset inputs. Preset inputs P_{1x} are enabled by a HIGH signal on PL_1 and Preset inputs P_{2x} are enabled by a HIGH signal on PL_2. The normal procedure for paralled entry of data consists of re-setting the flip-flops by applying a momentary HIGH signal to CL, followed by a HIGH signal on either PL_1, or PL_2, depending on which set of parallel data is desired. For serial operation the CL and both PL inputs must be LOW. Serial transfer is initiated by the rising edge of the clock.

ORDERING CODE: See Section 9

PKGS	PIN OUT	COMMERCIAL GRADE V_{CC} = +5.0 V ±5%, T_A = 0°C to +70°C	MILITARY GRADE V_{CC} = +5.0 V ±10%, T_A = -55°C to +125°C	PKG TYPE
Plastic DIP (P)	A	7494PC		9B
Ceramic DIP (D)	A	7494DC	5494DM	7B
Flatpak (F)	A	7494FC	5494FM	4L

INPUT LOADING/FAN-OUT: See Section 3 for U.L. definitions

PIN NAMES	DESCRIPTION	54/74 (U.L.) HIGH/LOW
P_{1A} — P_{1D}	Source 1 Parallel Data Inputs	1.0/1.0
P_{2A} — P_{2D}	Source 2 Parallel Data Inputs	1.0/1.0
PL_1	Asynchronous Parallel Load Input (Source 1)	4.0/4.0
PL_2	Asynchronous Parallel Load Input (Source 2)	4.0/4.0
D_S	Serial Data Input	1.0/1.0
CP	Clock Pulse Input (Active Rising Edge)	1.0/1.0
CL	Asynchronous Clear Input (Active HIGH)	1.0/1.0
Q_D	Serial Data Output	10/10

LOGIC SYMBOL

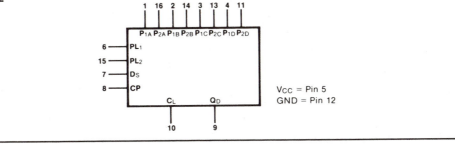

V_{CC} = Pin 5
GND = Pin 12

TRUTH TABLE

CP	CL	PL₁ • P₁D	PL₂ • P₂D	Q_D	RESPONSE
X	H	L	L	L	Clear
X	L	H	X	H	Preset
X	L	X	H	H	Preset
X	H	H	X	H	Indeterminate
X	H	X	H	H	Indeterminate
⌐	L	L	L	Q_C	Shift Right

NOTE: All four flip-flops respond in a similar manner.
H = HIGH Voltage Level
L = LOW Voltage Level
X = Immaterial

LOGIC DIAGRAM

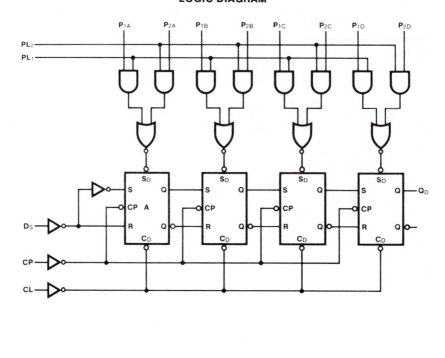

DC CHARACTERISTICS OVER OPERATING TEMPERATURE RANGE (unless otherwise specified)

SYMBOL	PARAMETER		54/74		UNITS	CONDITIONS
			Min	Max		
I_{CC}	Power Supply Current	XM		50	mA	V_{CC} = Max
		XC		58		

AC CHARACTERISTICS: V_{CC} = 5.0 V, T_A = 25°C (See Section 3 for waveforms and load configurations)

SYMBOL	PARAMETER	54/74		UNITS	CONDITIONS
		C_L = 15 pF R_L = 400 Ω			
		Min	Max		
f_{max}	Maximum Shift Frequency	10		MHz	Figs. 3-1, 3-8
t_{PLH} t_{PHL}	Propagation Delay CP to Q_D		40 40	ns	Figs. 3-1, 3-8
t_{PLH}	Propagation Delay, PL_n to Q_D		35	ns	Figs. 3-1, 3-17
t_{PHL}	Propagation Delay, CL to Q_D		40		

AC OPERATING REQUIREMENTS: V_{CC} = +5.0 V, T_A = +25°C

SYMBOL	PARAMETER	54/74		UNITS	CONDITIONS
		Min	Max		
t_s (H)	Setup Time HIGH, D_S to CP	35		ns	Fig. 3-6
t_h (H)	Hold Time HIGH, D_S to CP	0		ns	
t_s (L)	Setup Time LOW, D_S to CP	25		ns	Fig. 3-6
t_h (L)	Hold Time LOW, D_S to CP	0		ns	
t_w (H)	CP Pulse Width HIGH	35		ns	Fig. 3-8
t_w (H)	CL Pulse Width HIGH	30		ns	Fig. 3-16
t_w (H)	PL_n Pulse Width HIGH	30		ns	Fig. 3-16

54/74107
54LS/74LS107

DUAL JK FLIP-FLOP
(With Separate Clears and Clocks)

DESCRIPTION — The '107 dual JK master/slave flip-flops have a separate clock for each flip-flop. Inputs to the master section are controlled by the clock pulse. The clock pulse also regulates the state of the coupling transistors which connect the master and slave sections. The sequence of operation is as follows: 1) isolate slave from master; 2) enter information from J and K inputs to master; 3) disable J and K inputs; 4) transfer information from master to slave.

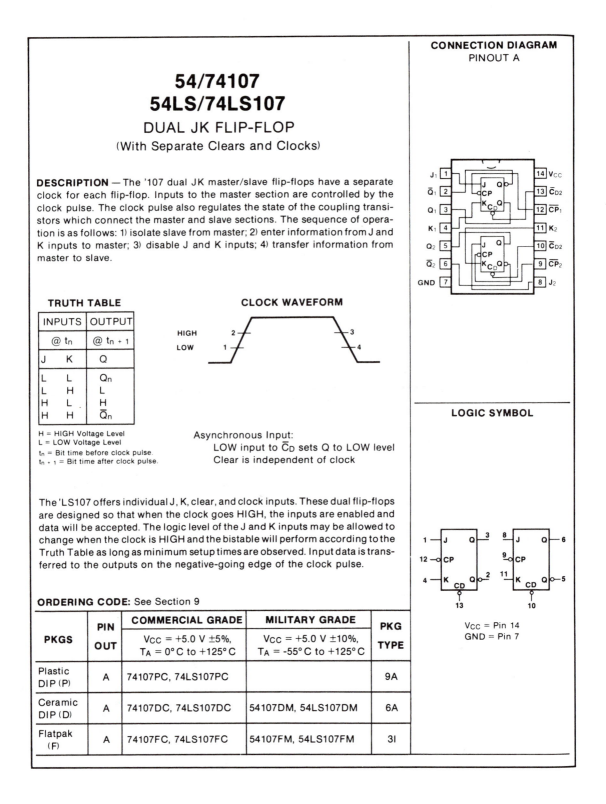

TRUTH TABLE

INPUTS		OUTPUT
@ t_n		@ t_{n+1}
J	K	Q
L	L	Q_n
L	H	L
H	L	H
H	H	$\overline{Q}_n$

H = HIGH Voltage Level
L = LOW Voltage Level
t_n = Bit time before clock pulse.
t_{n+1} = Bit time after clock pulse.

CLOCK WAVEFORM

Asynchronous Input:
LOW input to $\overline{C}_D$ sets Q to LOW level
Clear is independent of clock

The 'LS107 offers individual J, K, clear, and clock inputs. These dual flip-flops are designed so that when the clock goes HIGH, the inputs are enabled and data will be accepted. The logic level of the J and K inputs may be allowed to change when the clock is HIGH and the bistable will perform according to the Truth Table as long as minimum setup times are observed. Input data is transferred to the outputs on the negative-going edge of the clock pulse.

LOGIC SYMBOL

V_{CC} = Pin 14
GND = Pin 7

ORDERING CODE: See Section 9

PKGS	PIN OUT	COMMERCIAL GRADE V_{CC} = +5.0 V ±5%, T_A = 0°C to +125°C	MILITARY GRADE V_{CC} = +5.0 V ±10%, T_A = -55°C to +125°C	PKG TYPE
Plastic DIP (P)	A	74107PC, 74LS107PC		9A
Ceramic DIP (D)	A	74107DC, 74LS107DC	54107DM, 54LS107DM	6A
Flatpak (F)	A	74107FC, 74LS107FC	54107FM, 54LS107FM	3I

INPUT LOADING/FAN-OUT: See Section 3 for U.L. definitions

PIN NAMES	DESCRIPTION	54/74 (U.L.) HIGH/LOW	54/74LS (U.L.) HIGH/LOW
J_1, J_2, K_1, K_2	Data Inputs	1.0/1.0	0.5/0.25
$\overline{CP}_1$, $\overline{CP}_2$	Clock Pulse Inputs (Active Falling Edge)	2.0/2.0	2.0/0.5
$\overline{C}_{D1}$, $\overline{C}_{D2}$	Direct Clear Inputs (Active LOW)	2.0/2.0	1.5/0.5
Q_1, Q_2, $\overline{Q}_1$, $\overline{Q}_2$	Outputs	20/10	10/5.0 (2.5)

LOGIC DIAGRAM (one half shown)

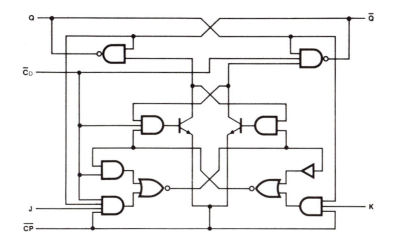

DC CHARACTERISTICS OVER OPERATING TEMPERATURE RANGE (unless otherwise specified)

SYMBOL	PARAMETER	54/74		54/74LS		UNITS	CONDITIONS
		Min	Max	Min	Max		
I_{CC}	Power Supply Current		40		8.0	mA	V_{CC} = Max, V_{CP} = 0 V

AC CHARACTERISTICS: V_{CC} = +5.0 V, T_A = +25° C (See Section 3 for waveforms and load configurations)

SYMBOL	PARAMETER	54/74 C_L = 15 pF R_L = 400 Ω		54/74LS C_L = 15 pF		UNITS	CONDITIONS
		Min	Max	Min	Max		
f_{max}	Maximum Clock Frequency	15		30		MHz	Figs. 3-1, 3-9
t_{PLH} t_{PHL}	Propagation Delay $\overline{CP}_n$ to Q_n or $\overline{Q}_n$		25 40		20 30	ns	Figs. 3-1, 3-9
t_{PLH} t_{PHL}	Propagation Delay $\overline{C}_{Dn}$ to Q_n or $\overline{Q}_n$		25 40		20 30	ns	Figs. 3-1, 3-10

AC OPERATING REQUIREMENTS: V_{CC} = +5.0 V, T_A = +25° C

SYMBOL	PARAMETER	54/74		54/74LS		UNITS	CONDITIONS
		Min	Max	Min	Max		
t_s (H)	Setup Time HIGH J_n or K_n to $\overline{CP}_n$	0		20		ns	Fig. 3-18 ('107) Fig. 3-7 ('LS107)
t_h (H)	Hold Time HIGH J_n or K_n to $\overline{CP}_n$	0		0		ns	
t_s (L)	Setup Time LOW J_n or K_n to $\overline{CP}_n$	0		20		ns	
t_h (L)	Hold Time LOW J_n or K_n to $\overline{CP}_n$	0		0		ns	
t_w (H) t_w (L)	$\overline{CP}_n$ Pulse Width	20 47		13.5 20		ns	Fig. 3-9
t_w (L)	$\overline{C}_{Dn}$ Pulse Width LOW	25		25		ns	Fig. 3-10

54/74121
MONOSTABLE MULTIVIBRATOR

CONNECTION DIAGRAM
PINOUT A

Q	1	14	V_{CC}
NC	2	13	NC
$\bar{A}_1$	3	12	NC
$\bar{A}_2$	4	11	$R_x C_x$
B	5	10	C_x
Q	6	9	R_{INT}
GND	7	8	NC

DESCRIPTION — The '121 features positive and negative dc level triggering inputs and complementary outputs. Input pin 5 directly activates a Schmitt circuit which provides temperature compensated level detection, increases immunity to positive-going noise and assures jitter-free response to slowly rising triggers.

When triggering occurs, internal feedback latches the circuit, prevents re-triggering while the output pulse is in progress and increases immunity to negative-going noise. Noise immunity is typically 1.2 V at the inputs and 1.5 V on V_{CC}.

Output pulse width stability is primarily a function of the external R_x and C_x chosen for the application. A 2 kΩ internal resistor is provided for optional use where output pulse width stability requirements are less stringent. Maximum duty cycle capability ranges from 67% with a 2 kΩ resistor to 90% with a 40 kΩ resistor. Duty cycles beyond this range tend to reduce the output pulse width. Otherwise, output pulse width follows the relationship:

$$t_w = 0.69\ R_x C_x$$

ORDERING CODE: See Section 9

PKGS	PIN OUT	COMMERCIAL GRADE $V_{CC} = +5.0\ V \pm 5\%$, $T_A = 0°C$ to $+70°C$	MILITARY GRADE $V_{CC} = +5.0\ V \pm 10\%$, $T_A = -55°C$ to $+125°C$	PKG TYPE
Plastic DIP (P)	A	74121PC		9A
Ceramic DIP (D)	A	74121DC	54121DM	6A
Flatpak (F)	A	74121FC	54121FM	3I

LOGIC SYMBOL

V_{CC} = Pin 14
GND = Pin 7
NC = Pins 2,8,12,13

INPUT LOADING/FAN-OUT: See Section 3 for U.L.definitions

PIN NAMES	DESCRIPTION	54/74 (U.L.) HIGH/LOW
$\bar{A}_1$, $\bar{A}_2$	Trigger Inputs (Active Falling Edge)	1.0/1.0
B	Schmitt Trigger Input (Active Rising Edge)	2.0/2.0
Q, $\bar{Q}$	Outputs	20/10

TRIGGERING TRUTH TABLE

INPUTS			RESPONSE
$\bar{A}_1$	$\bar{A}_2$	B	
H	H	⌐	No Trigger
L	X	⌐	Trigger
X	L	⌐	Trigger
⅃	L	X	No Trigger
⅃	X	L	No Trigger
⅃	H	H	Trigger
L	⅃	X	No Trigger
X	⅃	L	No Trigger
H	⅃	H	Trigger

NOTE:

Triggering occurs only when the $\bar{Q}$ output is HIGH (not in timing cycle) and one of the above triggering situations is satisfied.
H = HIGH Voltage Level
L = LOW Voltage Level
X = Immaterial

DC CHARACTERISTICS OVER OPERATING TEMPERATURE RANGE (unless otherwise specified)

SYMBOL	PARAMETER		54/74		UNITS	CONDITIONS
			Min	Max		
V_{T+}	Positive-going Threshold Voltage at $\bar{A}_n$ or B Inputs			2.0	V	V_{CC} = Min
V_{T-}	Negative-going Threshold Voltage at $\bar{A}_n$ or B Inputs		0.8		V	V_{CC} = Min
I_{OS}	Output Short Circuit Current	XM	−20	−55	mA	V_{CC} = Max
		XC	−18	−55		
I_{CC}	Power Supply Current	Quiescent State		25	mA	V_{CC} = Max
		Fired State		40		

AC CHARACTERISTICS: V_{CC} = +5.0 V, T_A = +25°C (See Section 3 for waveforms and load configurations)

SYMBOL	PARAMETER	54/74 C_L = 15 pF		UNITS	CONDITIONS	
		Min	Max			
t_{PLH}	Propagation Delay B to Q	15	55	ns		
t_{PLH}	Propagation Delay $\bar{A}_n$ to Q	25	70	ns	C_x = 80 pF Fig. 3-1, Fig. a	
t_{PHL}	Propagation Delay B to $\bar{Q}$	20	65	ns		
t_{PHL}	Propagation Delay $\bar{A}_n$ to $\bar{Q}$	30	80	ns		
t_w	Pulse Width Using Internal Timing Resistor	70	150	ns	C_x = 80 pF	R_x = Open Fig. 3-1 Fig. a Pin 9 = V_{CC}
t_w	Pulse Width with Zero Timing Capacitance	20	50	ns	C_x = 0 pF	
t_w	Pulse Width Using External Timing Resistor	600	800	ns	C_x = 100 pF	R_x = 10 kΩ Pin 9 = Open Fig. 3-1, a
		6.0	8.0	ms	C_x = 1.0 μF	
t_{HOLD}	Minimum Duration of Trigger Pulse		50	ns	C_x = 80 pF, R_x = Open Pin 9 = V_{CC}, Fig. a	

AC OPERATING REQUIREMENTS: V_{CC} = +5.0 V, T_A = +25° C

SYBMOL	PARAMETER		54/74		UNITS	CONDITIONS
			Min	Max		
V_{r-f}	Input Pulse Rise/Fall Slew Rate	@ A_n		1.0	V/μs	
		@ B		1.0	V/s	
R_X	External Timing Resistor	XC	1.4	40	kΩ	
		XM	1.4	30		
C_X	External Timing Capacitor		0	1000	μF	
t_W	Output Pulse Width			40	sec	Fig. a
	Duty Cycle	XM,XC		67	%	R_X = 2 kΩ
		XM		90		R_X = 30 kΩ
		XC		90		R_X = 40 kΩ

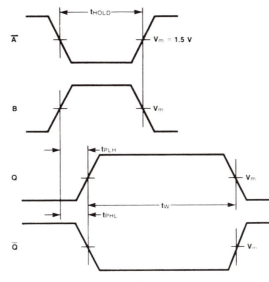

Fig. a

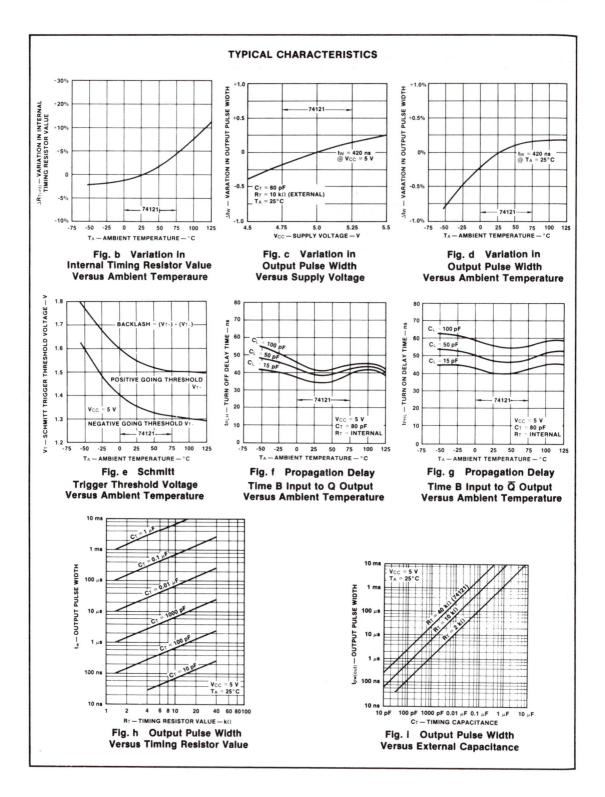

TYPICAL CHARACTERISTICS

Fig. b Variation in
Internal Timing Resistor Value
Versus Ambient Temperaure

Fig. c Variation in
Output Pulse Width
Versus Supply Voltage

Fig. d Variation in
Output Pulse Width
Versus Ambient Temperature

Fig. e Schmitt
Trigger Threshold Voltage
Versus Ambient Temperature

Fig. f Propagation Delay
Time B Input to Q Output
Versus Ambient Temperature

Fig. g Propagation Delay
Time B Input to $\overline{Q}$ Output
Versus Ambient Temperature

Fig. h Output Pulse Width
Versus Timing Resistor Value

Fig. i Output Pulse Width
Versus External Capacitance

54/74166
8-BIT SHIFT REGISTER

DESCRIPTION — The '166 is an 8-bit, serial- or parallel-in, serial-out shift register using edge triggered D-type flip-flops. Serial and parallel entry are synchronous, with state changes initiated by the rising edge of the clock. An asynchronous Master Reset overrides other inputs and clears all flip-flops. The circuit can be clocked from two sources or one CP input can be used to trigger the other.

- **35 MHz TYPICAL SHIFT FREQUENCY**
- **ASYNCHRONOUS MASTER RESET**
- **SYNCHRONOUS PARALLEL ENTRY**
- **GATED CLOCK INPUT CIRCUITRY**

CONNECTION DIAGRAM
PINOUT A

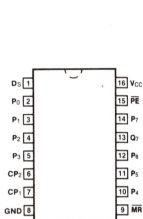

ORDERING CODE: See Section 9

PKGS	PIN OUT	COMMERCIAL GRADE $V_{CC} = +5.0$ V $\pm5\%$, $T_A = 0°$C to $+70°$C	MILITARY GRADE $V_{CC} = +5.0$ V $\pm10\%$, $T_A = -55°$C to $+125°$C	PKG TYPE
Plastic DIP (P)	A	74166PC		9B
Ceramic DIP (D)	A	74166DC	54166DM	7B
Flatpak (F)	A	74166FC	54166FM	4L

INPUT LOADING/FAN-OUT: See Section 3 for U.L. definitions

PIN NAMES	DESCRIPTION	54/74 (U.L.) HIGH/LOW
CP_1, CP_2	Clock Pulse Inputs (Active Rising Edge)	1.0/1.0
D_S	Serial Data Input	1.0/1.0
$\overline{PE}$	Parallel Enable Input (Active LOW)	1.0/1.0
$P_0 - P_7$	Parallel Data Inputs	1.0/1.0
$\overline{MR}$	Asynchronous Master Reset Input (Active LOW)	1.0/1.0
Q_7	Last Stage Output	20/10

LOGIC SYMBOL

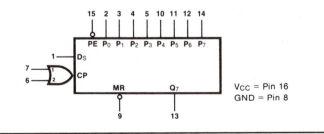

V_{CC} = Pin 16
GND = Pin 8

FUNCTIONAL DESCRIPTION — Operation is synchronous (except for Master Reset) and state changes are initiated by the rising edge of either clock input if the other clock input is LOW. When one of the clock inputs is used as an active HIGH clock inhibit, it should attain the HIGH state while the other clock is still in the HIGH state following the previous operation. When the Parallel Enable ($\overline{PE}$) input is LOW, data is loaded into the register from the Parallel Data ($P_0 - P_7$) inputs on the next rising edge of the clock. When $\overline{PE}$ is HIGH, information is shifted from the Serial Data (D_S) input to Q_0 and all data in the register is shifted one bit position (i.e., $Q_0 \rightarrow Q_1$, $Q_1 \rightarrow Q_2$, etc.) on the rising edge of the clock.

MODE SELECT TABLE

INPUTS				RESPONSE
$\overline{MR}$	$\overline{PE}$	CP_1	CP_2	
L	X	X	X	Asynchronous Reset; Q_n = LOW
H	X	H*	X	Hold
H	X	X	H*	
H	L	L	⌐	Parallel Load; $P_n \longrightarrow Q_n$
H	L	⌐	L	
H	H	L	⌐	Shift; $D_S \rightarrow Q_0$, $Q_0 \longrightarrow Q_1$, etc.
H	H	⌐	L	

*The HIGH signal on one CP input must be established while the other CP input is HIGH.
H = HIGH Voltage Level
L = LOW Voltage Level
X = Immaterial

LOGIC DIAGRAM

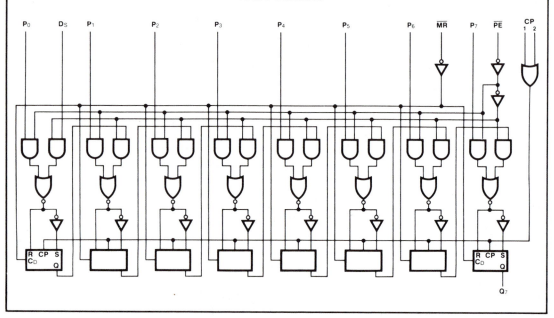

DC CHARACTERISTICS OVER OPERATING TEMPERATURE RANGE (unless otherwise specified)

SYMBOL	PARAMETER	54/74 Min	54/74 Max	UNITS	CONDITIONS
I_{CC}	Power Supply Current		127	mA	V_{CC} = Max, CP_1 = ⌐_ D_S = 4.5 V CP_2, $\overline{MR}$, $\overline{PE}$, P_n = Gnd

AC CHARACTERISTICS: V_{CC} = +5.0 V, T_A = +25° C (See Section 3 for waveforms and load configurations)

SYMBOL	PARAMETER	54/74 C_L = 15 pF R_L = 400 Ω Min	Max	UNITS	CONDITIONS
f_{max}	Maximum Clock Frequency	25		MHz	Figs. 3-1, 3-8
t_{PLH} t_{PHL}	Propagation Delay CP_n to Q_7		26 30	ns	
t_{PHL}	Propagation Delay $\overline{MR}$ to Q_7		35	ns	Figs. 3-1, 3-16

AC OPERATING REQUIREMENTS: V_{CC} = +5.0 V, T_A = +25° C

SYMBOL	PARAMETER	54/74 Min	54/74 Max	UNITS	CONDITIONS
t_s (H) t_s (L)	Setup Time HIGH or LOW D_S or P_n to CP_n	20 20		ns	
t_h (H) t_h (L)	Hold Time HIGH or LOW D_S or P_n to CP_n	0 0		ns	Fig. 3-6
t_s (H) t_s (L)	Setup Time HIGH or LOW $\overline{PE}$ to CP_n	30 30		ns	
t_h (H) t_h (L)	Hold Time HIGH or LOW $\overline{PE}$ to CP_n	0 0		ns	
t_w (H)	CP_n Pulse Width HIGH	20		ns	Fig. 3-8
t_w (L)	$\overline{MR}$ Pulse Width LOW	20		ns	Fig. 3-16

54LS/74LS181
4-BIT ARITHMETIC LOGIC UNIT

DESCRIPTION — The '181 is a 4-bit Arithmetic Logic Unit (ALU) which can perform all the possible 16 logic operations on two variables and a variety of arithmetic operations. For improved TTL, S-TTL and LP-TTL versions, please see the 9341 data sheet.

- **PROVIDES 16 ARITHMETIC OPERATIONS**
  ADD, SUBTRACT, COMPARE, DOUBLE, PLUS
  TWELVE OTHER ARITHMETIC OPERATIONS
- **PROVIDES ALL 16 LOGIC OPERATIONS OF TWO VARIABLES**
  EXCLUSIVE - OR, COMPARE, AND, NAND, OR, NOR,
  PLUS TEN OTHER LOGIC OPERATIONS
- **FULL LOOKAHEAD FOR HIGH SPEED ARITHMETIC**
  OPERATION ON LONG WORDS

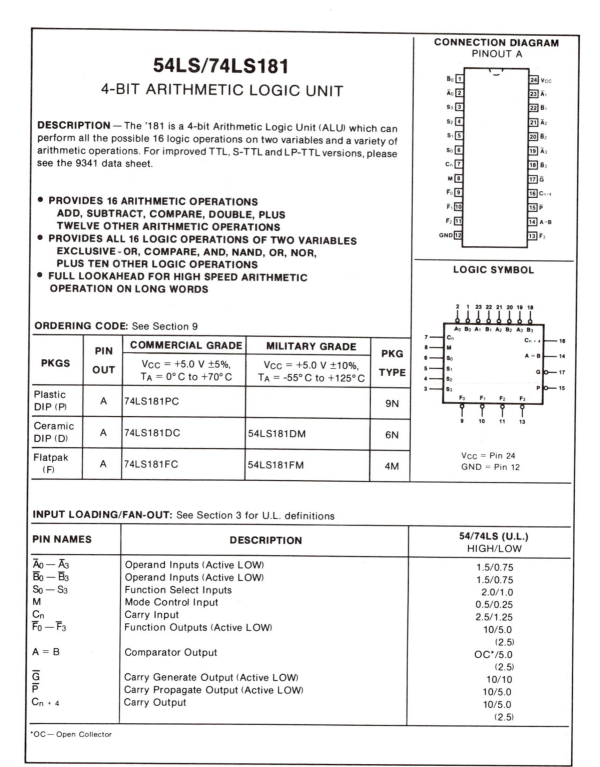

CONNECTION DIAGRAM
PINOUT A

LOGIC SYMBOL

V_{CC} = Pin 24
GND = Pin 12

ORDERING CODE: See Section 9

PKGS	PIN OUT	COMMERCIAL GRADE V_{CC} = +5.0 V ±5%, T_A = 0°C to +70°C	MILITARY GRADE V_{CC} = +5.0 V ±10%, T_A = -55°C to +125°C	PKG TYPE
Plastic DIP (P)	A	74LS181PC		9N
Ceramic DIP (D)	A	74LS181DC	54LS181DM	6N
Flatpak (F)	A	74LS181FC	54LS181FM	4M

INPUT LOADING/FAN-OUT: See Section 3 for U.L. definitions

PIN NAMES	DESCRIPTION	54/74LS (U.L.) HIGH/LOW
$\overline{A}_0 - \overline{A}_3$	Operand Inputs (Active LOW)	1.5/0.75
$\overline{B}_0 - \overline{B}_3$	Operand Inputs (Active LOW)	1.5/0.75
$S_0 - S_3$	Function Select Inputs	2.0/1.0
M	Mode Control Input	0.5/0.25
C_n	Carry Input	2.5/1.25
$\overline{F}_0 - \overline{F}_3$	Function Outputs (Active LOW)	10/5.0 (2.5)
A = B	Comparator Output	OC*/5.0 (2.5)
$\overline{G}$	Carry Generate Output (Active LOW)	10/10
$\overline{P}$	Carry Propagate Output (Active LOW)	10/5.0
$C_{n + 4}$	Carry Output	10/5.0 (2.5)

*OC — Open Collector

FUNCTIONAL DESCRIPTION — The 'LS181 is a 4-bit high speed parallel Arithmetic Logic Unit (ALU). Controlled by the four Function Select inputs ($S_0 - S_3$) and the Mode Control input (M), it can perform all the 16 possible logic operations or 16 different arithmetic operations on active HIGH or active LOW operands. The Function Table lists these operations.

When the Mode Control input (M) is HIGH, all internal carries are inhibited and the device performs logic operations on the individual bits as listed. When the Mode Control input is LOW, the carries are enabled and the device performs arithmetic operations on the two 4-bit words. The device incorporates full internal carry lookahead and provides for either ripple carry between devices using the C_{n+4} output, or for carry lookahead between packages using the signals $\overline{P}$ (Carry Propagate) and $\overline{G}$ (Carry Generate). In the ADD mode, $\overline{P}$ indicates that $\overline{F}$ is 15 or more, while $\overline{G}$ indicates that $\overline{F}$ is 16 or more. In the SUBTRACT mode, $\overline{P}$ indicates that $\overline{F}$ is zero or less, while $\overline{G}$ indicates that $\overline{F}$ is less than zero. $\overline{P}$ and $\overline{G}$ are not affected by carry in. When speed requirements are not stringent, it can be used in a simple ripple carry mode by connecting the Carry output (C_{n+4}) signal to the Carry input (C_n) of the next unit. For high speed operation the device is used in conjunction with the 9342 or 93S42 carry lookahead circuit. One carry lookahead package is required for each group of four 'LS181 devices. Carry lookahead can be provided at various levels and offers high speed capability over extremely long word lengths.

The A = B output from the device goes HIGH when all four $\overline{F}$ outputs are HIGH and can be used to indicate logic equivalence over four bits when the unit is in the subtract mode. The A = B output is open-collector and can be wired-AND with other A = B outputs to give a comparison for more than four bits. The A = B signal can also be used with the C_{n+4} signal to indicate A > B and A < B.

The Function Table lists the arithmetic operations that are performed without a carry in. An incoming carry adds a one to each operation. Thus, select code LHHL generates A minus B minus 1 (2s complement notation) without a carry in and generates A minus B when a carry is applied. Because subtraction is actually performed by complementary addition (1s complement), a carry out means borrow; thus a carry is generated when there is no underflow and no carry is generated when there is underflow. As indicated, this device can be used with either active LOW inputs producing active LOW outputs or with active HIGH inputs producing active HIGH outputs. For either case the table lists the operations that are performed to the operands labeled inside the logic symbol.

FUNCTION TABLE

MODE SELECT INPUTS				ACTIVE LOW OPERANDS & F_n OUTPUTS		ACTIVE HIGH OPERANDS & F_n OUTPUTS	
S_3	S_2	S_1	S_0	LOGIC (M = H)	ARITHMETIC** (M = L) (C_n = L)	LOGIC (M = H)	ARITHMETIC** (M = L) (C_n = H)
L	L	L	L	$\overline{A}$	A minus 1	$\overline{A}$	A
L	L	L	H	$\overline{AB}$	AB minus 1	$\overline{A+B}$	A + B
L	L	H	L	$\overline{A}+B$	$A\overline{B}$ minus 1	$\overline{A}B$	$A + \overline{B}$
L	L	H	H	Logic 1	minus 1	Logic 0	minus 1
L	H	L	L	$\overline{A+B}$	A plus $(A + \overline{B})$	$\overline{AB}$	A plus $A\overline{B}$
L	H	L	H	$\overline{B}$	AB plus $(A + \overline{B})$	$\overline{B}$	$(A + B)$ plus $A\overline{B}$
L	H	H	L	$A \oplus B$	A minus B minus 1	$A \oplus B$	A minus B minus 1
L	H	H	H	$A + \overline{B}$	$A + \overline{B}$	$A\overline{B}$	AB minus 1
H	L	L	L	$\overline{A}B$	A plus $(A + B)$	$\overline{A}+B$	A plus AB
H	L	L	H	$A \oplus B$	A plus B	$A \oplus B$	A plus B
H	L	H	L	B	$A\overline{B}$ plus $(A + B)$	B	$(A + \overline{B})$ plus AB
H	L	H	H	A + B	A + B	AB	AB minus 1
H	H	L	L	Logic 0	A plus A*	Logic 1	A plus A*
H	H	L	H	$A\overline{B}$	AB plus A	$A + \overline{B}$	$(A + B)$ plus A
H	H	H	L	AB	$A\overline{B}$ minus A	A + B	$(A + \overline{B})$ plus A
H	H	H	H	A	A	A	A minus 1

*each bit is shifted to the next more significant position
**arithmetic operations expressed in 2s complement notation

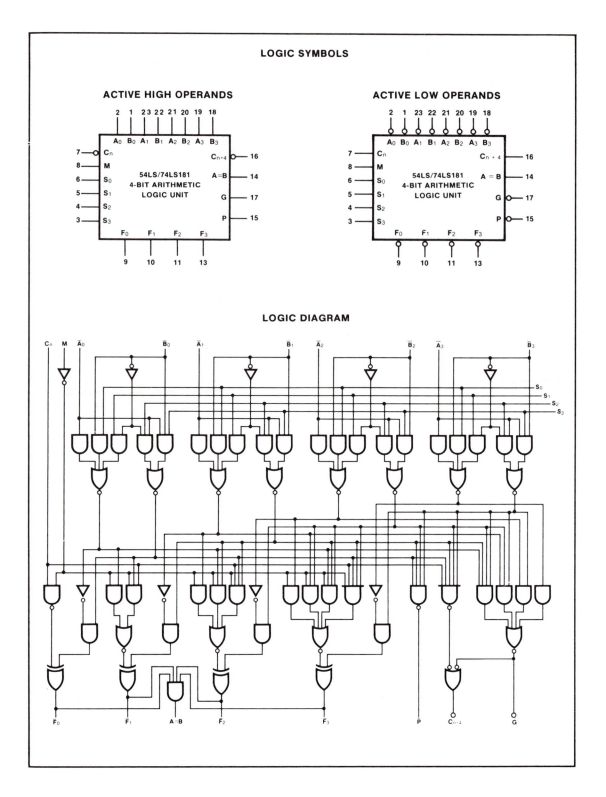

LOGIC SYMBOLS

ACTIVE HIGH OPERANDS

ACTIVE LOW OPERANDS

LOGIC DIAGRAM

DC CHARACTERISTICS OVER OPERATING TEMPERATURE RANGE (unless otherwise specified)

SYMBOL	PARAMETER		54/74LS		UNITS	CONDITIONS
			Min	Max		
I_{OH}	Output HIGH Current, A = B			100	μA	V_{CC} = Min, V_{OH} = 5.5 V
I_{CC}	Power Supply Current	XM		32	mA	V_{CC} = Max $\overline{B}_n$, C_n = Gnd S_n, M, $\overline{A}_n$ = 4.5 V
		XC		34		
		XM		35	mA	V_{CC} = Max $\overline{A}_n$, $\overline{B}_n$, C_n = Gnd M, S_n = 4.5 V
		XC		37		

AC CHARACTERISTICS: V_{CC} = +5.0 V, T_A = +25° C (See Section 3 for waveforms and load configurations)

SYMBOL	PARAMETER	54/74LS C_L = 15 pF		UNITS	CONDITIONS
		Min	Max		
t_{PLH} t_{PHL}	Propagation Delay C_n to C_{n+4}		27 20	ns	M = Gnd, Figs. 3-1, 3-5 Tables I & II
t_{PLH} t_{PHL}	Propagation Delay C_n to $\overline{F}$		26 20	ns	M = Gnd, Figs. 3-1, 3-5 Table I
t_{PLH} t_{PHL}	Propagation Delay $\overline{A}$ or $\overline{B}$ to $\overline{G}$		29 23	ns	M, S_1, S_2 = Gnd; S_1, S_3 = 4.5 V; Figs. 3-1, 3-5 Table I
t_{PLH} t_{PHL}	Propagation Delay $\overline{A}$ or $\overline{B}$ to $\overline{G}$		32 26	ns	M, S_0, S_3 = Gnd; S_1, S_2 = 4.5 V; Figs. 3-1, 3-4, 3-5; Table II
t_{PLH} t_{PHL}	Propagation Delay $\overline{A}$ or $\overline{B}$ to $\overline{P}$		30 30	ns	M, S_1, S_2 = Gnd; S_0, S_3 = 4.5 V; Figs. 3-1, 3-4; Table I
t_{PLH} t_{PHL}	Propagation Delay $\overline{A}$ or $\overline{B}$ to $\overline{P}$		30 33	ns	M, S_0, S_3 = Gnd; S_1, S_2 = 4.5 V; Figs. 3-1, 3-4, 3-5; Table II
t_{PLH} t_{PHL}	Propagation Delay $\overline{A}_i$ or $\overline{B}_i$ to $\overline{F}_i$		32 25	ns	M, S_1, S_2 = Gnd; S_0, S_3 = 4.5 V; Figs. 3-1, 3-5; Table I
t_{PLH} t_{PHL}	Propagation Delay $\overline{A}_i$ or $\overline{B}_i$ to $\overline{F}_i$		32 32	ns	M, S_0, S_3 = Gnd; S_1, S_2 = 4.5 V; Figs. 3-1, 3-4, 3-5; Table II
t_{PLH} t_{PHL}	Propagation Delay $\overline{A}$ or $\overline{B}$ to $\overline{F}$		33 29	ns	M = 4.5 V; Figs. 3-1, 3-5; Table III
t_{PLH} t_{PHL}	Propagation Delay $\overline{A}$ or $\overline{B}$ to C_{n+4}		38 38	ns	M, S_1, S_2 = Gnd; S_0, S_3 = 4.5 V; Figs. 3-1, 3-4; Table I

AC CHARACTERISTICS: $V_{CC} = +5.0$ V, $T_A = +25°$ C (Cont'd)

SYMBOL	PARAMETER	54/74LS $C_L = 15$ pF		UNITS	CONDITIONS
		Min	Max		
t_{PLH} t_{PHL}	Propagation Delay $\overline{A}$ or $\overline{B}$ to C_{n+4}		41 41	ns	M, S_0, S_3 = Gnd; S_1, S_2 = 4.5 V; Figs. 3-1, 3-4, 3-5; Table II
t_{PLH} t_{PHL}	Propagation Delay $\overline{A}$ or $\overline{B}$ to A = B		50 62	ns	M, S_0, S_3 = Gnd; S_1, S_2 = 4.5 V; R_L = 2 kΩ to 5.0 V; Figs. 3-2, 3-4, 3-5; Table II

SUM MODE TEST TABLE I   **FUNCTION INPUTS:** $S_0 = S_3 = 4.5$ V, $S_1 = S_2 = M = 0$ V

SYMBOL	INPUT UNDER TEST	OTHER INPUT SAME BIT		OTHER DATA INPUTS		OUTPUT UNDER TEST
		APPLY 4.5 V	APPLY GND	APPLY 4.5 V	APPLY GND	
t_{PLH} t_{PHL}	$\overline{A}_i$	$\overline{B}_i$	None	Remaining $\overline{A}$ and $\overline{B}$	C_n	$\overline{F}_i$
t_{PLH} t_{PHL}	$\overline{B}_i$	$\overline{A}_i$	None	Remaining $\overline{A}$ and $\overline{B}$	C_n	$\overline{F}_i$
t_{PLH} t_{PHL}	$\overline{A}$	$\overline{B}$	None	None	Remaining $\overline{A}$ and $\overline{B}$, C_n	$\overline{P}$
t_{PLH} t_{PHL}	$\overline{B}$	$\overline{A}$	None	None	Remaining $\overline{A}$ and $\overline{B}$, C_n	$\overline{P}$
t_{PLH} t_{PHL}	$\overline{A}$	None	$\overline{B}$	Remaining $\overline{B}$	Remaining $\overline{A}$, C_n	$\overline{G}$
t_{PLH} t_{PHL}	$\overline{B}$	None	$\overline{A}$	Remaining $\overline{B}$	Remaining $\overline{A}$, C_n	$\overline{G}$
t_{PLH} t_{PHL}	$\overline{A}$	None	$\overline{B}$	Remaining $\overline{B}$	Remaining $\overline{A}$, C_n	C_{n+4}
t_{PLH} t_{PHL}	$\overline{B}$	None	$\overline{A}$	Remaining $\overline{B}$	Remaining $\overline{A}$, C_n	C_{n+4}
t_{PLH} t_{PHL}	C_n	None	None	All $\overline{A}$	All $\overline{B}$	Any $\overline{F}$ or C_{n+4}

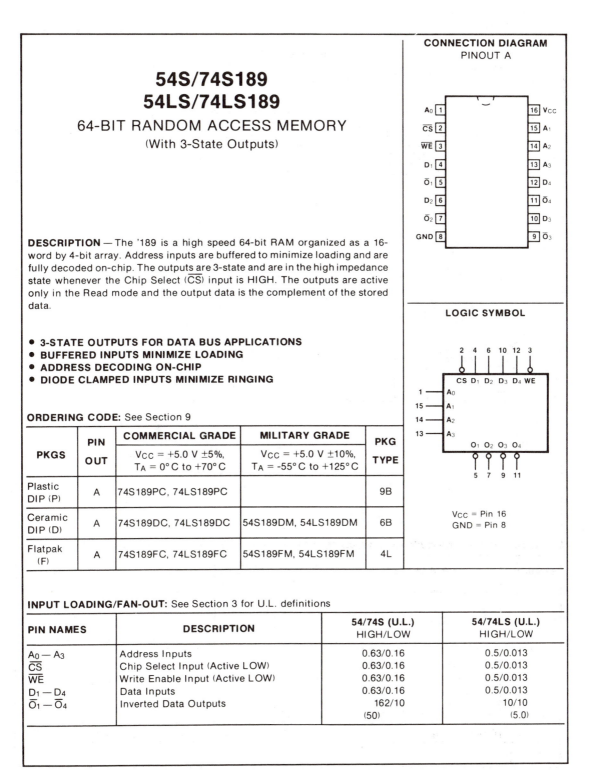

54S/74S189
54LS/74LS189

64-BIT RANDOM ACCESS MEMORY
(With 3-State Outputs)

CONNECTION DIAGRAM
PINOUT A

A_0	1	16	V_{CC}
$\overline{CS}$	2	15	A_1
$\overline{WE}$	3	14	A_2
D_1	4	13	A_3
$\overline{O}_1$	5	12	D_4
D_2	6	11	$\overline{O}_4$
$\overline{O}_2$	7	10	D_3
GND	8	9	$\overline{O}_3$

DESCRIPTION — The '189 is a high speed 64-bit RAM organized as a 16-word by 4-bit array. Address inputs are buffered to minimize loading and are fully decoded on-chip. The outputs are 3-state and are in the high impedance state whenever the Chip Select ($\overline{CS}$) input is HIGH. The outputs are active only in the Read mode and the output data is the complement of the stored data.

- 3-STATE OUTPUTS FOR DATA BUS APPLICATIONS
- BUFFERED INPUTS MINIMIZE LOADING
- ADDRESS DECODING ON-CHIP
- DIODE CLAMPED INPUTS MINIMIZE RINGING

LOGIC SYMBOL

V_{CC} = Pin 16
GND = Pin 8

ORDERING CODE: See Section 9

PKGS	PIN OUT	COMMERCIAL GRADE V_{CC} = +5.0 V ±5%, T_A = 0°C to +70°C	MILITARY GRADE V_{CC} = +5.0 V ±10%, T_A = -55°C to +125°C	PKG TYPE
Plastic DIP (P)	A	74S189PC, 74LS189PC		9B
Ceramic DIP (D)	A	74S189DC, 74LS189DC	54S189DM, 54LS189DM	6B
Flatpak (F)	A	74S189FC, 74LS189FC	54S189FM, 54LS189FM	4L

INPUT LOADING/FAN-OUT: See Section 3 for U.L. definitions

PIN NAMES	DESCRIPTION	54/74S (U.L.) HIGH/LOW	54/74LS (U.L.) HIGH/LOW
$A_0 - A_3$	Address Inputs	0.63/0.16	0.5/0.013
$\overline{CS}$	Chip Select Input (Active LOW)	0.63/0.16	0.5/0.013
$\overline{WE}$	Write Enable Input (Active LOW)	0.63/0.16	0.5/0.013
$D_1 - D_4$	Data Inputs	0.63/0.16	0.5/0.013
$\overline{O}_1 - \overline{O}_4$	Inverted Data Outputs	162/10 (50)	10/10 (5.0)

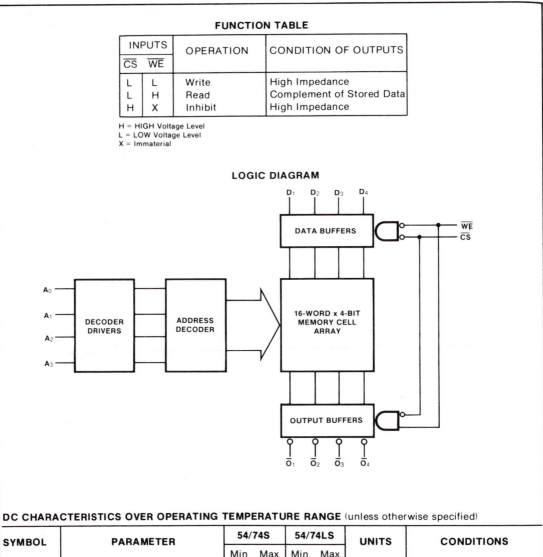

FUNCTION TABLE

INPUTS		OPERATION	CONDITION OF OUTPUTS
$\overline{CS}$	$\overline{WE}$		
L	L	Write	High Impedance
L	H	Read	Complement of Stored Data
H	X	Inhibit	High Impedance

H = HIGH Voltage Level
L = LOW Voltage Level
X = Immaterial

LOGIC DIAGRAM

DC CHARACTERISTICS OVER OPERATING TEMPERATURE RANGE (unless otherwise specified)

SYMBOL	PARAMETER		54/74S		54/74LS		UNITS	CONDITIONS
			Min	Max	Min	Max		
V_{OL}	Output LOW Voltage	XM		0.5		0.4	V	V_{CC} = Min I_{OL} = 16 mA ('S189) I_{OL} = 8.0 mA (54LS189) I_{OL} = 16 mA (74LS189)
		XC		0.45		0.5		
V_{OH}	Output HIGH Voltage	XM		2.4		2.8	V	V_{CC} = Min I_{OH} = 2.0 mA (54S189) I_{OH} = 6.5 mA (74S189) I_{OH} = 0.4 mA ('LS189)
		XC		2.4		2.8		
I_{OS}	Output Short Circuit Current		-30	-100	-80*		mA	V_{CC} = Max
I_{CC}	Power Supply Current			110		40	mA	V_{CC} = Max; $\overline{WE}$, $\overline{CS}$, Gnd

*Typical Value

AC CHARACTERISTICS OVER RECOMMENDED V$_{CC}$ AND T$_A$ RANGE (unless otherwise specified)

SYMBOL	PARAMETER		54/74S $C_L = 30$ pF $R_L = 300$ Ω		54/74LS $C_L = 15$ pF		UNITS	CONDITIONS
			Min	Max	Min	Max		
t$_{PLH}$	Access Time, HIGH or	XM		50	37*		ns	Figs. 3-1, 3-20
t$_{PHL}$	LOW, A$_n$ to $\overline{O}_n$	XC		35	37*			
t$_{PZH}$	Access Time, HIGH or	XM		32	10*		ns	Figs. 3-3, 3-11, 3-12
t$_{PZL}$	LOW, $\overline{CS}$ to $\overline{O}_n$	XC		22	10*			R$_L$ = 2 kΩ ('LS189)
t$_{PHZ}$	Disable Time $\overline{CS}$ to $\overline{O}_n$	XM		25			ns	Figs. 3-3, 3-11, 3-12 R$_L$ = 2 kΩ ('LS189) C$_L$ = 5 pF
		XC		25				
t$_{PLZ}$	Disable Time $\overline{CS}$ to $\overline{O}_n$	XM		25				
		XC		17				
t$_{PZH}$	Access Time, HIGH or	XM		40			ns	Figs. 3-3, 3-11, 3-12
t$_{PZL}$	LOW, $\overline{WE}$ to $\overline{O}_n$	XC		30				R$_L$ = 2 kΩ ('LS189)
t$_{PHZ}$	Disable Time $\overline{WE}$ to $\overline{O}_n$	XM		30			ns	Figs. 3-3, 3-11, 3-12 R$_L$ = 2 kΩ ('LS189) C$_L$ = 5 pF
		XC		20				
t$_{PLZ}$	Disable Time $\overline{WE}$ to $\overline{O}_n$	XM		32				
		XC		20				

AC OPERATING REQUIREMENTS OVER RECOMMENDED V$_{CC}$ AND T$_A$ RANGE (unless otherwise specified)

SYMBOL	PARAMETER	54/74S		54/74LS		UNITS	CONDITIONS
		Min	Max	Min	Max		
t$_s$ (H)	Setup Time HIGH or LOW	0		10*		ns	Fig. 3-21
t$_s$ (L)	A$_n$ to $\overline{WE}$	0		10*			
t$_h$ (H)	Hold Time HIGH or LOW	0		0*		ns	
t$_h$ (L)	A$_n$ to $\overline{WE}$	0		0*			
t$_s$ (H)	Setup Time HIGH or LOW	20		25*		ns	Fig. 3-13
t$_s$ (L)	D$_n$ to $\overline{WE}$	20		25*			
t$_h$ (H)	Hold Time HIGH or LOW	0		0*		ns	
t$_h$ (L)	D$_n$ to $\overline{WE}$	0		0*			
t$_s$ (L)	Setup Time LOW $\overline{CS}$ to $\overline{WE}$	0				ns	Fig. 3-14
t$_h$ (L)	Hold Time LOW $\overline{CS}$ to $\overline{WE}$	0				ns	Fig. 3-13
t$_w$ (L)	$\overline{WE}$ Pulse Width LOW	20		25*		ns	Fig. 3-14

*Typical Value

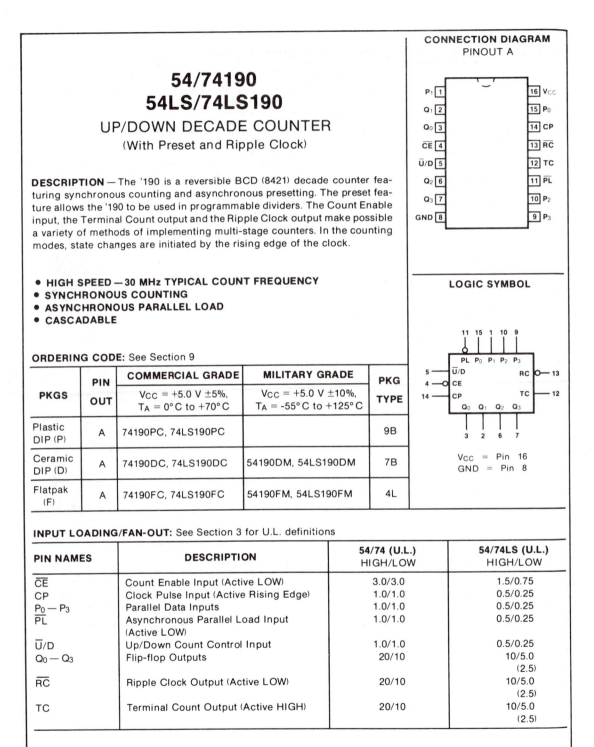

CONNECTION DIAGRAM
PINOUT A

54/74190
54LS/74LS190

UP/DOWN DECADE COUNTER
(With Preset and Ripple Clock)

DESCRIPTION — The '190 is a reversible BCD (8421) decade counter featuring synchronous counting and asynchronous presetting. The preset feature allows the '190 to be used in programmable dividers. The Count Enable input, the Terminal Count output and the Ripple Clock output make possible a variety of methods of implementing multi-stage counters. In the counting modes, state changes are initiated by the rising edge of the clock.

- **HIGH SPEED — 30 MHz TYPICAL COUNT FREQUENCY**
- **SYNCHRONOUS COUNTING**
- **ASYNCHRONOUS PARALLEL LOAD**
- **CASCADABLE**

LOGIC SYMBOL

V_{CC} = Pin 16
GND = Pin 8

ORDERING CODE: See Section 9

PKGS	PIN OUT	COMMERCIAL GRADE V_{CC} = +5.0 V ±5%, T_A = 0°C to +70°C	MILITARY GRADE V_{CC} = +5.0 V ±10%, T_A = -55°C to +125°C	PKG TYPE
Plastic DIP (P)	A	74190PC, 74LS190PC		9B
Ceramic DIP (D)	A	74190DC, 74LS190DC	54190DM, 54LS190DM	7B
Flatpak (F)	A	74190FC, 74LS190FC	54190FM, 54LS190FM	4L

INPUT LOADING/FAN-OUT: See Section 3 for U.L. definitions

PIN NAMES	DESCRIPTION	54/74 (U.L.) HIGH/LOW	54/74LS (U.L.) HIGH/LOW
$\overline{CE}$	Count Enable Input (Active LOW)	3.0/3.0	1.5/0.75
CP	Clock Pulse Input (Active Rising Edge)	1.0/1.0	0.5/0.25
$P_0 — P_3$	Parallel Data Inputs	1.0/1.0	0.5/0.25
$\overline{PL}$	Asynchronous Parallel Load Input (Active LOW)	1.0/1.0	0.5/0.25
$\overline{U}/D$	Up/Down Count Control Input	1.0/1.0	0.5/0.25
$Q_0 — Q_3$	Flip-flop Outputs	20/10	10/5.0 (2.5)
$\overline{RC}$	Ripple Clock Output (Active LOW)	20/10	10/5.0 (2.5)
TC	Terminal Count Output (Active HIGH)	20/10	10/5.0 (2.5)

LOGIC DIAGRAM

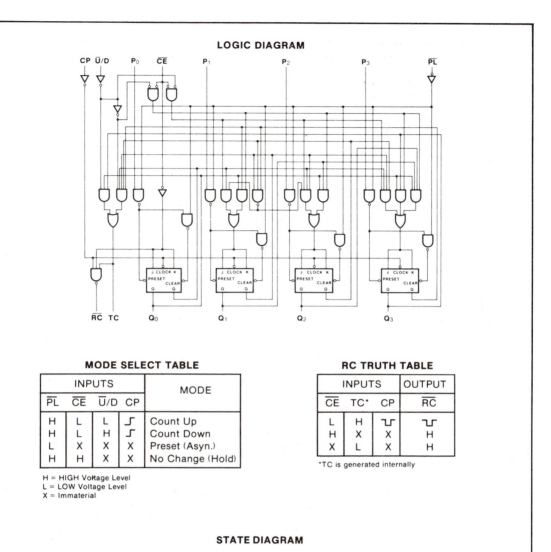

MODE SELECT TABLE

INPUTS				MODE
$\overline{PL}$	$\overline{CE}$	$\overline{U}/D$	CP	
H	L	L	⌐_	Count Up
H	L	H	⌐_	Count Down
L	X	X	X	Preset (Asyn.)
H	H	X	X	No Change (Hold)

H = HIGH Voltage Level
L = LOW Voltage Level
X = Immaterial

RC TRUTH TABLE

INPUTS			OUTPUT
$\overline{CE}$	TC*	CP	$\overline{RC}$
L	H	⊓	⊓
H	X	X	H
X	L	X	H

*TC is generated internally

STATE DIAGRAM

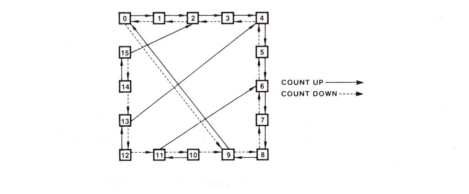

COUNT UP ——————▶
COUNT DOWN -----▶

FUNCTIONAL DESCRIPTION — The '190 is a synchronous up/down BCD decade counter and the '191 is a synchronous up/down 4-bit binary counter. The operating modes of the '190 decade counter and the '191 binary counter are identical, with the only difference being the count sequences as noted in the state diagrams. Each circuit contains four master/slave flip-flops, with internal gating and steering logic to provide individual preset, count-up and count-down operations.

Each circuit has an asynchronous parallel load capability permitting the counter to be preset to any desired number. When the Parallel Load ($\overline{PL}$) input is LOW, information present on the Parallel Data inputs ($P_0 — P_3$) is loaded into the counter and appears on the Q outputs. This operation overrides the counting functions, as indicated in the Mode Select Table.

A HIGH signal on the $\overline{CE}$ input inhibits counting. When $\overline{CE}$ is LOW, internal state changes are initiated synchronously by the LOW-to-HIGH transition of the clock input. The direction of counting is determined by the $\overline{U}/D$ input signal, as indicated in the Mode Select Table. When counting is to be enabled, the $\overline{CE}$ signal can be made LOW when the clock is in either state. However, when counting is to be inhibited, the LOW-to-HIGH $\overline{CE}$ transition must occur only while the clock is HIGH. Similarly, the $\overline{U}/D$ signal should only be changed when either $\overline{CE}$ or the clock is HIGH. These restrictions do not apply to the 'LS190 and 'LS191; $\overline{CE}$ and $\overline{U}/D$ can be changed with the clock in either state, provided only that the recommended setup and hold times are observed.

Two types of outputs are provided as overflow/underflow indicators. The Terminal Count (TC) output is normally LOW and goes HIGH when a circuit reaches zero in the count-down mode or reaches maximum (9 for the '190, 15 for the '191) in the count-up mode. The TC output will then remain HIGH until a state change occurs, whether by counting or presetting or until $\overline{U}/D$ is changed. The TC output should not be used as a clock signal because it is subject to decoding spikes.

The TC signal is also used internally to enable the Ripple Clock ($\overline{RC}$) output. The $\overline{RC}$ output is normally HIGH. When $\overline{CE}$ is LOW and TC is HIGH, the $\overline{RC}$ output will go LOW when the clock next goes LOW and will stay LOW until the clock goes HIGH again. This feature simplifies the design of multi-stage counters, as indicated in *Figures a and b*. In *Figure a*, each $\overline{RC}$ output is used as the clock input for the next higher stage. This configuration is particularly advantageous when the clock source has a limited drive capability, since it drives only the first stage. To prevent counting in all stages it is only necessary to inhibit the first stage, since a HIGH signal on $\overline{CE}$ inhibits the $\overline{RC}$ output pulse, as indicated in the $\overline{RC}$ Truth Table. A disadvantage of this configuration, in some applications, is the timing skew between state changes in the first and last stages. This represents the cumulative delay of the clock as it ripples through the preceding stages.

A method of causing state changes to occur simultaneously in all stages in shown in *Figure b*. All clock inputs are driven in parallel and the $\overline{RC}$ outputs propagate the carry/borrow signals in ripple fashion. In this configuration the LOW state duration of the clock must be long enough to allow the negative-going edge of the carry/borrow signal to ripple through to the last stage before the clock goes HIGH. There is no such restriction on the HIGH state duration of the clock, since the $\overline{RC}$ output of any package goes HIGH shortly after its CP input goes HIGH.

The configuration shown in *Figure c* avoids ripple delays and their associated restrictions. The $\overline{CE}$ input for a given stage is formed by combining the TC signals from all the preceding stages. Note that in order to inhibit counting an enable signal must be included in each carry gate. The simple inhibit scheme of *Figures a and b* doesn't apply, because the TC output of a given stage is not affected by its own $\overline{CE}$.

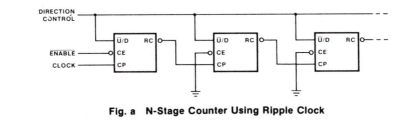

Fig. a N-Stage Counter Using Ripple Clock

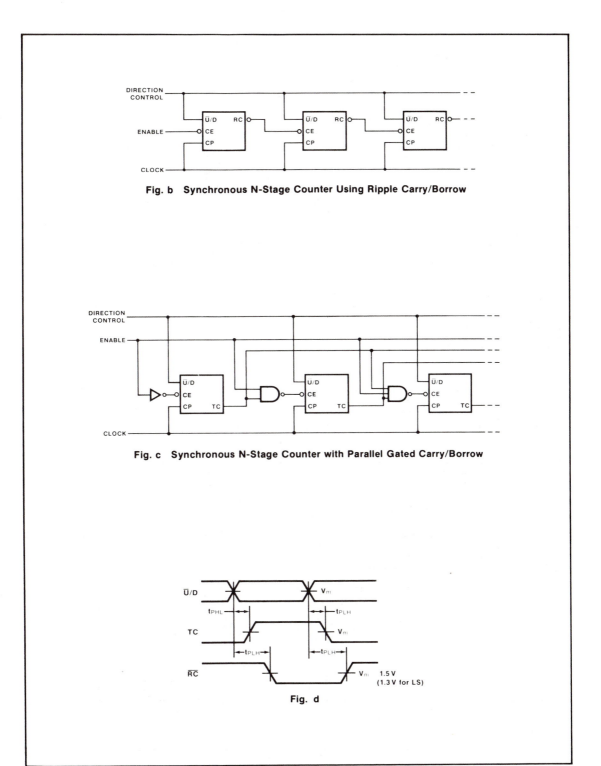

Fig. b   Synchronous N-Stage Counter Using Ripple Carry/Borrow

Fig. c   Synchronous N-Stage Counter with Parallel Gated Carry/Borrow

Fig. d

DC CHARACTERISTICS OVER OPERATING TEMPERATURE RANGE (unless otherwise specified)

SYMBOL	PARAMETER		54/74		54/74LS		UNITS	CONDITIONS
			Min	Max	Min	Max		
I_{CC}	Power Supply Current	XM		99		35	mA	V_{CC} = Max
		XC		105		35		All Inputs = Gnd

AC CHARACTERISTICS: V_{CC} = +5.0 V, T_A = +25°C (See Section 3 for waveforms and load configurations)

SYMBOL	PARAMETER	54/74 C_L = 15 pF R_L = 400 Ω		54/74LS C_L = 15 pF		UNITS	CONDITIONS
		Min	Max	Min	Max		
f_{max}	Maximum Count Frequency	20		20		MHz	
t_{PLH} t_{PHL}	Propagation Delay CP to Q_n		24 36		24 36	ns	Figs. 3-1, 3-8
t_{PLH} t_{PHL}	Propagation Delay CP to TC		42 52		42 52	ns	
t_{PLH} t_{PHL}	Propagation Delay CP to $\overline{RC}$		20 24		20 24	ns	
t_{PLH} t_{PHL}	Propagation Delay P_n to Q_n		22 50		22 50	ns	Figs. 3-1, 3-5
t_{PLH} t_{PHL}	Propagation Delay $\overline{CE}$ to $\overline{RC}$		33 33		33 33	ns	
t_{PLH} t_{PHL}	Propagation Delay $\overline{PL}$ to Q_n		33 50		33 50	ns	Figs. 3-1, 3-16
t_{PLH} t_{PHL}	Propagation Delay $\overline{U}/D$ to $\overline{RC}$		45 45		45 45	ns	Fig. 3-1, Fig. d
t_{PLH} t_{PHL}	Propagation Delay $\overline{U}/D$ to $\overline{TC}$		33 33		33 33	ns	

AC OPERATING REQUIREMENTS: V_{CC} = +5.0 V, T_A = +25°C

SYMBOL	PARAMETER	54/74		54/74LS		UNITS	CONDITIONS
		Min	Max	Min	Max		
t_s (H) t_s (L)	Setup Time HIGH or LOW P_n to $\overline{PL}$	20 20		20 20		ns	Fig. 3-13
t_h (H) t_h (L)	Hold Time HIGH or LOW P_n to $\overline{PL}$	0 0		5.0 5.0		ns	
t_s (L)	Setup Time LOW $\overline{CE}$ to CP	20		20		ns	Fig. 3-6
t_h (L)	Hold Time LOW $\overline{CE}$ to CP	0		0		ns	
t_w (L)	CP Pulse Width LOW	25		20		ns	Fig. 3-8
t_w (L)	$\overline{PL}$ Pulse Width LOW	35		35		ns	Fig. 3-16
t_{rec}	Recovery Time $\overline{PL}$ to CP	20		20		ns	Fig. 3-16

54LS/74LS247

BCD TO 7-SEGMENT DECODER/DRIVER
(With Open-Collector Outputs)

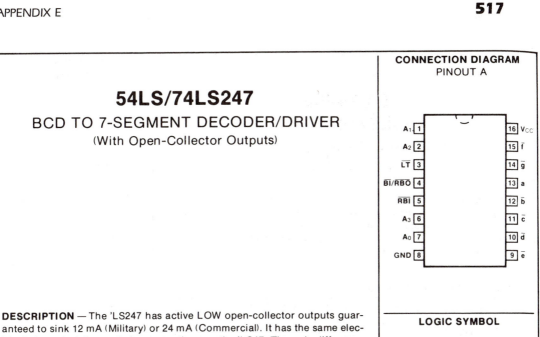

CONNECTION DIAGRAM
PINOUT A

A_1	1	16	V_{CC}
A_2	2	15	$\bar{f}$
$\overline{LT}$	3	14	$\bar{g}$
$\overline{BI/RBO}$	4	13	$\bar{a}$
$\overline{RBI}$	5	12	$\bar{b}$
A_3	6	11	$\bar{c}$
A_0	7	10	$\bar{d}$
GND	8	9	$\bar{e}$

DESCRIPTION — The 'LS247 has active LOW open-collector outputs guaranteed to sink 12 mA (Military) or 24 mA (Commercial). It has the same electrical characteristics and pin connections as the 'LS47. The only difference is that the 'LS247 will light the top bar (segment a) for numeral 6 and the bottom bar (segment d) for numeral 9. For detailed description and specifications please refer to the 'LS47 data sheet.

LOGIC SYMBOL

Inputs: 7 1 2 6 3 5 → A_0 A_1 A_2 A_3 LT RBI

Outputs: a b c d e f g / BI/RBO → 13 12 11 10 9 15 14 / 4

V_{CC} = Pin 16
GND = Pin 8

ORDERING CODE: See Section 9

PKGS	PIN OUT	COMMERCIAL GRADE V_{CC} = +5.0 V ±5%, T_A = 0°C to +70°C	MILITARY GRADE V_{CC} = +5.0 V ±10%, T_A = -55°C to +125°C	PKG TYPE
Plastic DIP (P)	A	74LS247PC		9B
Ceramic DIP (D)	A	74LS247DC	54LS247DM	6B
Flatpak (F)	A	74LS247FC	54LS247FM	4L

INPUT LOADING/FAN-OUT: See Section 3 for U.L. definitions

PIN NAMES	DESCRIPTION	54/74LS (U.L.) HIGH/LOW
$A_0 - A_3$	BCD Inputs	0.5/0.25
$\overline{RBI}$	Ripple Blanking Input (Active LOW)	0.5/0.25
$\overline{LT}$	Lamp Test Input (Active LOW)	0.5/0.25
$\overline{BI/RBO}$	Blanking Input (Active LOW) or	0.5/0.25
	Ripple Blanking Output (Active LOW)	1.25/2.0
		(1.0)
$\bar{a} - \bar{g}$	Segment Outputs (Active LOW)	OC*/15
		(7.5)

*OC — Open Collector

54/74283
54LS/74LS283
4-BIT BINARY FULL ADDER
(With Fast Carry)

DESCRIPTION — The '283 high speed 4-bit binary full adders with internal carry lookahead accept two 4-bit binary words (A_0 — A_3, B_0 — B_3) and a Carry input (C_0). They generate the binary Sum outputs (S_0 — S_3) and the Carry output (C_4) from the most significant bit. They operate with either active HIGH or active LOW operands (positive or negative logic).

CONNECTION DIAGRAM
PINOUT A

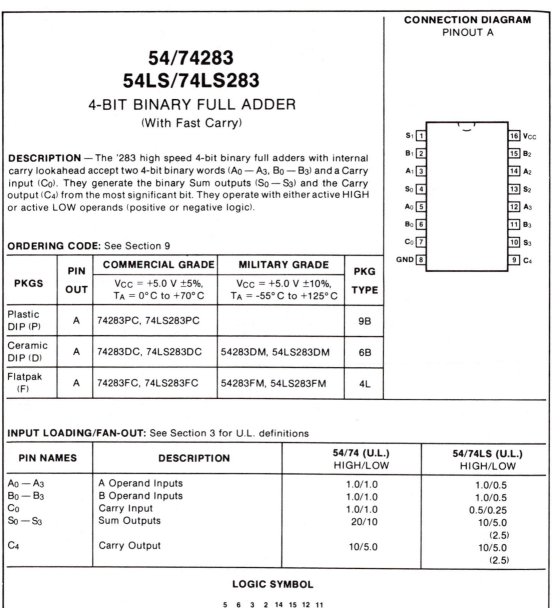

S_1	1	16	V_{CC}
B_1	2	15	B_2
A_1	3	14	A_2
S_0	4	13	S_2
A_0	5	12	A_3
B_0	6	11	B_3
C_0	7	10	S_3
GND	8	9	C_4

ORDERING CODE: See Section 9

PKGS	PIN OUT	COMMERCIAL GRADE V_{CC} = +5.0 V ±5%, T_A = 0°C to +70°C	MILITARY GRADE V_{CC} = +5.0 V ±10%, T_A = -55°C to +125°C	PKG TYPE
Plastic DIP (P)	A	74283PC, 74LS283PC		9B
Ceramic DIP (D)	A	74283DC, 74LS283DC	54283DM, 54LS283DM	6B
Flatpak (F)	A	74283FC, 74LS283FC	54283FM, 54LS283FM	4L

INPUT LOADING/FAN-OUT: See Section 3 for U.L. definitions

PIN NAMES	DESCRIPTION	54/74 (U.L.) HIGH/LOW	54/74LS (U.L.) HIGH/LOW
A_0 — A_3	A Operand Inputs	1.0/1.0	1.0/0.5
B_0 — B_3	B Operand Inputs	1.0/1.0	1.0/0.5
C_0	Carry Input	1.0/1.0	0.5/0.25
S_0 — S_3	Sum Outputs	20/10	10/5.0 (2.5)
C_4	Carry Output	10/5.0	10/5.0 (2.5)

LOGIC SYMBOL

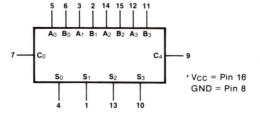

5	6	3	2	14	15	12	11
A_0	B_0	A_1	B_1	A_2	B_2	A_3	B_3

7 — C_0 C_4 — 9

S_0 S_1 S_2 S_3

4   1   13   10

V_{CC} = Pin 16
GND = Pin 8

FUNCTIONAL DESCRIPTION — The '283 adds two 4-bit binary words (A plus B) plus the incoming carry C_0. The binary sum appears on the Sum (S_0 — S_3) and outgoing carry (C_4) outputs. The binary weight of the various inputs and outputs is indicated by the subscript numbers, representing powers of two.

$$2^0\ (A_0 + B_0 + C_0) + 2^1\ (A_1 + B_1) + 2^2\ (A_2 + B_2) + 2^3\ (A_3 + B_3) = S_0 + 2S_1 + 4S_2 + 8S_3 + 16C_4$$
$$\text{Where (+) = plus}$$

Interchanging inputs of equal weight does not affect the operation. Thus C_0, A_0, B_0 can be arbitrarily assigned to pins 5, 6 and 7. Due to the symmetry of the binary add function, the '283 can be used either with all inputs and outputs active HIGH (positive logic) or with all inputs and outputs active LOW (negative logic). Note that if C_0 is not used it must be tied LOW for active HIGH logic or tied HIGH for active LOW logic.

Example:

	C_0	A_0	A_1	A_2	A_3	B_0	B_1	B_2	B_3	S_0	S_1	S_2	S_3	C_4
Logic Levels	L	L	H	L	H	H	L	L	H	H	H	L	L	H
Active HIGH	0	0	1	0	1	1	0	0	1	1	1	0	0	1
Active LOW	1	1	0	1	0	0	1	1	0	0	0	1	1	0

Active HIGH: 0 + 10 + 9 = 3 + 16   Active LOW: 1 + 5 + 6 = 12 + 0

Due to pin limitations, the intermediate carries of the '283 are not brought out for use as inputs or outputs. However, other means can be used to effectively insert a carry into, or bring a carry out from, an intermediate stage. *Figure a* shows a way of making a 3-bit adder. Tying the operand inputs of the fourth adder (A_3, B_3) LOW makes S_3 dependent only on, and equal to, the carry from the third adder. Using somewhat the same principle, *Figure b* shows a way of dividing the '283 into a 2-bit and a 1-bit adder. The third stage adder (A_2, B_2, S_2) is used merely as a means of getting a carry (C_{10}) signal into the fourth stage (via A_2 and B_2) and bringing out the carry from the second stage on S_2. Note that as long as A_2 and B_2 are the same, whether HIGH or LOW, they do not influence S_2. Similarly, when A_2 and B_2 are the same the carry into the third stage does not influence the carry out of the third stage. *Figure c* shows a method of implementing a 5-input encoder, where the inputs are equally weighted. The outputs S_0, S_1 and S_2 present a binary number equal to the number of inputs I_1 — I_5 that are true. *Figure d* shows one method of implementing a 5-input majority gate. When three or more of the inputs I_1 — I_5 are true, the output M_5 is true.

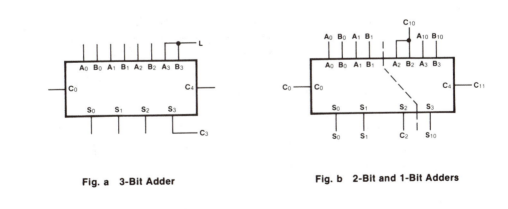

Fig. a   3-Bit Adder   **Fig. b   2-Bit and 1-Bit Adders**

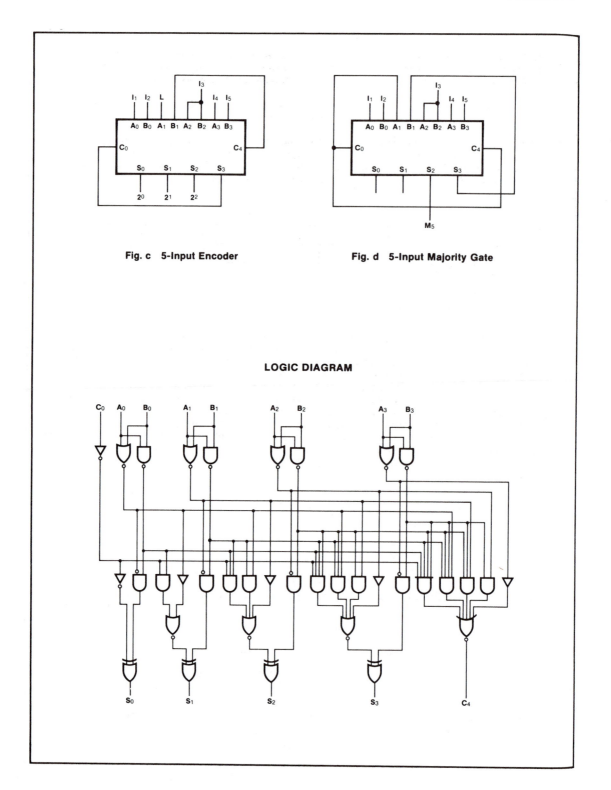

Fig. c   5-Input Encoder

Fig. d   5-Input Majority Gate

LOGIC DIAGRAM

DC CHARACTERISTICS OVER OPERATING TEMPERATURE RANGE (unless otherwise specified)

SYMBOL	PARAMETER		54/74 Min	54/74 Max	54/74LS Min	54/74LS Max	UNITS	CONDITIONS
I_{OS}	Output Short Circuit Current at S_n	XM	-20	-55	-20	-100	mA	V_{CC} = Max
		XC	-18	-55	-20	-100		
I_{OS}	Output Short Circuit Current at C_4	XM	-20	-70	-20	-100	mA	V_{CC} = Max
		XC	-18	-70	-20	-100		
I_{CC}	Power Supply Current	XM		99		39	mA	V_{CC} = Max, Inputs = Gnd ('LS283) Inputs = 4.5 V ('283)
		XC		110		39		
		XM, XC				34	mA	V_{CC} = Max Inputs = 4.5 V ('LS283)

AC CHARACTERISTICS: V_{CC} = 5.0 V, T_A = 25°C (See Section 3 for waveforms and load configurations)

SYMBOL	PARAMETER	54/74 C_L = 15 pF R_L = 400 Ω Min	Max	54/74LS C_L = 15 pF Min	Max	UNITS	CONDITIONS
t_{PLH}	Propagation Delay C_0 to S_n		21		24	ns	Figs. 3-1, 3-20
t_{PHL}			21		24		
t_{PLH}	Propagation Delay A_n or B_n to S_n		24		24	ns	Figs. 3-1, 3-20
t_{PHL}			24		24		
t_{PLH}	Propagation Delay C_0 to C_4		14		17	ns	Figs. 3-1, 3-5 R_L = 780 Ω ('283)
t_{PHL}			16		17		
t_{PLH}	Propagation Delay A_n or B_n to C_4		14		17	ns	Figs. 3-1, 3-5 R_L = 780Ω ('283)
t_{PHL}			16		17		

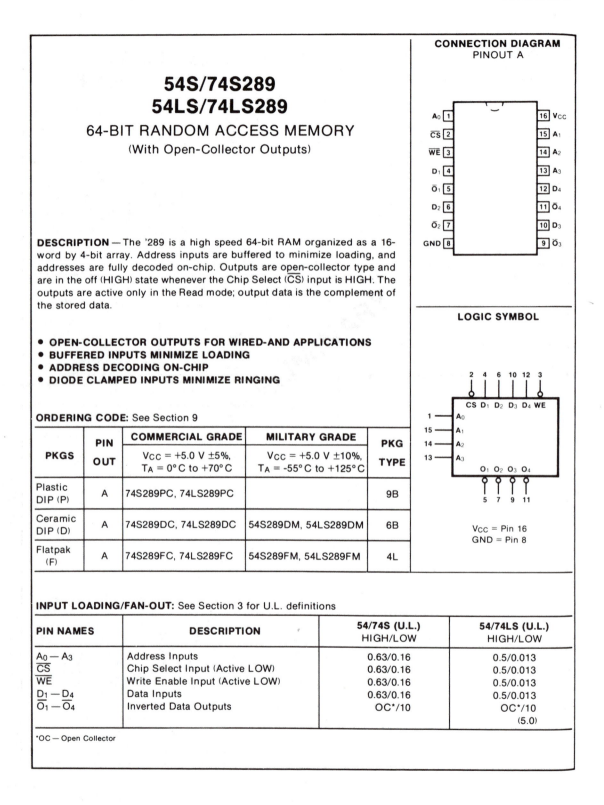

54S/74S289
54LS/74LS289

64-BIT RANDOM ACCESS MEMORY
(With Open-Collector Outputs)

CONNECTION DIAGRAM
PINOUT A

A_0	1	16	V_{CC}
$\overline{CS}$	2	15	A_1
$\overline{WE}$	3	14	A_2
D_1	4	13	A_3
$\overline{O}_1$	5	12	D_4
D_2	6	11	$\overline{O}_4$
$\overline{O}_2$	7	10	D_3
GND	8	9	$\overline{O}_3$

DESCRIPTION — The '289 is a high speed 64-bit RAM organized as a 16-word by 4-bit array. Address inputs are buffered to minimize loading, and addresses are fully decoded on-chip. Outputs are open-collector type and are in the off (HIGH) state whenever the Chip Select ($\overline{CS}$) input is HIGH. The outputs are active only in the Read mode; output data is the complement of the stored data.

* **OPEN-COLLECTOR OUTPUTS FOR WIRED-AND APPLICATIONS**
* **BUFFERED INPUTS MINIMIZE LOADING**
* **ADDRESS DECODING ON-CHIP**
* **DIODE CLAMPED INPUTS MINIMIZE RINGING**

LOGIC SYMBOL

V_{CC} = Pin 16
GND = Pin 8

ORDERING CODE: See Section 9

PKGS	PIN OUT	COMMERCIAL GRADE V_{CC} = +5.0 V ±5%, T_A = 0°C to +70°C	MILITARY GRADE V_{CC} = +5.0 V ±10%, T_A = -55°C to +125°C	PKG TYPE
Plastic DIP (P)	A	74S289PC, 74LS289PC		9B
Ceramic DIP (D)	A	74S289DC, 74LS289DC	54S289DM, 54LS289DM	6B
Flatpak (F)	A	74S289FC, 74LS289FC	54S289FM, 54LS289FM	4L

INPUT LOADING/FAN-OUT: See Section 3 for U.L. definitions

PIN NAMES	DESCRIPTION	54/74S (U.L.) HIGH/LOW	54/74LS (U.L.) HIGH/LOW
$A_0 - A_3$	Address Inputs	0.63/0.16	0.5/0.013
$\overline{CS}$	Chip Select Input (Active LOW)	0.63/0.16	0.5/0.013
$\overline{WE}$	Write Enable Input (Active LOW)	0.63/0.16	0.5/0.013
$D_1 - D_4$	Data Inputs	0.63/0.16	0.5/0.013
$\overline{O}_1 - \overline{O}_4$	Inverted Data Outputs	OC*/10	OC*/10 (5.0)

*OC — Open Collector

FUNCTION TABLE

INPUTS		OPERATION	CONDITION OF OUTPUTS
$\overline{CS}$	$\overline{WE}$		
L	L	Write	Off (HIGH)
L	H	Read	Complement of Stored Data
H	X	Inhibit	Off (HIGH)

H = HIGH Voltage Level
L = LOW Voltage Level
X = Immaterial

LOGIC DIAGRAM

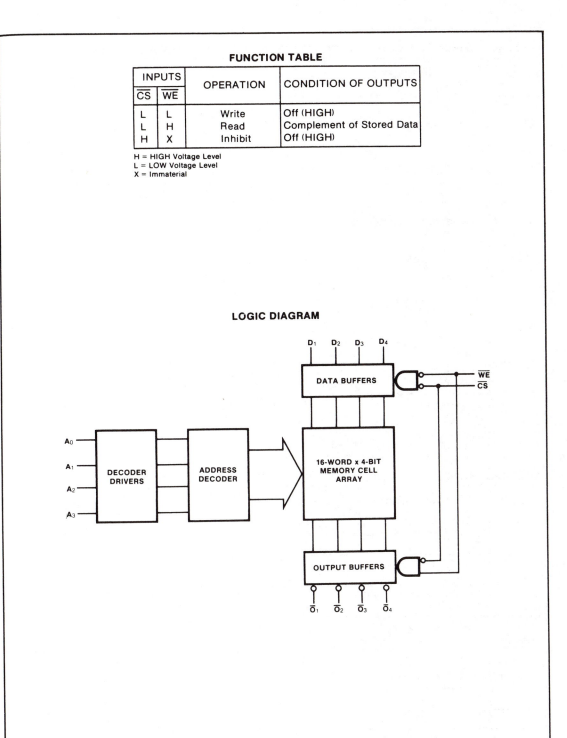

DC CHARACTERISTICS OVER OPERATING TEMPERATURE RANGE (unless otherwise specified)

SYMBOL	PARAMETER		54/74S Min Max	54/74LS Min Max	UNITS	CONDITIONS
V_{OL}	Output LOW Voltage	XM	0.5	0.4	V	V_{CC} = Min
		XC	0.45	0.5		I_{OL} = 16 mA ('S289)
						I_{OL} = 8.0 mA (54LS289)
						I_{OL} = 16 mA (74LS289)
I_{OH}	Output HIGH Current		40	20	μA	V_{OH} = 2.4 V, V_{CC} = Min
			100	100		V_{OH} = 5.5 V
I_{CC}	Power Supply Current		105	40	mA	V_{CC} = Max

AC CHARACTERISTICS OVER RECOMMENDED V_{CC} AND T_A RANGE (unless otherwise specified)

SYMBOL	PARAMETER		54/74S C_L = 30 pF R_L = * Min Max	54/74LS C_L = 15 pF R_L = 2 kΩ Min Max	UNITS	CONDITIONS
t_{PLH} t_{PHL}	Access Time, HIGH or LOW, A_n to $\overline{O}_n$	XM XC	50 35	37** 37**	ns	Figs. 3-2, 3-20
t_{PHL}	Access Time $\overline{CS}$ to $\overline{O}_n$	XM XC	25 17	10** 10**	ns	Figs. 3-2, 3-5
t_{PLH}	Disable Time $\overline{CS}$ to $\overline{O}_n$	XM XC	20 17		ns	
t_{PHL}	Recovery Time $\overline{WE}$ to $\overline{O}_n$	XM XC	40 35	30** 30**	ns	Figs. 3-2, 3-4
t_{PLH}	Disable Time $\overline{WE}$ to $\overline{O}_n$	XM XC	30 25		ns	

AC OPERATING REQUIREMENTS OVER RECOMMENDED V_{CC} AND T_A RANGE (unless otherwise specified)

SYMBOL	PARAMETER	54/74S Min Max	54/74LS Min Max	UNITS	CONDITIONS
t_s (H) t_s (L)	Setup Time, HIGH or LOW A_n to $\overline{WE}$	0 0	10** 10**	ns	Fig. 3-21
t_h (H) t_h (L)	Hold Time, HIGH or LOW A_n to $\overline{WE}$	0 0	0** 0**	ns	
t_s (H) t_s (L)	Setup Time, HIGH or LOW D_n to $\overline{WE}$	20 20	25** 25**	ns	Fig. 3-13
t_h (H) t_h (L)	Hold Time HIGH or LOW D_n to WE	0 0	0* 0*	ns	
t_s (L)	Setup Time LOW $\overline{CS}$ to $\overline{WE}$	0		ns	Fig. 3-14
t_h (L)	Hold Time LOW $\overline{CS}$ to $\overline{WE}$	0		ns	Fig. 3-13
t_w (L)	$\overline{WE}$ Pulse Width LOW	20	25**	ns	Fig. 3-14

*R_L = 300 Ω to V_{CC} and 600 Ω to Gnd.
**Typical Value

APPENDIX

*Data Sheets for CMOS Logic Devices

*Reprinted by permission of the National Semiconductor Corp.

National Semiconductor

MM54C30/MM74C30 8-Input NAND Gate

General Description

The logic gate employs complementary MOS (CMOS) to achieve wide power supply operating range, low power consumption and high noise immunity. Function and pin out compatibility with series 54/74 devices minimizes design time for those designers familiar with the standard 54/74 logic family.

All inputs are protected from damage due to static discharge by diode clamps to V_{CC} and GND.

Features

■ Wide supply voltage range   3.0V to 15V

■ Guaranteed noise margin   1.0V

■ High noise immunity $0.45\,V_{CC}$ (typ.)

■ Low power   fan out of 2
  TTL compatibility   driving 74L

Logic and Connection Diagrams

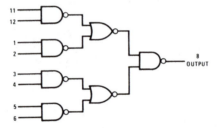

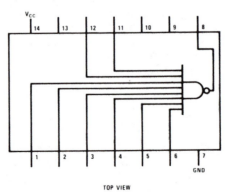

TOP VIEW

Absolute Maximum Ratings (Note 1)

Voltage at Any Pin	$-0.3V$ to $V_{CC} + 0.3V$
Operating Temperature Range	
MM54C30	$-55°C$ to $+125°C$
MM74C30	$-40°C$ to $+85°C$
Storage Temperature Range	$-65°C$ to $+150°C$
Package Dissipation	500 mW
Operating V_{CC} Range	3.0V to 15V
Absolute Maximum V_{CC}	18V
Lead Temperature (Soldering, 10 seconds)	300°C

DC Electrical Characteristics

Min/max limits apply across temperature range unless otherwise noted.

Parameter		Conditions	Min.	Typ.	Max.	Units
CMOS to CMOS						
$V_{IN(1)}$	Logical "1" Input Voltage	$V_{CC} = 5.0V$	3.5			V
		$V_{CC} = 10V$	8.0			V
$V_{IN(0)}$	Logical "0" Input Voltage	$V_{CC} = 5.0V$			1.5	V
		$V_{CC} = 10V$			2.0	V
$V_{OUT(1)}$	Logical "1" Output Voltage	$V_{CC} = 5.0V, I_O = -10\mu A$	4.5			V
		$V_{CC} = 10V, I_O = -10\mu A$	9.0			V
$V_{OUT(0)}$	Logical "0" Output Voltage	$V_{CC} = 5.0V, I_O = +10\mu A$			0.5	V
		$V_{CC} = 10V, I_O = +10\mu A$			1.0	V
$I_{IN(1)}$	Logical "1" Input Current	$V_{CC} = 15V, V_{IN} = 15V$		0.005	1.0	μA
$I_{IN(0)}$	Logical "0" Input Current	$V_{CC} = 15V, V_{IN} = 0V$	-1.0	-0.005		μA
I_{CC}	Supply Current	$V_{CC} = 15V$		0.01	15	μA
CMOS/LPTTL Interface						
$V_{IN(1)}$	Logical "1" Input Voltage	54C, $V_{CC} = 4.5V$	$V_{CC} - 1.5$			V
		74C, $V_{CC} = 4.75V$	$V_{CC} - 1.5$			V
$V_{IN(0)}$	Logical "0" Input Voltage	54C, $V_{CC} = 4.5V$			0.8	V
		74C, $V_{CC} = 4.75V$			0.8	V
$V_{OUT(1)}$	Logical "1" Output Voltage	54C, $V_{CC} = 4.5V, I_O = -360\mu A$	2.4			V
		74C, $V_{CC} = 4.75V, I_O = -360\mu A$	2.4			V
$V_{OUT(0)}$	Logical "0" Output Voltage	54C, $V_{CC} = 4.5V, I_O = 360\mu A$			0.4	V
		74C, $V_{CC} = 4.75V, I_O = 360\mu A$			0.4	V
Output Drive (See 54C/74C Family Characteristics Data Sheet) (short circuit current)						
I_{SOURCE}	Output Source Current (P-Channel)	$V_{CC} = 5.0V, V_{OUT} = 0V$ $T_A = 25°C$	-1.75	-3.3		mA
I_{SOURCE}	Output Source Current (P-Channel)	$V_{CC} = 10V, V_{OUT} = 0V$ $T_A = 25°C$	-8.0	-15		mA
I_{SINK}	Output Sink Current (N-Channel)	$V_{CC} = 5.0V, V_{OUT} = V_{CC}$ $T_A = 25°C$	1.75	3.6		mA
I_{SINK}	Output Sink Current (N-Channel)	$V_{CC} = 10V, V_{OUT} = V_{CC}$ $T_A = 25°C$	8.0	16		mA

AC Electrical Characteristics

$T_A = 25°C$, $C_L = 50\,pF$, unless otherwise specified.

	Parameter	Conditions	Min.	Typ.	Max.	Units
t_{pd}	Propagation Delay Time to Logical "1" or "0"	$V_{CC} = 5.0\,V$ $V_{CC} = 10\,V$		125 55	180 90	ns ns
C_{IN}	Input Capacitance	(Note 2)		4.0		pF
C_{PD}	Power Dissipation Capacitance	(Note 3) Per Gate		26		pF

Note 1: "Absolute Maximum Ratings" are those values beyond which the safety of the device cannot be guaranteed. Except for "Operating Temperature Range" they are not meant to imply that the devices should be operated at these limits. The table of "Electrical Characteristics" provides conditions for actual device operation.

Note 2: Capacitance is guaranteed by periodic testing.

Note 3: C_{PD} determines the no load ac power consumption of any CMOS device. For complete explanation see 54C/74C Family Characteristics application note — AN–90.

Typical Performance Characteristics

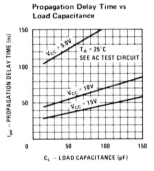

Propagation Delay Time vs Load Capacitance

Switching Time Waveforms

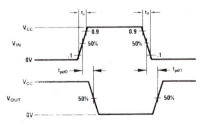

NOTE: DELAYS MEASURED WITH INPUT t_r, t_f = 20 ns.

AC Test Circuit

National Semiconductor

MM54C42/MM74C42 BCD-to-Decimal Decoder

General Description

The MM54C42/MM74C42 one-of-ten decoder is a monolithic complementary MOS (CMOS) integrated circuit constructed with N- and P-channel enhancement transistors. This decoder produces a logical "0" at the output corresponding to a four bit binary input from zero to nine, and a logical "1" at the other outputs. For binary inputs from ten to fifteen all outputs are logical "1".

- High noise immunity $0.45\,V_{CC}$ (typ.)
- Low power   50 nW (typ.)
- Medium speed operation   10 MHz (typ.) with 10 V V_{CC}

Features

- Supply voltage range   3V to 15V
- Tenth power TTL compatible   drive 2 LPTTL loads

Applications

- Automotive
- Data terminals
- Instrumentation
- Medical electronics
- Alarm systems
- Industrial electronics
- Remote metering
- Computers

Schematic Diagram

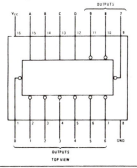

Connection Diagram

Truth Table

INPUTS				OUTPUTS									
D	C	B	A	0	1	2	3	4	5	6	7	8	9
0	0	0	0	0	1	1	1	1	1	1	1	1	1
0	0	0	1	1	0	1	1	1	1	1	1	1	1
0	0	1	0	1	1	0	1	1	1	1	1	1	1
0	0	1	1	1	1	1	0	1	1	1	1	1	1
0	1	0	0	1	1	1	1	0	1	1	1	1	1
0	1	0	1	1	1	1	1	1	0	1	1	1	1
0	1	1	0	1	1	1	1	1	1	0	1	1	1
0	1	1	1	1	1	1	1	1	1	1	0	1	1
1	0	0	0	1	1	1	1	1	1	1	1	0	1
1	0	0	1	1	1	1	1	1	1	1	1	1	0
1	0	1	0	1	1	1	1	1	1	1	1	1	1
1	0	1	1	1	1	1	1	1	1	1	1	1	1
1	1	0	0	1	1	1	1	1	1	1	1	1	1
1	1	0	1	1	1	1	1	1	1	1	1	1	1
1	1	1	0	1	1	1	1	1	1	1	1	1	1
1	1	1	1	1	1	1	1	1	1	1	1	1	1

Absolute Maximum Ratings (Note 1)

Voltage at Any Pin (Note 1)	-0.3V to $V_{CC} + 0.3$V
Operating Temperature Range	
MM54C42	$-55°$C to $+125°$C
MM74C42	$-40°$C to $+85°$C
Storage Temperature Range	$-65°$C to $+150°$C

Package Dissipation	500 mW
Operating V_{CC} Range	3.0 V to 15 V
Absolute Maximum V_{CC}	18 V
Lead Temperature (Soldering, 10 seconds)	300°C

DC Electrical Characteristics Min./max. limits apply across temperature range unless otherwise noted.

	Parameter	Conditions	Min.	Typ.	Max.	Units
	CMOS to CMOS					
$V_{IN(1)}$	Logical "1" Input Voltage	$V_{CC} = 5.0$V	3.5			V
		$V_{CC} = 10$V	8.0			V
$V_{IN(0)}$	Logical "0" Input Voltage	$V_{CC} = 5.0$V			1.5	V
		$V_{CC} = 10$V			2.0	V
$V_{OUT(1)}$	Logical "1" Output Voltage	$V_{CC} = 5.0$V, $I_O = -10\,\mu$A	4.5			V
		$V_{CC} = 10$V, $I_O = -10\,\mu$A	9.0			V
$V_{OUT(0)}$	Logical "0" Output Voltage	$V_{CC} = 5.0$V, $I_O = 10\,\mu$A			0.5	V
		$V_{CC} = 10$V, $I_O = 10\,\mu$A			1.0	V
$I_{IN(1)}$	Logical "1" Input Current	$V_{CC} = 15$V, $V_{IN} = 15$V			1.0	μA
$I_{IN(0)}$	Logical "0" Input Current	$V_{CC} = 15$V, $V_{IN} = 0$V	-1.0		300	μA
I_{CC}	Supply Current	$V_{CC} = 15$V		0.05	300	μA
	CMOS/LPTTL Interface					
$V_{IN(1)}$	Logical "1" Input Voltage	54C, $V_{CC} = 4.5$V	$V_{CC} - 1.5$			V
		74C, $V_{CC} = 4.75$V	$V_{CC} - 1.5$			V
$V_{IN(0)}$	Logical "0" Input Voltage	54C, $V_{CC} = 4.5$V			0.8	V
		74C, $V_{CC} = 4.75$V			0.8	V
$V_{OUT(1)}$	Logical "1" Output Voltage	54C, $V_{CC} = 4.5$V, $I_O = -360\,\mu$A	2.4			V
		74C, $V_{CC} = 4.75$V, $I_O = -360\,\mu$A	2.4			V
$V_{OUT(0)}$	Logical "0" Output Voltage	54C, $V_{CC} = 4.5$V, $I_O = 360\,\mu$A			0.4	V
		74C, $V_{CC} = 4.75$V, $I_O = 360\,\mu$A			0.4	V
	Output Drive (See 54C/74C Family Characteristics Data Sheet) $T_A = 25°$C (short circuit current)					
I_{SOURCE}	Output Source Current	$V_{CC} = 5.0$V, $V_{IN(0)} = 0$V, $V_{OUT} = 0$V	-1.75			mA
I_{SOURCE}	Output Source Current	$V_{CC} = 10$V, $V_{IN(0)} = 0$V, $V_{OUT} = 0$V	-8.0			mA
I_{SINK}	Output Sink Current	$V_{CC} = 5.0$V, $V_{IN(1)} = 5.0$V, $V_{OUT} = V_{CC}$	1.75			mA
I_{SINK}	Output Sink Current	$V_{CC} = 10$V, $V_{IN(1)} = 10$V, $V_{OUT} = V_{CC}$	8.0			mA

AC Electrical Characteristics $T_A = 25°$C, $C_L = 50$ pF, unless otherwise specified.

	Parameter	Conditions	Min.	Typ.	Max.	Units
t_{pd}	Propagation Delay Time to	$V_{CC} = 5.0$V		200	300	ns
	Logical "0" or "1"	$V_{CC} = 10$V		90	140	ns
C_{IN}	Input Capacitance	(See note 2)		5		pF
C_{PD}	Power Dissipation Capacitance	(See note 3)		50		pF

Note 1: "Absolute Maximum Ratings" are those values beyond which the safety of the device cannot be guaranteed. Except for "Operating Temperature Range" they are not meant to imply that the devices should be operated at these limits. The table of "Electrical Characteristics" provides conditions for actual device operation.

Note 2: Capacitance is guaranteed by periodic testing.

Note 3: C_{PD} determines the no load ac power consumption of any CMOS device. For complete explanation see 54C/74C Family Characteristics application note — AN-90.

![National Semiconductor logo] **National Semiconductor**

MM54C85/MM74C85 4–Bit Magnitude Comparator

General Description

The MM54C85/MM74C85 is a four-bit magnitude comparator which will perform comparison of straight binary or BCD codes. The circuit consists of eight comparing inputs (A0, A1, A2, A3, B0, B1, B2, B3), three cascading inputs ($A > B$, $A < B$ and $A = B$), and three outputs ($A > B$, $A < B$ and $A = B$). This device compares two four-bit words (A and B) and determines whether they are "greater than," "less than," or "equal to" each other by a high level on the appropriate output. For words greater than four-bits, units can be cascaded by connecting the outputs ($A > B$, $A < B$, and $A = B$) of the least significant stage to the cascade inputs ($A > B$, $A < B$ and $A = B$) of the next-significant stage. In addition the least significant stage must have a high level voltage ($V_{IN(1)}$ applied to the $A = B$ input and low level voltages ($V_{IN(0)}$) applied to $A > B$ and $A < B$ inputs.

Features

- Wide supply voltage range   3.0V to 15V
- Guaranteed noise margin   1.0V
- High noise immunity   0.45 V_{CC} (typ.)
- Low power   fan out of 2
  TTL compatibility   driving 74L
- Expandable to 'N' stages
- Applicable to binary or BCD
- The MM54C85/MM74C85 follows the DM54LS85/DM74LS85 Pinout.

Logic Diagrams

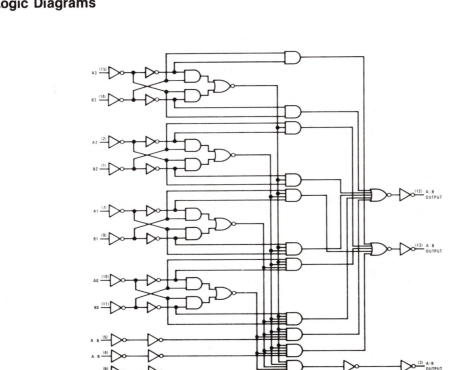

Absolute Maximum Ratings (Note 1)

Voltage at Any Pin	-0.3 V to $V_{CC} + 0.3$ V
Operating Temperature Range	
MM54C85	$-55°C$ to $+125°C$
MM74C85	$-40°C$ to $+85°C$
Storage Temperature Range	$-65°C$ to $+150°C$

Package Dissipation	500 mW
Operating V_{CC} Range	3.0 V to 15 V
V_{CC}	18 V
Lead Temperature (Soldering, 10 seconds)	300°C

DC Electrical Characteristics Min./max. limits apply across temperature range unless otherwise noted.

Parameter		Conditions	Min.	Typ.	Max.	Units
CMOS to CMOS						
$V_{IN(1)}$	Logical "1" Input Voltage	$V_{CC} = 5.0$ V	3.5			V
		$V_{CC} = 10$ V	8.0			V
$V_{IN(0)}$	Logical "0" Input Voltage	$V_{CC} = 5.0$ V			1.5	V
		$V_{CC} = 10$ V			2.0	V
$V_{OUT(1)}$	Logical "1" Output Voltage	$V_{CC} = 5.0$ V, $I_O = -10\,\mu A$	4.5			V
		$V_{CC} = 10$ V, $I_O = -10\,\mu A$	9.0			V
$V_{OUT(0)}$	Logical "0" Output Voltage	$V_{CC} = 5.0$ V, $I_O = +10\,\mu A$			0.5	V
		$V_{CC} = 10$ V, $I_O = +10\,\mu A$			1.0	V
$I_{IN(1)}$	Logical "1" Input Current	$V_{CC} = 15$ V, $V_{IN} = 15$ V		0.005	1.0	μA
$I_{IN(0)}$	Logical "0" Input Current	$V_{CC} = 15$ V, $V_{IN} = 0$ V	-1.0	-0.005		μA
I_{CC}	Supply Current	$V_{CC} = 15$ V		0.05	300	μA
CMOS/LPTTL Interface						
$V_{IN(1)}$	Logical "1" Input Voltage	54C, $V_{CC} = 4.5$ V	$V_{CC} - 1.5$			V
		74C, $V_{CC} = 4.75$ V	$V_{CC} - 1.5$			V
$V_{IN(0)}$	Logical "0" Input Voltage	54C, $V_{CC} = 4.5$ V			0.8	V
		74C, $V_{CC} = 4.75$ V			0.8	V
$V_{OUT(1)}$	Logical "1" Output Voltage	54C, $V_{CC} = 4.5$ V, $I_O = -360\,\mu A$	2.4			V
		74C, $V_{CC} = 4.75$ V, $I_O = -360\,\mu A$	2.4			V
$V_{OUT(0)}$	Logical "0" Output Voltage	54C, $V_{CC} = 4.5$ V, $I_O = 360\,\mu A$			0.4	V
		74C, $V_{CC} = 4.75$ V, $I_O = 360\,\mu A$			0.4	V
Output Drive (See 54C/74C Family Characteristics Data Sheet) (short circuit current)						
I_{SOURCE}	Output Source Current (P-Channel)	$V_{CC} = 5.0$ V, $V_{OUT} = 0$ V $T_A = 25°C$	-1.75	-3.3		mA
I_{SOURCE}	Output Source Current (P-Channel)	$V_{CC} = 10$ V, $V_{OUT} = 0$ V $T_A = 25°C$	-8.0	-15		mA
I_{SINK}	Output Sink Current (N-Channel)	$V_{CC} = 5.0$ V, $V_{OUT} = V_{CC}$ $T_A = 25°C$	1.75	3.6		mA
I_{SINK}	Output Sink Current (N-Channel)	$V_{CC} = 10$ V, $V_{OUT} = V_{CC}$ $T_A = 25°C$	8.0	16		mA

AC Electrical Characteristics $T_A = 25°C$, $C_L = 50$ pF, unless otherwise specified.

Parameter		Conditions	Min.	Typ.	Max.	Units
t_{pd}	Propagation Delay from any A or B Data Input to any Data Output	$V_{CC} = 5.0$ V		250	600	ns
		$V_{CC} = 10$ V		100	300	ns
t_{pd}	Propagation Delay Time from any Cascade Input to any Output	$V_{CC} = 5.0$ V		200	500	ns
		$V_{CC} = 10$ V		100	250	ns
C_{IN}	Input Capacitance	Any Input		5.0		pF
C_{PD}	Power Dissipation Capacitance	(Note 3) Per Package		45		pF

Note 1: "Absolute Maximum Ratings" are those values beyond which the safety of the device cannot be guaranteed. Except for "Operating Temperature Range" they are not meant to imply that the devices should be operated at these limits. The table of "Electrical Characteristics" provides conditions for actual device operation.

Note 2: Capacitance is guaranteed by periodic testing.

Note 3: C_{PD} determines the no load ac power consumption of any CMOS device. For complete explanation see 54C/74C Family Characteristics application note — AN-90.

Typical Applications

Four Digit Comparator

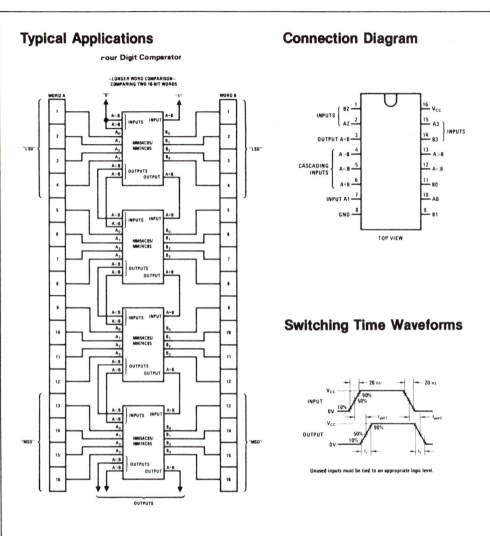

Connection Diagram

Switching Time Waveforms

Unused inputs must be tied to an appropriate logic level.

Truth Table

COMPARING INPUTS				CASCADING INPUTS			OUTPUTS		
A3, B3	A2, B2	A1, B1	A0, B0	A > B	A < B	A = B	A > B	A < B	A = B
A3 > B3	X	X	X	X	X	X	H	L	L
A3 < B3	X	X	X	X	X	X	L	H	L
A3 = B3	A2 > B2	X	X	X	X	X	H	L	L
A3 = B3	A2 < B2	X	X	X	X	X	L	H	L
A3 = B3	A2 = B2	A1 > B1	X	X	X	X	H	L	L
A3 = B3	A2 = B2	A1 < B1	X	X	X	X	L	H	L
A3 = B3	A2 = B2	A1 = B1	A0 > B0	X	X	X	H	L	L
A3 = B3	A2 = B2	A1 = B1	A0 < B0	X	X	X	L	H	L
A3 = B3	A2 = B2	A1 = B1	A0 = B0	H	L	L	H	L	L
A3 = B3	A2 = B2	A1 = B1	A0 = B0	L	H	L	L	H	L
A3 = B3	A2 = B2	A1 = B1	A0 = B0	L	L	H	L	L	H
A3 = B3	A2 = B2	A1 = B1	A0 = B0	L	H	H	L	H	H
A3 = B3	A2 = B2	A1 = B1	A0 = B0	H	L	H	H	L	H
A3 = B3	A2 = B2	A1 = B1	A0 = B0	H	H	H	H	H	H
A3 = B3	A2 = B2	A1 = B1	A0 = B0	H	H	L	H	H	L
A3 = B3	A2 = B2	A1 = B1	A0 = B0	L	L	L	L	L	L

H = high level, L = low level, X = irrelevant

⚡ National
Semiconductor

MM54C150/MM74C150 16-Line to 1-Line Multiplexer
MM72C19/MM82C19 TRI-STATE® 16-Line to 1-Line Multiplexer

General Description

The MM54C150/MM74C150 and MM72C19/MM82C19 multiplex 16 digital lines to 1 output. A 4-bit address code determines the particular 1-of-16 inputs which is routed to the output. The data is inverted from input to output.

A strobe override places the output of MM54C150/MM74C150 in the logical "1" state and the output of MM72C19/MM82C19 in the high-impedance state.

All inputs are protected from damage due to static discharge by diode clamps to V_{CC} and GND.

Features

■ Wide supply voltage range   3.0 V to 15 V
■ Guaranteed noise margin   1.0 V
■ High noise immunity   0.45 V_{CC} (typ.)
■ TTL compatibility   Drive 1 TTL Load

Connection Diagram

Absolute Maximum Ratings (Note 1)

Voltage at Any Pin	-0.3 V to $V_{CC}+0.3$ V
Operating Temperature Range	
MM54C150, MM72C19	$-55°C$ to $+125°C$
MM74C150, MM82C19	$-40°C$ to $+85°C$
Storage Temperature Range	$-65°C$ to $+150°C$
Package Dissipation	500 mW
Operating V_{CC} Range	3.0 V to 15 V
V_{CC}	18 V
Lead Temperature (Soldering, 10 sec.)	300°C

DC Electrical Characteristics Max./min. limits apply across temperature range, unless otherwise noted.

Parameter		Conditions	Min.	Typ.	Max.	Units
CMOS to CMOS						
$V_{IN(1)}$	Logical "1" Input Voltage	$V_{CC}=5.0$ V	3.5			V
		$V_{CC}=10$ V	8.0			V
$V_{IN(0)}$	Logical "0" Input Voltage	$V_{CC}=5.0$ V			1.5	V
		$V_{CC}=10$ V			2.0	V
$V_{OUT(1)}$	Logical "1" Output Voltage	$V_{CC}=5.0$ V, $I_O=-10\,\mu A$	4.5			V
		$V_{CC}=10$ V, $I_O=-10\,\mu A$	9.0			V
$V_{OUT(0)}$	Logical "0" Output Voltage	$V_{CC}=5.0$ V, $I_O=+10\,\mu A$			0.5	V
		$V_{CC}=10$ V, $I_O=+10\,\mu A$			1.0	V
$I_{IN(1)}$	Logical "1" Input Current	$V_{CC}=15$ V, $V_{IN}=15$ V		0.005	1.0	μA
$I_{IN(0)}$	Logical "0" Input Current	$V_{CC}=15$ V, $V_{IN}=0$ V	-1.0	-0.005		μA
I_{OZ}	Output Current in High Impedance State					
	MM73C19/MM82C19	$V_{CC}=15$ V, $V_O=15$ V		0.005	1.0	μA
		$V_{CC}=15$ V, $V_O=0$ V	-1.0	-0.005		μA
I_{CC}	Supply Current	$V_{CC}=15$ V		0.05	300	μA
TTL Interface						
$V_{IN(1)}$	Logical "1" Input Voltage	54C, 72C $V_{CC}=4.5$ V	$V_{CC}-1.5$			V
		74C, 82C $V_{CC}=4.75$ V	$V_{CC}-1.5$			V
$V_{IN(0)}$	Logical "0" Input Voltage	54C, 72C $V_{CC}=4.5$ V			0.8	V
		74C, 82C $V_{CC}=4.75$ V			0.8	V
$V_{OUT(1)}$	Logical "1" Output Voltage	54C, 72C $V_{CC}=4.5$ V, $I_O=-1.6$ mA	2.4			V
		74C, 82C $V_{CC}=4.75$ V, $I_O=-1.6$ mA	2.4			V
$V_{OUT(0)}$	Logical "0" Output Voltage	54C, 72C $V_{CC}=4.5$ V, $I_O=1.6$ mA			0.4	V
		74C, 82C $V_{CC}=4.75$ V, $I_O=1.6$ mA			0.4	V
Output Drive (Short Circuit Current)						
I_{SOURCE}	Output Source Current (P-Channel)	$V_{CC}=5.0$ V, $V_{OUT}=0$ V, $T_A=25°C$	-4.35	-8		mA
I_{SOURCE}	Output Source Current (P-Channel)	$V_{CC}=10$ V, $V_{OUT}=0$ V, $T_A=25°C$	-20	-40		mA
I_{SINK}	Output Sink Current (N-Channel)	$V_{CC}=5.0$ V, $V_{OUT}=V_{CC}$, $T_A=25°C$	4.35	8		mA
I_{SINK}	Output Sink Current (N-Channel)	$V_{CC}=10$ V, $V_{OUT}=V_{CC}$, $T_A=25°C$	20	40		mA

AC Electrical Characteristics
$T_A = 25°C$, $C_L = 50\,pF$, unless otherwise noted.

Parameter		Conditions	Min.	Typ.	Max.	Units
t_{pd0}, t_{pd1}	Propagation Delay Time to a Logical "0" or Logical "1" from Data Inputs to Output	$V_{CC} = 5.0\,V$ $V_{CC} = 10\,V$ $V_{CC} = 5.0\,V$, $C_L = 150\,pF$ $V_{CC} = 10\,V$, $C_L = 150\,pF$		250 110 290 120	600 300 650 330	ns ns ns ns
t_{pd0}, t_{pd1}	Propagation Delay Time to a Logical "0" or Logical "1" from Data Select Inputs to Output	$V_{CC} = 5.0\,V$ $V_{CC} = 10\,V$		290 120	650 330	ns ns
t_{pd0}, t_{pd1}	Propagation Delay Time to a Logical "0" or Logical "1" from Strobe to Output MM54C150/MM74C150	$V_{CC} = 5.0\,V$ $V_{CC} = 10V$		120 55	300 150	ns ns
t_{1H}, t_{0H}	Delay from Strobe to High Impedance State MM72C19/MM82C19	$V_{CC} = 5.0\,V$, $R_L = 10\,k$, $C_L = 5\,pF$ $V_{CC} = 10\,V$, $R_L = 10\,k$, $C_L = 5\,pF$		80 60	200 150	ns ns
t_{H1}, t_{H0}	Delay from Strobe to Logical "1" Level or to Logical "0" Level (from High Impedance State) MM72C19/MM82C19	$V_{CC} = 5.0\,V$, $R_L = 10\,k$, $C_L = 5\,pF$ $V_{CC} = 10\,V$, $R_L = 10\,k$, $C_L = 5\,pF$		80 30	250 120	ns ns
C_{IN}	Input Capacitance	Any Input, (Note 2)		5.0		pF
C_{OUT}	Output Capacitance MM72C19/MM82C19	(Note 2)		11.0		pF
C_{PD}	Power Dissipation Capacitance	(Note 3)		100		pF

Note 1: "Absolute Maximum Ratings" are those values beyond which the safety of the device cannot be guaranteed. Except for "Operating Temperature Range" they are not meant to imply that the devices should be operated at these limits. The table of "Electrical Characteristics" provides conditions for actual device operation.

Note 2: Capacitance is guaranteed by periodic testing.

Note 3: C_{PD} determines the no load AC power consumption of any CMOS device. For complete explanation see 54C/74C Family Characteristics application note AN-90.

Truth Table

MM54C150/MM74C150

| | | | INPUTS | | | | | | | | | | | | | | | | | | | OUTPUT |
D	C	B	A	STROBE	E0	E1	E2	E3	E4	E5	E6	E7	E8	E9	E10	E11	E12	E13	E14	E15	W
X	X	X	X	1	X	X	X	X	X	X	X	X	X	X	X	X	X	X	X	X	1 *
0	0	0	0	0	0	X	X	X	X	X	X	X	X	X	X	X	X	X	X	X	1
0	0	0	0	0	1	X	X	X	X	X	X	X	X	X	X	X	X	X	X	X	0
0	0	0	1	0	X	0	X	X	X	X	X	X	X	X	X	X	X	X	X	X	1
0	0	0	1	0	X	1	X	X	X	X	X	X	X	X	X	X	X	X	X	X	0
0	0	1	0	0	X	X	0	X	X	X	X	X	X	X	X	X	X	X	X	X	1
0	0	1	0	0	X	X	1	X	X	X	X	X	X	X	X	X	X	X	X	X	0
0	0	1	1	0	X	X	X	0	X	X	X	X	X	X	X	X	X	X	X	X	1
0	0	1	1	0	X	X	X	1	X	X	X	X	X	X	X	X	X	X	X	X	0
0	1	0	0	0	X	X	X	X	0	X	X	X	X	X	X	X	X	X	X	X	1
0	1	0	0	0	X	X	X	X	1	X	X	X	X	X	X	X	X	X	X	X	0
0	1	0	1	0	X	X	X	X	X	0	X	X	X	X	X	X	X	X	X	X	1
0	1	0	1	0	X	X	X	X	X	1	X	X	X	X	X	X	X	X	X	X	0
0	1	1	0	0	X	X	X	X	X	X	0	X	X	X	X	X	X	X	X	X	1
0	1	1	0	0	X	X	X	X	X	X	1	X	X	X	X	X	X	X	X	X	0
0	1	1	1	0	X	X	X	X	X	X	X	0	X	X	X	X	X	X	X	X	1
0	1	1	1	0	X	X	X	X	X	X	X	1	X	X	X	X	X	X	X	X	0
1	0	0	0	0	X	X	X	X	X	X	X	X	0	X	X	X	X	X	X	X	1
1	0	0	0	0	X	X	X	X	X	X	X	X	1	X	X	X	X	X	X	X	0
1	0	0	1	0	X	X	X	X	X	X	X	X	X	0	X	X	X	X	X	X	1
1	0	0	1	0	X	X	X	X	X	X	X	X	X	1	X	X	X	X	X	X	0
1	0	1	0	0	X	X	X	X	X	X	X	X	X	X	0	X	X	X	X	X	1
1	0	1	0	0	X	X	X	X	X	X	X	X	X	X	1	X	X	X	X	X	0
1	0	1	1	0	X	X	X	X	X	X	X	X	X	X	X	0	X	X	X	X	1
1	0	1	1	0	X	X	X	X	X	X	X	X	X	X	X	1	X	X	X	X	0
1	1	0	0	0	X	X	X	X	X	X	X	X	X	X	X	X	0	X	X	X	1
1	1	0	0	0	X	X	X	X	X	X	X	X	X	X	X	X	1	X	X	X	0
1	1	0	1	0	X	X	X	X	X	X	X	X	X	X	X	X	X	0	X	X	1
1	1	0	1	0	X	X	X	X	X	X	X	X	X	X	X	X	X	1	X	X	0
1	1	1	0	0	X	X	X	X	X	X	X	X	X	X	X	X	X	X	0	X	1
1	1	1	0	0	X	X	X	X	X	X	X	X	X	X	X	X	X	X	1	X	0
1	1	1	1	0	X	X	X	X	X	X	X	X	X	X	X	X	X	X	X	0	1
1	1	1	1	0	X	X	X	X	X	X	X	X	X	X	X	X	X	X	X	1	0

*For MM72C19/MM82C19 this would be Hi-Z, everything else is the same.

Switching Time Waveforms

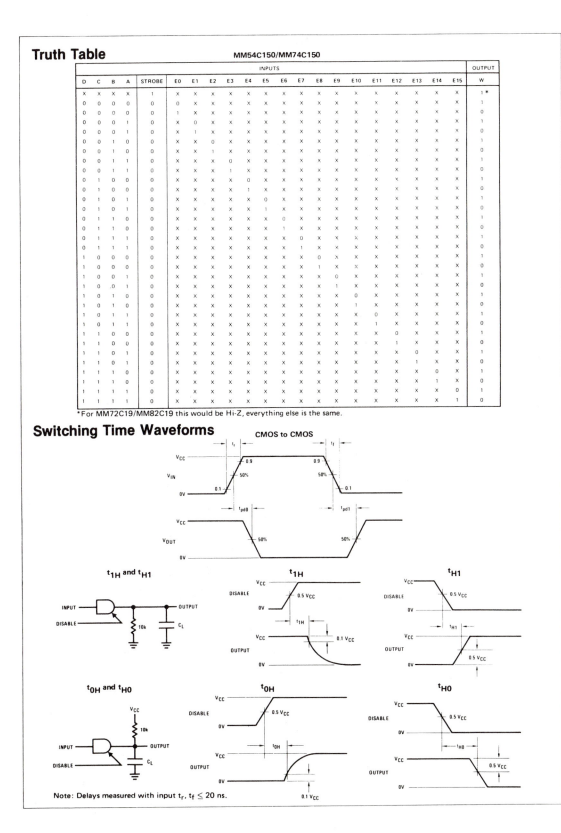

t_{1H} and t_{H1}

t_{0H} and t_{H0}

Note: Delays measured with input t_r, $t_f \leq 20$ ns.

Logic Diagram

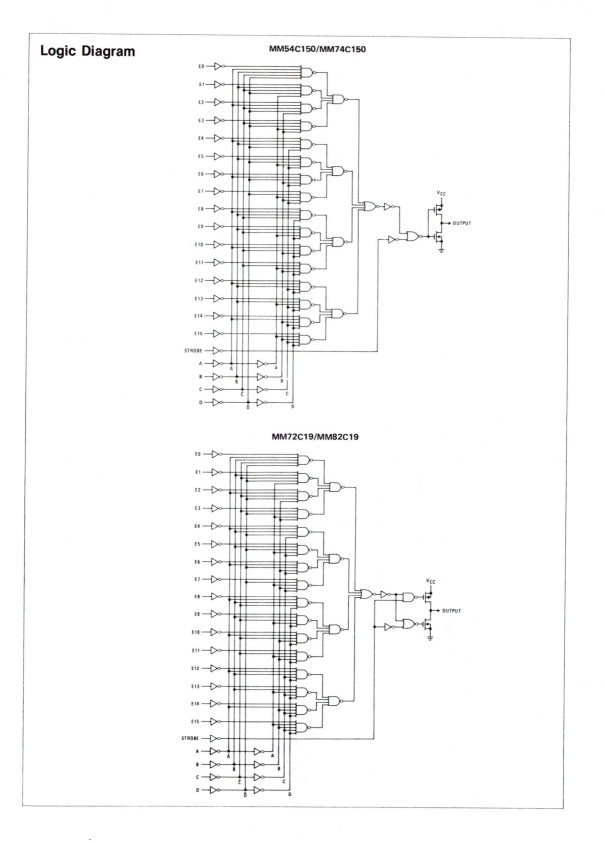

MM54C150/MM74C150

MM72C19/MM82C19

National Semiconductor

MM54C151/MM74C151 8-Channel Digital Multiplexer

General Description

The MM54C151/MM74C151 multiplexer is a monolithic complementary MOS (CMOS) integrated circuit constructed with N- and P-channel enhancement transistors.

This data selector/multiplexer contains on-chip binary decoding. Two outputs provide true (output Y) and complement (output W) data. A logical "1" on the strobe input forces W to a logical "1" and Y to a logical "0".

All inputs are protected against electrostatic effects.

Features

- Supply voltage range — 3 V to 15 V
- Tenth power TTL compatible — drive 2 LPTTL loads
- High noise immunity — 0.45 V_{CC} (typ.)
- Low power — 50 nW (typ.)

Applications

- Automotive
- Data terminals
- Instrumentation
- Medical electronics
- Alarm systems
- Industrial electronics
- Remote metering
- Computers

Logic and Connection Diagrams

Absolute Maximum Ratings (Note 1)

Voltage at Any Pin	$-0.3\,V$ to $V_{CC}+0.3\,V$
Operating Temperature Range	
MM54C151	$-55°C$ to $+125°C$
MM74C151	$-40°C$ to $+85°C$
Storage Temperature Range	$-65°C$ to $+150°C$
Maximum V_{CC} Voltage	$18\,V$
Package Dissipation	$500\,mW$
Operating V_{CC} Range	$3\,V$ to $15\,V$
Lead Temperature (Soldering, 10 sec.)	$300°C$

DC Electrical Characteristics Max./min. limits apply across temperature range, unless otherwise noted.

Parameter		Conditions	Min.	Typ.	Max.	Units
CMOS to CMOS						
$V_{IN(1)}$	Logical "1" Input Voltage	$V_{CC} = 5.0\,V$ $V_{CC} = 10\,V$	3.5 8			V V
$V_{IN(0)}$	Logical "0" Input Voltage	$V_{CC} = 5.0\,V$ $V_{CC} = 10\,V$			1.5 2	V V
$V_{OUT(1)}$	Logical "1" Output Voltage	$V_{CC} = 5.0\,V, I_O = -10\,\mu A$ $V_{CC} = 10\,V, I_O = -10\,\mu A$	4.5 9.0			V V
$V_{OUT(0)}$	Logical "0" Output Voltage	$V_{CC} = 5.0\,V, I_O = +10\,\mu A$ $V_{CC} = 10\,V, I_O = +10\,\mu A$			0.5 1.0	V V
$I_{IN(1)}$	Logical "1" Input Current	$V_{CC} = 15\,V, V_{IN} = 15\,V$			1.0	μA
$I_{IN(0)}$	Logical "0" Input Current	$V_{CC} = 15\,V, V_{IN} = 0\,V$	-1.0			μA
I_{CC}	Supply Current	$V_{CC} = 15\,V$		0.05	300	μA
CMOS to LPTTL Interface						
$V_{IN(1)}$	Logical "1" Input Voltage	54C $V_{CC} = 4.5\,V$ 74C $V_{CC} = 4.75\,V$	$V_{CC} - 1.5$ $V_{CC} - 1.5$			V mA
$V_{IN(0)}$	Logical "0" Input Voltage	54C $V_{CC} = 4.5\,V$ 74C $V_{CC} = 4.75\,V$			0.8 0.8	V V
$V_{OUT(1)}$	Logical "1" Output Voltage	54C $V_{CC} = 4.5\,V, I_O = -360\,\mu A$ 74C $V_{CC} = 4.75\,V, I_O = -360\,\mu A$	2.4 2.4			V V
$V_{OUT(O)}$	Logical "0" Output Voltage	54C $V_{CC} = 4.5\,V, I_O = 360\,\mu A$ 74C $V_{CC} = 4.75\,V, I_O = 360\,\mu A$			0.4 0.4	V V
Output Drive (See 54C/74C Family Characteristics Data Sheet) (Short Circuit Current)						
I_{SOURCE}	Output Source Current	$V_{CC} = 5.0\,V, V_{IN(0)} = 0\,V$ $T_A = 25°C, V_{OUT} = 0\,V$	-1.75			mA
I_{SOURCE}	Output Source Current	$V_{CC} = 10\,V, V_{IN(0)} = 0\,V$ $T_A = 25°C, V_{OUT} = 0\,V$	-8.0			mA
I_{SINK}	Output Sink Current	$V_{CC} = 5.0\,V, V_{IN(1)} = 5.0\,V$ $T_A = 25°C, V_{OUT} = V_{CC}$	1.75			mA
I_{SINK}	Output Sink Current	$V_{CC} = 10\,V, V_{IN(1)} = 10\,V$ $T_A = 25°C, V_{OUT} = V_{CC}$	8.0			mA

AC Electrical Characteristics $T_A = 25°C$, $C_L = 50\,pF$, unless otherwise noted.

	Parameter	Conditions	Min.	Typ.	Max.	Units
t_{pd0}, t_{pd1}	Propagation Delay Time to a Logical "0" or Logical "1" from Data to Y	$V_{CC} = 5.0\,V$, $V_{CC} = 10\,V$		170 80	270 130	ns ns
t_{pd0}, t_{pd1}	Propagation Delay Time to a Logical "0" or Logical "1" from Data to W	$V_{CC} = 5.0\,V$, $V_{CC} = 10\,V$		200 90	300 140	ns ns
t_{pd0}, t_{pd1}	Propagation Delay Time to a Logical "0" or Logical "1" from Strobe or Data Select to Y	$V_{CC} = 5.0\,V$, $V_{CC} = 10\,V$		240 110	360 170	ns ns
C_{IN}	Input Capacitance	(Note 2)		5.0		pF
C_{PD}	Power Dissipation Capacitance	(Note 3)		50		pF

Note 1: "Absolute Maximum Ratings" are those values beyond which the safety of the device cannot be guaranteed. Except for "Operating Temperature Range" they are not meant to imply that the devices should be operated at these limits. The table of "Electrical Characteristics" provides conditions for actual device operation.

Note 2: Capacitance is guaranteed by periodic testing.

Note 3: C_{PD} determines the no load AC power consumption of any CMOS device. For complete explanation see 54C/74C Family Characteristics application note AN–90.

Switching Time Waveforms

CMOS to CMOS (t_{pd1} & t_{pd0})

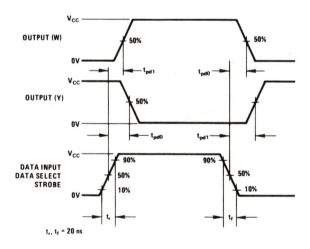

AC Test Circuit

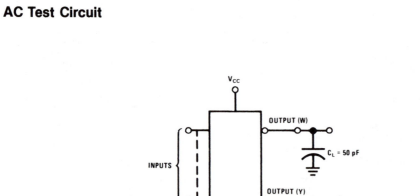

Truth Table

INPUTS												OUTPUTS	
C	B	A	STROBE	D_0	D_1	D_2	D_3	D_4	D_5	D_6	D_7	Y	W
X	X	X	1	X	X	X	X	X	X	X	X	0	1
0	0	0	0	0	X	X	X	X	X	X	X	0	1
0	0	0	0	1	X	X	X	X	X	X	X	1	0
0	0	1	0	X	0	X	X	X	X	X	X	0	1
0	0	1	0	X	1	X	X	X	X	X	X	1	0
0	1	0	0	X	X	0	X	X	X	X	X	0	1
0	1	0	0	X	X	1	X	X	X	X	X	1	0
0	1	1	0	X	X	X	0	X	X	X	X	0	1
0	1	1	0	X	X	X	1	X	X	X	X	1	0
1	0	0	0	X	X	X	X	0	X	X	X	0	1
1	0	0	0	X	X	X	X	1	X	X	X	1	0
1	0	1	0	X	X	X	X	X	0	X	X	0	1
1	0	1	0	X	X	X	X	X	1	X	X	1	0
1	1	0	0	X	X	X	X	X	X	0	X	0	1
1	1	0	0	X	X	X	X	X	X	1	X	1	0
1	1	1	0	X	X	X	X	X	X	X	0	0	1
1	1	1	0	X	X	X	X	X	X	X	1	1	0

National Semiconductor

MM54C174/MM74C174 Hex D Flip-Flop

General Description

The MM54C174/MM74C174 hex D flip-flop is a monolithic complementary MOS (CMOS) integrated circuit constructed with N- and P-channel enhancement transistors. All have a direct clear input. Information at the D inputs meeting the setup time requirements is transferred to the Q outputs on the positive-going edge of the clock pulse. Clear is independent of clock and accomplished by a low level at the clear input. All inputs are protected by diodes To V_{CC} and GND.

Features

- Wide supply voltage range   3.0 V to 15 V
- Guaranteed noise margin   1.0 V
- High noise immunity   0.45 V_{CC} (typ.)
- Low power TTL compatibility   fan out of 2
                                                  driving 74L

Logic Diagrams

Connection Diagram

TOP VIEW

Truth Table

	INPUTS		OUTPUT
CLEAR	CLOCK	D	Q
L	X	X	L
H	↑	H	H
H	↑	L	L
H	L	X	Q

Absolute Maximum Ratings (Note 1)

Voltage at Any Pin	$-0.3\,V$ to $V_{CC} + 0.3\,V$
Operating Temperature Range	
MM54C174	$-55°C$ to $+125°C$
MM74C174	$-40°C$ to $+85°C$
Storage Temperature Range	$-65°C$ to $+150°C$
Package Dissipation	$500\,mW$
Operating V_{CC} Range	$3.0\,V$ to $15\,V$
Absolute Maximum V_{CC}	$18\,V$
Lead Temperature (Soldering, 10 sec.)	$300°C$

DC Electrical Characteristics Max./min. limits apply across temperature range, unless otherwise noted.

Parameter		Conditions	Min.	Typ.	Max.	Units
CMOS to CMOS						
$V_{IN(1)}$	Logical "1" Input Voltage	$V_{CC} = 5.0\,V$ $V_{CC} = 10\,V$	3.5 8.0			V V
$V_{IN(0)}$	Logical "0" Input Voltage	$V_{CC} = 5.0\,V$ $V_{CC} = 10\,V$			1.5 2.0	V V
$V_{OUT(1)}$	Logical "1" Output Voltage	$V_{CC} = 5.0\,V$, $I_O = -10\,\mu A$ $V_{CC} = 10\,V$, $I_O = -10\,\mu A$	4.5 9.0			V V
$V_{OUT(0)}$	Logical "0" Output Voltage	$V_{CC} = 5.0\,V$, $I_O = +10\,\mu A$ $V_{CC} = 10\,V$, $I_O = +10\,\mu A$			0.5 1.0	V V
$I_{IN(1)}$	Logical "1" Input Current	$V_{CC} = 15\,V$, $V_{IN} = 15\,V$		0.005	1.0	μA
$I_{IN(0)}$	Logical "0" Input Current	$V_{CC} = 15\,V$, $V_{IN} = 0\,V$	-1.0	-0.005		μA
I_{CC}	Supply Current	$V_{CC} = 15\,V$		0.05	300	μA
CMOS/LPTTL Interface						
$V_{IN(1)}$	Logical "1" Input Voltage	54C $V_{CC} = 4.5\,V$ 74C $V_{CC} = 4.75\,V$	$V_{CC} - 1.5$ $V_{CC} - 1.5$			V v
$V_{IN(0)}$	Logical "0" Input Voltage	54C $V_{CC} = 4.5\,V$ 74C $V_{CC} = 4.75\,V$			0.8 0.8	V V
$V_{OUT(1)}$	Logical "1" Output Voltage	54C $V_{CC} = 4.5\,V$, $I_O = -360\,\mu A$ 74C $V_{CC} = 4.75\,V$, $I_O = -360\,\mu A$	2.4 2.4			V V
$V_{OUT(0)}$	Logical "0" Output Voltage	54C $V_{CC} = 4.5\,V$, $I_O = 360\,\mu A$ 74C $V_{CC} = 4.75\,V$, $I_O = 360\,\mu A$			0.4 0.4	V V
Output Drive (See 54C/74C Family Characteristics Data Sheet) (Short Circuit Current)						
I_{SOURCE}	Output Source Current (P-Channel)	$V_{CC} = 5.0\,V$ $T_A = 25°C$, $V_{OUT} = 0\,V$	-1.75	-3.3		mA
I_{SOURCE}	Output Source Current (P-Channel)	$V_{CC} = 10\,V$ $T_A = 25°C$, $V_{OUT} = 0\,V$	-8.0	-15		mA
I_{SINK}	Output Sink Current (N-Channel)	$V_{CC} = 5.0\,V$ $T_A = 25°C$, $V_{OUT} = V_{CC}$	1.75	3.6		mA
I_{SINK}	Output Sink Current (N-Channel)	$V_{CC} = 10\,V$ $T_A = 25°C$, $V_{OUT} = V_{CC}$	8.0	16		mA

AC Electrical Characteristics $T_A = 25°C$, $C_L = 50\,pF$, unless otherwise noted.

	Parameter	Conditions	Min.	Typ.	Max.	Units
t_{pd}	Propagation Delay Time to a Logical "0" or Logical "1" from Clock to Q	$V_{CC} = 5.0\,V$ $V_{CC} = 10\,V$		150 70	300 110	ns ns
t_{pd}	Propagation Delay Time to a Logical "0" from Clear	$V_{CC} = 5.0\,V$ $V_{CC} = 10\,V$		110 50	300 110	ns ns
t_{S1}, t_{S0}	Time Prior to Clock Pulse that Data must be Present	$V_{CC} = 5.0\,V$ $V_{CC} = 10\,V$	75 25			ns ns
t_{H1}, t_{H0}	Time after Clock Pulse that Data must be Held	$V_{CC} = 5.0\,V$ $V_{CC} = 10\,V$	0 0	−10 −5.0		ns ns
t_W	Minimum Clock Pulse Width	$V_{CC} = 5.0\,V$ $V_{CC} = 10\,V$		50 35	250 100	ns ns
t_W	Minimum Clear Pulse Width	$V_{CC} = 5.0\,V$ $V_{CC} = 10\,V$		65 35	140 70	ns ns
t_r, t_f	Maximum Clock Rise and Fall Time	$V_{CC} = 5.0\,V$ $V_{CC} = 10\,V$	15 5.0	>1200 >1200		μs μs
f_{MAX}	Maximum Clock Frequency	$V_{CC} = 5.0\,V$ $V_{CC} = 10\,V$	2.0 5.0	6.5 12		MHz MHz
C_{IN}	Input Capacitance	Clear Input (Note 2) Any Other Input		11 5.0		pF pF
C_{PD}	Power Dissipation Capacitance	Per Package (Note 3)		95		pF

Note 1: "Absolute Maximum Ratings" are those values beyond which the safety of the device cannot be guaranteed. Except for "Operating Temperature Range" they are not meant to imply that the devices should be operated at these limits. The table of "Electrical Characteristics" provides conditions for actual device operation.

Note 2: Capacitance is guaranteed by periodic testing.

Note 3: C_{PD} determines the no load AC power consumption of any CMOS device. For complete explanation see 54C/74C Family Characteristics application note AN-90.

Switching Time Waveforms

CMOS to CMOS

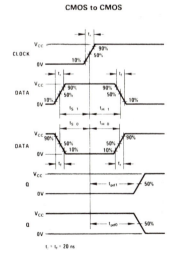

AC Test Circuit

Answers to Odd-Numbered Problems

Chapter 1

1. Continuous.

3. Origin is uncertain, since both European and Asian historical records mention the abacus. The word *abacus* meant dust.

5. Watch, voltmeter, thermometer.

7. Increased resolution, ease of interfacing with computers.

Chapter 2

1. Jack may own a bicycle.

3.

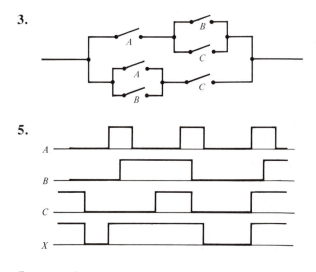

5.

7. $x = 1$.

9.

Voltage Levels			Logic Levels		
A	B	X	A	B	X
0	0	0	1	1	1
0	5	0	1	0	1
5	0	0	0	1	1
5	5	5	0	0	0

11.

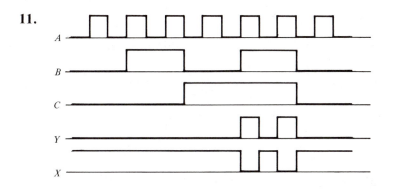

13. N = neutral, BS = brake set, ST = switch in "start" position.

N	BS	ST	X
0	0	0	0
0	0	1	0
0	1	0	0
0	1	1	0
1	0	0	0
1	0	1	0
1	1	0	0
1	1	1	1

15.

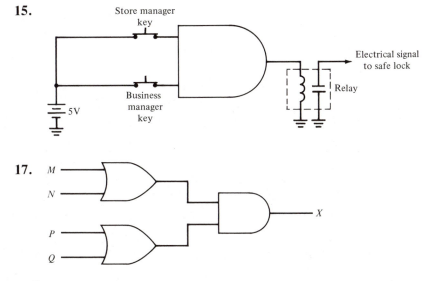

17.

M
N
P
Q
X

Chapter 3

1. (a) Distributive laws, (b) De Morgan's theorem, (c) distributive laws.

3. $A + \overline{A}B = A + B$

A B	$A + \overline{A}B =$ __	$A + B =$ __
0 0	$0 + \overline{0}0 = 0$	$0 + 0 = 0$
0 1	$0 + \overline{0}1 = 1$	$0 + 1 = 1$
1 0	$1 + \overline{1}0 = 1$	$1 + 0 = 1$
1 1	$1 + \overline{1}1 = 1$	$1 + 1 = 1$

5. (a) $X = A(B + C)$, (b) $X = B$, (c) $X = A$.

7. $X = A + \bar{B} + \bar{C}$

9. (a) $X = \bar{A}C + \bar{B}C$, (b) $X = \overline{A\bar{B}\bar{C}}$, (c) $X = \bar{A}BC$.

11. $X = A + B$.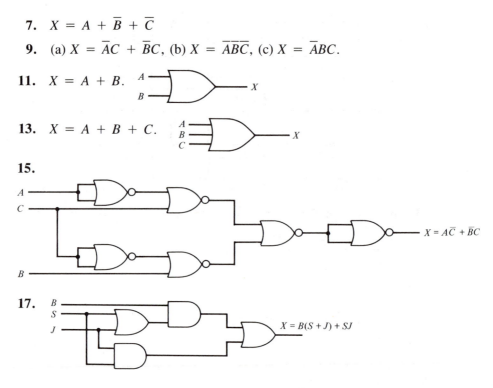

13. $X = A + B + C$.

15.

17.

19.

Chapter 4

1. $SOP = \bar{A}B\bar{C} + \bar{A}BC + A\bar{B}\bar{C} + A\bar{B}C + ABC$
$POS = (A + B + C)(A + B + \bar{C})(\bar{A} + B + C)(\bar{A} + \bar{B} + C)$

3.

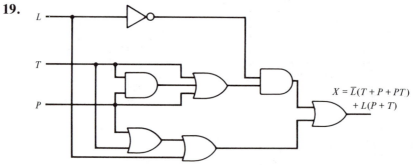

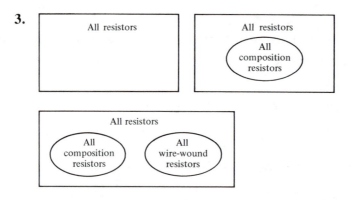

5. (a) $X = A + B + C$, (b) $X = 1$.

7. (a) $X = A + \bar{B} + \bar{C}$, (b) $X = \bar{A} + \bar{B} + C$.

9. (a) $X = A + \bar{B} + CD$, (b) $X = \overline{AB} + \bar{A}C + A\bar{D} + \overline{CD}$

11. (a)

A	B	C	X
0	0	0	0
0	0	1	0
0	1	0	0
0	1	1	1
1	0	0	0
1	0	1	1
1	1	0	0
1	1	1	1

(b)

A	B	C	X
0	0	0	0
0	0	1	0
0	1	0	0
0	1	1	1
1	0	0	0
1	0	1	0
1	1	0	0
1	1	1	1

13.

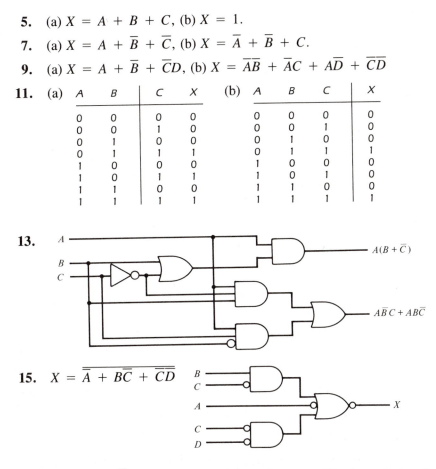

15. $X = \overline{\bar{A} + B\bar{C}} + \overline{CD}$

17. $X = AB + \bar{A}C$ using both the Karnaugh map and Boolean algebra theorems.

19. $X = JC\bar{M} + JW\bar{M} + \bar{J}CM + \bar{J}WM$.

21. $X = AC(B + D) + BD(A + C)$.

Chapter 5

1. (a) Advanced micro devices, (b) high speed, (c) 2-input OR gate, (d) ceramic DIP.

3. Low power TTL.

5. Low power Schottky.

7. (a) Device B, (b) Device B, (c) Device C, (d) Device A, (e) Device B, (f) Device B, (g) Device C.

9. Rise time ≈ 1.4 μs; fall time ≈ 1.6 μs.

11. 0.8 V.

13. 12.8 mA.

15. 36 ns.

17.

A	B	Enable	X
0	0	0	1
0	1	1	1
1	0	0	0
1	1	1	1
0	0	1	1
0	1	0	1
1	0	0	1
1	1	0	1

21. $A \approx 0.3$ V, $B \approx 0.2$ V, $C \approx 5.0$ V, $D \approx 4.3$ V, $E \approx 3.6$ V.

23. $P_T = 1580$ μW.

27. (a) $X = \overline{A \cdot B}$, (b) $X = \overline{A} \cdot B$, (c) $X = \overline{A \cdot B}$.

29. $X = 1$.

31. Propagation delay $= 4.16$ μs.

Chapter 6

1. (a) 5, (b) 6, (c) 4, (d) 7, (e) 9.

3. (a) 20.75, (b) 53.25, (c) 115.375, (d) 61.875, (e) 97.0625.

5. (a) 11110110, (b) 1111011, (c) 111011111, (d) 10101000, (e) 101000110.

7. (a) 10001001, (b) 11111110, (c) 101111101, (d) 100100111, (e) 110110000.

9. (a) 10010, (b) 11000, (c) 1100000, (d) 11010100.

11. (a) 01, (b) 10, (c) 10, (d) 11, (e) 101.

13. (a) 0101, (b) 0010, (c) 01001, (d) 00100, (e) 010010, (f) 011010.

15. (a) 0011, (b) 0111, (c) 01010, (d) 00110, (e) 00111, (f) 010011.

17. (a) 1001, (b) 1010, (c) 11110, (d) 110010, (e) 10001111, (f) 1001110110.

19. (a) 12, (b) 30, (c) 39, (d) 61, (e) 158, (f) 794.

21. (a) 35_8, (b) 104_8, (c) 175_8, (d) 363.527_8, (e) 643.2631_8, (f) 1172.450_8.

23. (a) 262_8, (b) 255_8, (c) 165_8, (d) 655_8, (e) 27.5_8, (f) 72.1_8.

25. (a) 00100111, (b) 101000100001, (c) 010111001101, (d) 010011110010.1100, (e) 011010110011.1110, (f) 111111101111.10110010.

27. (a) 57, (b) 180, (c) 27882, (d) 60022, (e) 63209, (f) 60039.

29. (a) $5F_{16}$, (b) FE_{16}, (c) $8E3_{16}$, (d) $F03_{16}$.

31. (a) 92, (b) 207, (c) 469.

Chapter 7

1. (a) Sum $= A\overline{B} + \overline{A}B$, carry $= AB$, (b) sum $= A\overline{B} + \overline{A}B$, carry $= AB$, (c) sum $= A\overline{B} + \overline{A}B$, carry $= AB$.

3.

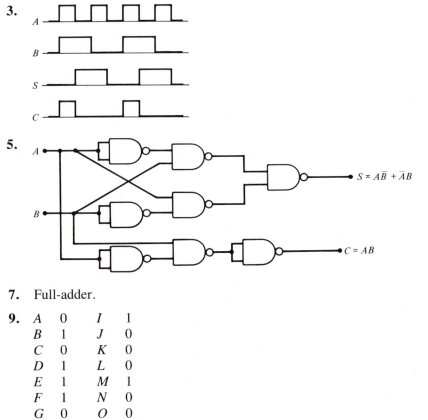

5.

$S = A\bar{B} + \bar{A}B$

$C = AB$

7. Full-adder.

9.

A	0	I	1
B	1	J	0
C	0	K	0
D	1	L	0
E	1	M	1
F	1	N	0
G	0	O	0
H	1	P	0

11.

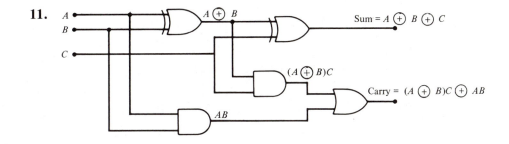

Sum $= A \oplus B \oplus C$

$(A \oplus B)C$

Carry $= (A \oplus B)C \oplus AB$

AB

Chapter 8

1.

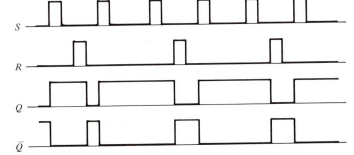

3.

5.

7.

9.

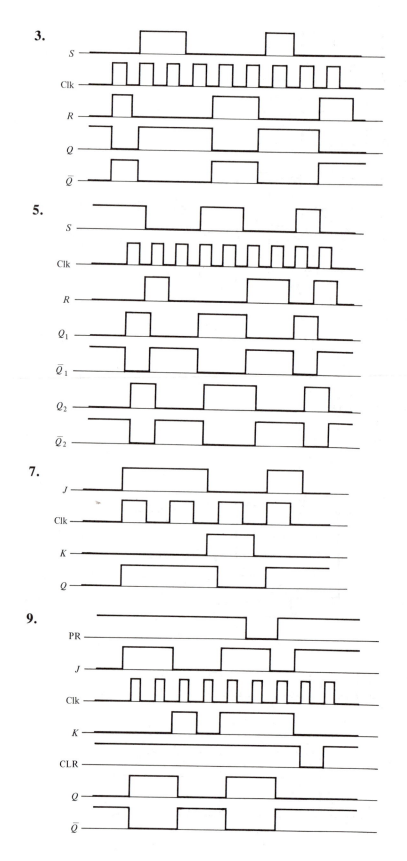

11.   1000 ns.

13.

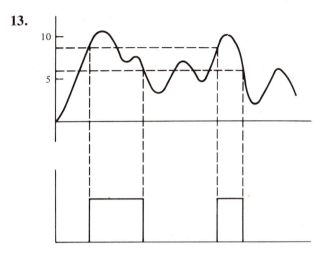

15. $t = 6.3$ μs.

17.

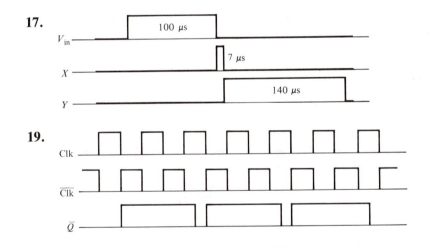

19.

21. $R_B = 2.97$ kΩ.

Chapter 9

1.   (a) Two flip-flops, (b) three flip-flops, (c) four flip-flops, (d) five flip-flops.

3. $n = 7$.

5. $f_{out} = 37.5$ kHz.

7.   00111.

9.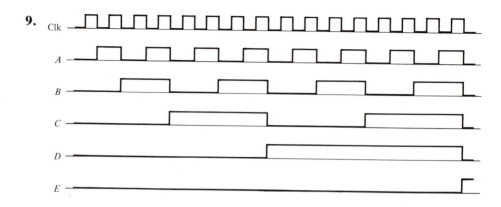

11. Mod-11.

13.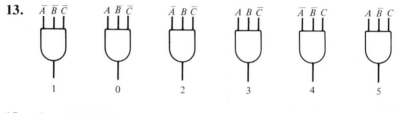

15. $f = 12.5$ MHz.

Chapter 10

3. $t_D = 27.5$ μs.

5. $t_D = 61.5$ μs.

7. Data must be shifted three positions further into the register.

9.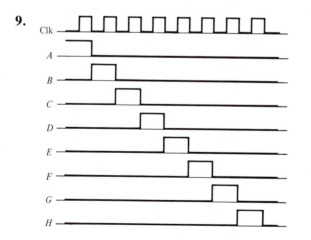

11. Set G, E, and B flip-flops HIGH.

13. Final register contents $= 21_{10}$.

15. Johnson counter costs more.

Chapter 11

1. (a) 00010010 *BCD*, (b) 00110111 *BCD*, (c) 10001001 *BCD*, (d) 000100110110 *BCD*, (e) 000101101000 *BCD*, (f) 001001000011 *BCD*.

3. (a) 0101, (b) 0111, (c) 00010011, (d) 00010100, (e) 1000 1001, (f) 0001 1000 0010.

5. (a) 2, (b) 4, (c) 37, (d) 01, (e) 162, (f) 783.

7. (a) 0101, (b) 0110, (c) 0101 0011, (d) 0110 1000, (e) 0111 1010, (f) 0111 0111 0100.

9. (a) 10110, (b) 110101, (c) 1000100, (d) 10110001, (e) 110100110, (f) 1011001110.

11. Eight rings are required.

13. (a) ECL = 1100 0101 1100 0011 1101 0011,
    (b) DIP = 1100 0100 1100 1001 1101 0111,
    (c) CMOS = 1100 0011 1101 0100 1101 0110 1110 0010.

15. Error in third row down and third column from left.

Chapter 12

1.

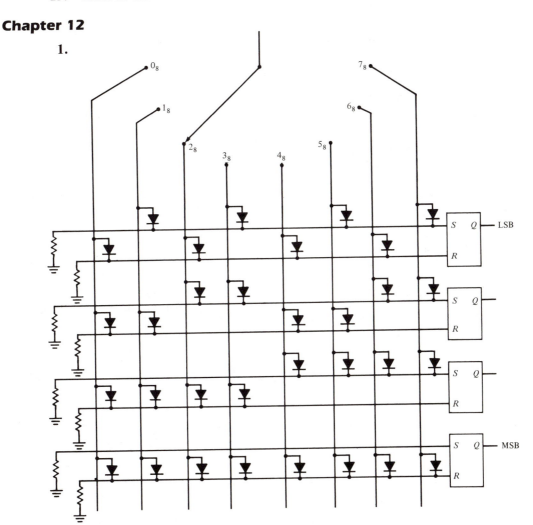

3.

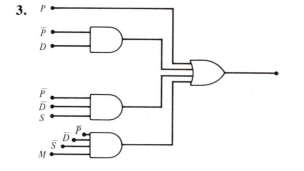

5. (a) 1101 (invalid BCD code group), (b) 0110, (c) 1001, (d) 1000.

7.

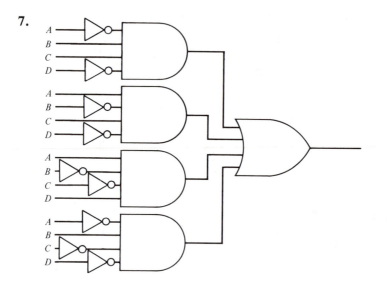

9.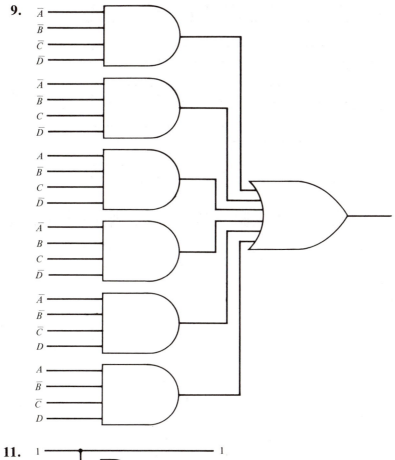

11.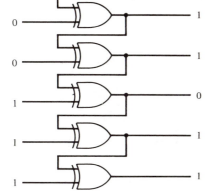

13.

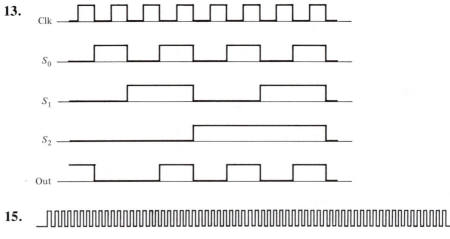

Clk

S_0

S_1

S_2

Out

15.

$|\!\!\leftarrow\!\!-F\!-\!\!\rightarrow\!\!|\!\!\leftarrow\!\!-O\!-\!\!\rightarrow\!\!|\!\!\leftarrow\!\!-E\!-\!\!\rightarrow\!\!|\!\!\leftarrow\!\!-F\!-\!\!\rightarrow\!\!|\!\!\leftarrow\!\!-W\!-\!\!\rightarrow\!\!|\!\!\leftarrow\!\!-N\!-\!\!\rightarrow\!\!|\!\!\leftarrow\!\!-O\!-\!\!\rightarrow\!\!|\!\!\leftarrow\!\!-T\!-\!\!\rightarrow\!\!|\!\!\leftarrow\!\!-O\!-\!\!\rightarrow\!\!|$

Chapter 13

  1. 0.125 μs/adder.

  3. -1111.

  5. 9 μs.

  9.

Data Lines		Select Lines	Mode
$A_3\ A_2\ A_1\ A_0$	$B_3\ B_2\ B_1\ B_0$	$S_3\ S_2\ S_1\ S_0$	M
1 1 0 0	0 1 0 1	0 1 1 0	1

Chapter 14

  1. $V_o = -105$ mV.

  3. $V_o = 3V$.

  5. $V_o = -11V$.

  7. $R_2 = 62.5$ kΩ, $R_3 = 31.25$ kΩ, $R_4 = 15.625$ kΩ.

  9. Percent Res $= 0.024\%$.

 11. Number of bits $= 13$.

 13. $f = 2400$ pps.

 15. (a) 256 μs, (b) 128 μs, (c) 3906 conversions/sec.

Chapter 15

  3. Storage capacity $= 4K$.

  5. Number of cells $= 262,144$.

7. (a) 8, (b) 8192 cores, (c) 32X and 32Y lines, (d) 8 sense lines, (e) 8 inhibit lines.

9. Inhibit current = 30 mA.

11. 8192 bytes.

13. Eight data input and eight data output.

15. 0101.

21. Tape B.

Chapter 16

1.

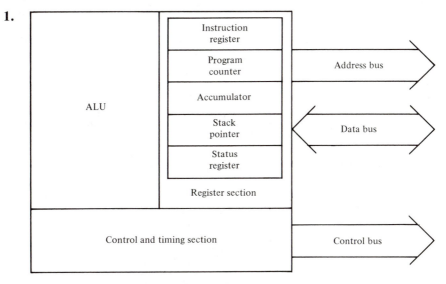

3. The microprocessor serves as the CPU for the microcomputer.

5. 65,536 words of memory.

7. 16 registers: 10 principal plus 6 general purpose.

9. The data bus lines are on pins 7 through 10 and 12 through 14.

11. The control and data buses.

13. Buffered outputs allow data transfer on the bidirectional bus.

Index